U.S. Marine Corps Physical Readiness Training for Combat

Plus

Jungle Operations

Department of Defense

Pentagon
Publishing

UNITED STATES MARINE CORPS
Marine Corps Combat Development Command
Quantico, Virginia 22134-5001

29 January 1988

FLEET MARINE FORCE REFERENCE PUBLICATION............0-1B

MARINE PHYSICAL READINESS TRAINING FOR COMBAT

1. PURPOSE

Fleet Marine Force Reference Publication (FMFRP) 0-1Marine Physical Readiness Training for Combat is intended for use by all Marines. It provides the information and references necessary to establish and conduct physical conditioning programs to prepare Marines for the physical demands of combat.

2. APPLICABILITY

Although not directive, this handbook is intended for use as a reference by all Marines in developing unit physical readiness training for combat. This FMFRP does not establish policy for the conduct of physical fitness training. Refer to Marine Corps Order 6300.3_ for applicable Marine Corps policy on physical fitness. It may also be used as a basis for instruction conducted by unit leaders or as a reference for instruction or training.

3. RECOMMENDATIONS

To improve this manual, the MAGTF Warfighting Center seeks comments and recommendations from the user. Officers and noncommissioned officers with constructive comments are encouraged to respond. Lack of response will be considered as concurrence. The User Suggestion Form which follows can be detached and forwarded to - -

Commanding General
Marine Corps Combat Development Command (WF 06)
Quantico, Virginia 22134-5001

4. DISTRIBUTION

Initial distribution of this FMFRP 0-1B is shown below. Addressees on this list should ensure that appropriate subordinate organizations are provided copies.

JOHN I. HOPKINS
Brigadier General, U.S. Marine Corps
Assistant Chief of Staff, Warfighting
Marine Corps Combat Development Command
Quantico, Virginia

DISTRIBUTION: TDJ

From:

To: Commanding General, Marine Corps Combat Development Command
(WF 06), Quantico, Virginia 22134-5001

Subj: FMFRP 0-1B, MARINE PHYSICAL READINESS TRAINING FOR COMBAT, RECOMMENDATIONS CONCERNING

1. In accordance with the Promulgation Page to FMFRP 0-1B, which invites individuals to submit suggestions concerning this FMFRP directly to the above addressee, the following unclassified recommendation is forwarded:

Page_________Article/Para No.__________Line No._________Figure/Table No._________

Nature of Change:_____Add_____Delete_____Change_____Correct

2. Proposed New Verbatim Text: (Verbatim, double spaced; continue on additional pages as necessary.)

3. Justification/Source: (Need not be double spaced.)

NOTE: Only one recommendation per page.

PREFACE

MARINE PHYSICAL READINESS TRAINING FOR COMBAT

Chapter 1
PHYSICAL READINESS LEADERSHIP
Section I. Role of Physical Fitness in Combat

1101. PURPOSE OF PHYSICAL FITNESS TRAINING

Physical fitness training in the Marine Corps has one purpose: to prepare Marines to physically withstand the rigors of combat. All other goals of physical fitness training are subordinate to and must support attainment of this goal. The idea that only infantry or reconnaissance units and their attachments normally face physically demanding combat is wrong. This error must not influence the priority commanders of combat support, combat service support, aviation, and headquarters units give to physical fitness for combat. Physical fitness for combat has a high priority for all Marines. A sound, effective unit program, requiring limited time and material, offers a greater payoff in combat than many more expensive and time-consuming training programs. This handbook describes the concept and provides the ingredients of a program for commanders to use in carrying out one of their most serious responsibilities: ensuring that their Marines are physically ready for combat.

1102. SCOPE OF THE MANUAL

This handbook provides guidance for all leaders, trainers, and planners of physical training programs. It describes unit physical fitness for combat training in the following chapters:

a. **Physical Readiness Leadership.** Chapter 1 provides guidance to leaders in the conduct of physical combat readiness training.

b. **Physical Readiness Training Programs.** Chapter 2 discusses how to structure programs to meet physical readiness goals in a variety of combat training situations.

c. **Physical Conditioning Activities.** Chapter 3 discusses the primary physical conditioning activities which commanders may use to attain readiness for combat goals. Sections in this chapter address foot marches under load, strength-building activities, and activities which build confidence and the aggressive spirit needed in combat.

d. **Combat Water Survival.** As an amphibious force, Marines cannot ignore the importance of combat water survival training. Chapter 4 describes a program for commanders to conduct this training.

e. **Competitive Conditioning Activities.** Chapter 5 describes competitive activities which stress the relationship to combat actions, featuring team-building types of competition.

f. **Evaluation of Performance During Training.** Chapter 6 describes tests which measure individual and unit physical fitness for combat.

g. **The Human Body.** Chapter 7 covers the structure and functioning of the body.

1103. COMMANDER'S ROLE

Major General Commandant John A. Lejeune in the 1921 edition of Marine Corps Manual expressed our philosophy of leadership. Among other things, General Lejeune stated that "it will be necessary for officers ... to endeavor to enlist the interest of their men in building up and maintaining their bodies in the finest physical condition." It is significant that the Commandant prefaced these writings in 1921 with the statement that his thoughts were greatly influenced by the experience of World War I. We have never varied as a Corps from this belief that one of a commander's most serious responsibilities is to have Marines physically prepared for combat. Senior commanders and their staffs carry out this role primarily by providing command emphasis, including planning, support, and supervision. Commanders at the battalion/squadron and company/battery level execute a program of physical fitness for combat by close supervision and, most importantly, leadership by example.

1104. HISTORY OF MILITARY PHYSICAL READINESS

Every war has revealed our military physical deficiencies during the initial periods of mobilization. This realization followed the Civil War and has recurred regularly with each national emergency.

a. Training programs in each war were geared to the physical need of the era. Success was dependent upon the amount of time available during training to physically prepare Marines for battle conditions. Frequently, casualties in initial engagements were attributed to the inability of our Marines to physically withstand the rigors of combat over rugged terrain and under unfavorable climatic conditions. With adequate preparation, Marines have always handled the test of battle.

b. During World War II, the first physical conditioning doctrine that could be scientifically justified by testing was introduced. As the war progressed, this program was effective in the physical conditioning of millions of Marines for combat.

c. Postwar periods have traditionally been a time of consolidation. Unfortunately, some leaders considered the conditioning phase of training to be a wartime tool. With such a philosophy prevalent between wars, physical readiness was relegated to secondary importance resulting in a serious lowering of combat effectiveness. The initial commitment of Marines in Korea dramatically displayed this failure to recognize the extreme physical demands of warfare. Call-up of large numbers of reserve Marines and pressing them into the war within weeks after call-up mainly caused the loss in effectiveness.

d. Costly lessons learned from our military experiences over a period of years and the course of several wars led to an increasing interest in the physical conditioning of the individual Marine. No longer can we afford to emphasize physical fitness during wartime and de-emphasize it during peacetime. It is evident that, in spite of increased mechanization and modern weapons, physical readiness retains a vital place in the life of each individual Marine and in every unit within the Marine Corps regardless of mission.

e. Commanders are well aware of the need for rugged and well conditioned Marines. The daily demands of housekeeping, maintenance, support, training, operations, and other time-consuming tasks make it necessary for commanders to set aside time for frequent, regular, and vigorous exercise periods.

1105. PHYSICAL DEMANDS OF COMBAT

There are three primary elements of effective physical fitness for combat: lower body strength and stamina; upper body strength and stamina; and a competitive, combative spirit. Cardiovascular functioning is not treated as a separate element since it is a necessary foundation for the other three. Exercises and activities designed to improve Marines' condition in the three primary elements will foster good cardiovascular and respiratory condition.

a. Lower Body Strength and Stamina. Some common demands that combat places on lower body strength and endurance are marching long distances under load and functioning effectively at the destination; moving quickly and evasively under fire; and carrying wounded Marines to safety.

b. Upper Body Strength and Stamina. Some common demands that combat places on upper body strength and stamina are rapidly emplacing crew-served weapons; handling large-caliber ammunition for extended periods; climbing walls, cliffs, and other high obstacles; and performing field maintenance on air craft or heavy machinery.

c. Competitive, Combative Spirit. While part of this critical aspect of combat readiness is mental and emotional, robust physical condition and a training program which promotes physical aggressiveness greatly enhance a competitive and combative spirit. Activities which force Marines to overcome natural physical fear, which directly leads to fatigue, are particularly valuable.

1106. PHYSICAL EFFECTS OF COMBAT STRESS

It has been thoroughly documented that the added uncertainty and stress of combat have a major physical effect on Marines as well as the acknowledged psychological effect. In other words, in combat, fear equals fatigue. In training, we cannot easily reproduce this combat stress which reduces the effectiveness of individuals and units. However, we can produce fatigue and psychological doubt in training situations by developing a challenging physical training program which stresses the participants. This compels Marines to exercise their ability to continue to function under difficult and trying conditions. Such activities as long foot marches under load and difficult confidence courses train Marines to overcome their own fear and fatigue. Rugged competitive activities such as martial arts and pugil stick fighting prepare Marines psychologically to overcome an opponent as well as their own fear and fatigue. This sort of physically demanding training builds self-confidence and unit morale. The adage that "the unit that sweats more in peacetime will bleed less in battle" certainly applies to its physical training program. Athletic coaches have long followed this approach by such practices as having teams shoot foul shots after wind sprints in order to strengthen their mental conditioning and toughness. Although pushing Marines to their limit is necessary, commanders often set lower standards initially than ultimately desired. The program should then call for steadily increasing the difficulty. A good guide for a starting point is a standard that will be physically and psychologically demanding for no less than three-fourths of the unit. Expressed another way, no more than one-fourth of the unit should be working at less than their limit. The program should progress quickly, augmented, if needed, by remedial training for those unable to meet the unit standards, until the training is challenging for all but the most exceptional Marines in the command. Commanders must never forget that Marines expect to be challenged and thrive on being pushed to their limit.

1107. LIMITS OF PHYSICAL READINESS

A well-conditioned Marine unit has a significant advantage in combat, but this does not mean that physical conditioning can substitute for technical knowledge or good planning. For instance, units must, as discussed in chapter 3, pursue a continuing and challenging program of marches under loads. This docs not disprove the axiom that fighting men, if required to carry over one-third of body weight, will generally become ineffective in battle. Leaders and planners must decide what is essential to be carried and must use an effective logistic distribution system, such as unit trains, to make remaining items available when they are needed. In a similar vein, the valid requirement to foster a competitive and combative spirit in Marines does not excuse leaders from the responsibility to practice tactics which will not needlessly risk lives. Physical readiness for combat is a vital goal, but it is only one part of a unit's overall readiness for the lest of combat.

Section II. Fundamentals of Physical Fitness

1201. FITNESS FOR MARINES

Total combat readiness includes technical proficiency and mental, emotional, and physical fitness. If any one of these attributes is lacking, combat effectiveness suffers proportionately. Without technical fitness, Marines lack the knowledge and skill to fight; without mental and emotional fitness, they lack the incentive and will to fight; and without physical fitness, they lack the physical ability and confidence to fight. Physical fitness in a Marine means a whole and healthy body, the capacity for skillful and sustained performance, the ability to recover from exertion rapidly, the desire to complete a designated task, and the confidence to face any eventuality.

1202. COMPONENTS OF PHYSICAL FITNESS

A sound body, free of disease and defect, does not in itself constitute physical fitness. Before an untrained Marine can be considered physically fit for combat, the following traits must be developed:

a. Strength. Every Marine needs enough strength to perform the heaviest task encountered in routine and emergency activities. The basic areas where strength is required are in the arm and shoulder girdle, abdomen, back, and legs. Muscles increase in size, strength, and firmness with regular and strenuous exercises. Without work, they grow flabby and weak.

b. Endurance. Each Marine needs sustaining power to maintain the maximum ability without undue fatigue. There are two types of endurance:

(1) Muscular Endurance. The Marine needs muscular endurance to fight the enemy under the most tiring combat conditions. Muscular endurance is the ability to perform continuous work over long periods of time. Endurance depends on the bloodstream's ability to deliver large amounts of oxygen and nutrition to the muscle masses and then carry away the waste products quickly.

(2) Cardiovascular-Respiratory Endurance. The development of cardiovascular-respiratory endurance ("wind") is necessary to maintain muscular endurance. Cardiovascular-respiratory endurance depends on the efficiency of the blood vessels, lungs, and heart. The maximum effort a Marine can exert over a period of time is limited by the capacity to absorb oxygen and expel carbon dioxide. The average Marine's cardiovascular-respiratory capacity can be greatly increased by exercise.

c. Agility. A Marine must be able to change direction quickly and as faultlessly as possible. The ability to react instantly and to maintain orientation during rapid changes of body position is important to survival. This important ability may be developed by conditioning exercises that require varied and rapid changes of body position on the ground and in the air, such as obstacle courses.

d. Coordination. Coordination is the ability to move all parts of the body in a smooth, efficient, and concerted effort (commonly called timing). A well-coordinated individual does not make useless movements. An individual who moves with precision and accuracy saves energy. Coordination is best developed by practicing diversified muscular activities and skills affecting all body parts.

1203. TYPES OF EXERCISE

Basically, there are two forms of exercise: isotonic and isometric. Both forms are contained in the activities found in this manual.

a. Isotonic. Isotonic exercises are those in which the expenditure of energy is regulated and released during concentrated efforts. The regulated expenditure of energy is controlled by both the mode of exercise and the individual's effort. This type of effort is common to most exercise and sports. To develop endurance, coordination, and flexibility, isotonic exercise must be used. Strength can also be developed or increased through this type of exercise.

b. Isometric. Isometric exercises are those in which maximum effort is applied and held until the engaged muscle energy is depleted during a single contraction effort. The individual exerts full force against an immovable object for a relatively short period of time (6 to 10 seconds) and then repeats the contraction several times with a short period of rest between each contraction. This type of exercise develops only strength; therefore, it has limited application.

1204. PRINCIPLES OF PHYSICAL CONDITIONING

Men and women vary in their physical makeup. Physiological (body) function and reaction also vary in proportion to heavy demands placed upon the body. To attain the maximum benefit without sacrificing Marines' welfare, fitness must grow with a careful program of conditioning. To allow for adjustments in body functioning as the conditioning program progresses and to ensure attainment of objectives, the following principles of physical conditioning must be applied:

a. Overload. As strength and endurance increase, the physical load must be increased until the desired level of fitness is reached.

b. Progression. In beginning stages, the load must be moderate. Gradual progression from this low state of fitness to a higher state is possible through application of a progressive program.

c. Balance. An effective program utilizes various types of activities and provides for the concurrent development of strength, endurance, and coordination as well as basic physical skills.

d. Variety. Some programs fail because the routine becomes boring. The most successful programs always include conditioning activities, competitive events, and military physical skill development.

e. Regularity. There is no easy or occasional way to develop physical fitness. Regularity of exercise is a must, with daily exercise preferred.

1205. THREE STAGES OF PHYSICAL CONDITIONING

Unconditioned or poorly conditioned Marines pass through the following stages in reaching the desired state of physical condition.

a. Toughening Stage. This stage is approximately 2 weeks long and is usually characterized by muscular stiffness and soreness followed by recovery.

b. Slow Improvement Stage. This stage is approximately 6 to 10 weeks long and is characterized by slow and steady improvement until the desired level, or a high level, of fitness is attained.

c. Sustaining Stage. This stage goes on indefinitely in order to maintain the level of conditioning achieved by passage through the previous stages.

1206. EFFECTS OF CLIMATIC CONDITIONS

Temperature, both atmospheric and body, affects the physical performance of Marines. Proper maintenance of body temperature through warm up exercise, proper dress in cold weather, and removal or adjustment of clothing in hot weather is necessary for effective performance and health. Climatic factors to be considered are the following:

a. Exercise In High Temperatures and High Humidity. Marines can endure strenuous physical activity in extremely hot temperatures if they are given an opportunity to become acclimated and if they take enough salt and water. It is essential to continue physical training programs in hot climates. Marines can better withstand high temperatures when they are well-conditioned. High humidity combined with high temperatures presents a serious danger. These conditions prevent the natural cooling of the body by the evaporation of perspiration. Training schedules should conform to the provisions of the base commander in respect to wet bulb conditions. Those who conduct training under these conditions should monitor weight loss and be careful to make fluid replacement adjustments.

b. Exercise at High Altitudes. Certain problems are encountered in conditioning Marines stationed in high altitudes because the heart undergoes greater exertion during exercise. It is particularly important that only light exercise be given initially at such altitudes. A Marine's body gradually adjusts to high altitudes within a few weeks. After this adjustment, progressively greater amounts of exercise are possible.

c. Exercise in Arctic Regions. Military duty in the arctic is so arduous that a high level of physical conditioning is essential. Because of the difficulties of carrying on physical conditioning exercises in extreme cold, Marines should be conditioned to the highest level possible before they arrive. A sustaining program will then maintain that level. When exercising in cold weather, Marines should be required to remove excess clothing to prevent them from becoming damp with perspiration.

1207. WARMING UP AND COOLING OFF

It is a fundamental physiological principle that Marines should warm up gradually before taking strenuous exercise. Such action speeds up the circulation to prepare the body to take an overload and helps to prevent injury to muscles and joints after exercising. Marines should be kept mildly active, walking, stretching, or performing some other mild muscular activity until breathing and temperature have returned to normal. Marines should never be allowed to cool off too rapidly; in cool or cold weather, they should put on additional clothing during the cooling-off period.

1208. PHYSICAL ACTIVITY AS AGE INCREASES

Combat makes severe physical demands on Marines. All Marines, regardless of age, must be physically ready to meet these demands. There is no physiological reason to cease exercise or exertion with age. Increased age usually brings increased responsibility which, in many instances, leads to a routine that can become almost devoid of physical activity. The key to fitness with increased age is to continue exercising at a reasonable level and to include exercise of a vigorous type in the daily routine. Older persons who have not regularly maintained a reasonable state of physical fitness require more time than younger persons to become fit. Such individuals usually require a longer period of time to recover from physical effort than younger Marines. If general health is good, evidence shows that older Marines can develop and maintain a degree of fitness which permits vigorous activity and proper performance of duties. It is both the individual's and the commander's responsibility that all Marines continue a daily sustaining exercise program.

Section III. Goals of Physical Readiness Training

1301. NECESSITY FOR PHYSICAL READINESS TRAINING

A very important objective of training is attainment and maintenance of operational readiness. Marines must be physically ready for operations at any time and under all conditions of climate and environment. A combination of training to develop proficiency in physical skills and conditioning to improve strength and endurance results in physical readiness for combat. The degree of physical fitness required of Marines can be acquired through physical exercises in a normal well-coordinated and closely monitored training routine. The performance of purely military duties, such as drills and marching, is not enough to build all the desired areas of fitness. Few recruits are physically fit for the arduous duties ahead of them. The softening influences of our mechanized civilization and the difficulties of conditioning Marines make physical fitness more important than ever before. If Marines are to be fully developed to and maintained at the desired standard of physical fitness, a well-conceived plan of physical readiness training must be a basic part of every training program. Marines cannot be adequately prepared in any other way for the hard work and arduous demands required on the battlefield.

1302. OBJECTIVE OF PHYSICAL READINESS TRAINING

The overall objective of the physical readiness training program is to develop individuals and units who are physically able and ready to perform their duty assignments or missions during training and in combat. To attain the objective of physical readiness, exercise activities must develop the following areas:

a. Strength and Endurance. Developing adequate strength to perform required duties and adequate endurance to sustain activity over a long period of time.

b. Muscle Tone. Developing muscle tone adequate to maintain proper posture and reasonable weight control.

c. Skills. Developing military physical skills which are essential to personal safety and effective combat performance. As skill is developed, agility and coordination are attained. The essential skills are-

(1) **Marching Under Load.** Marching with individual and unit weapons and equipment.

(2) **Running.** Distance and sprint running on roads and cross-country.

(3) **Jumping.** Broad jumping and vertical jumping downward from a height.

(4) **Dodging.** Changing body direction rapidly while running.

(5) **Climbing and Traversing.** Vertical climbing of ropes, poles, walls, and cargo nets. Traversing horizontal objects such as ropes, pipes, and ladders.

(6) **Crawling.** High crawling and low crawling for speed and stealth.

(7) **Throwing.** Propelling objects, such as grenades, for a distance with accuracy.

(8) **Vaulting.** Surmounting low objects, such as fences and barriers, by use of hand assists.

(9) **Carrying.** Carrying objects and employing man-carries.

(10) **Balancing.** Maintaining proper body balance on narrow walkways and at heights above normal.

(11) **Falling.** Contacting the ground from standing, running, and jumping postures.

(12) Surviving in Water (Or Other Specialized Situations) Using water survival techniques.

d. Character Traits. Jnstilling character traits which help accomplish military missions to include-

(1) Confidence. Confidence develops through achieving progressively more difficult tasks as physical ability grows.

(2) Aggressiveness. Participation in combative activities and contests develops desire and willingness to overcome an opponent.

(3) Reaction Under Pressure. Competitive contests and game situations train Marines to think and to act quickly while under pressure.

(4) Teamwork. Teamwork develops through competitive events in which a number of Marines must coordinate their efforts to accomplish a physical task.

1303. BENEFITS OF EXERCISE

The benefits of exercise are not always understood. Some of the more important results of exercise are listed below:

a. Improved Muscle Tone. Muscular tone improves and, at the same time, muscular strength and endurance are built up.

b. Cardiovascular-respiratory Endurance. Cardiovascular-respiratory endurance, or wind, improves through a process of opening up dormant lung capacity to absorb greater amounts of oxygen.

c. Circulation. Circulation of the blood speeds up and extends to a greater portion of the body as exercise forces the blood to service all parts of the body. The efficiency and effectiveness of the heart, lungs, and blood vessels improve.

d. Flexibility. A wider range of muscular movement is possible and rapidity in physical skills grows.

e. Elimination of Body Waste. Bending and twisting the body and the general speedup of body processes caused by exercise regulate and help eliminate body wastes.

f. Tension. Working off excess nervous energy and relief from daily worries and cares relieve tension. Participation in exercises leaves little time for worry.

g. Sleep. Sleep improves because muscles are healthfully tired after about of exercise. A by-product of sound sleep is relief of tension.

h. Obesity Control.. Control of obesity (fat) is made possible by using up excessive amounts of fat-producing food elements.

i. Injury Susceptibility. Susceptibility to injury is reduced through exercise. Muscles, tendons, and joints are strengthened. Injuries such as hernia, back strain, and joint sprains are less likely to occur if muscles are maintained in proper tone.

Section IV. Leadership Roles

1401. PSYCHOLOGICAL LEADERSHIP

The full development of a Marine's resources is not all physical. To be effective in developing physical readiness, leaders must realize that mind and attitude are also important to success. The more important psychological considerations are to—

a. Promote Understanding of the Value of Physical Readiness. A desire to be physically ready should be created in all Marines. Motivation is increased and Marines take greater interest in their individual physical fitness if they understand the value and benefits of vigorous exercise. When Marines realize their efforts are an investment in their own personal welfare, it should not be difficult to obtain their cooperation. Marines should understand the objectives, the benefits, and the value of each type of exercise activity in their program. They should also understand the relation of physical readiness to survival in combat.

b. Maintain a Positive Approach. Physical readiness training for combat is strenuous and demanding. It is a responsibility of leadership to create an atmosphere where all desire to participate fully. This attitude should be fostered. A negative approach must not be identified with physical readiness training, even with those having difficulty. Only in unusual cases should fear of punishment be the motivating factor behind good performance. For those few who cannot keep up or attempt to malinger, an effective remedial program is essential.

c. Seek Cooperation and Develop Morale. In a program placing maximum physical stress upon individuals, it is necessary to gain their cooperation. Favorable reaction is enhanced by proper planning and organization, challenging requirements, use of competition, and application of a progressive program resulting in physical fitness. As physical fitness grows, morale also grows.

1402. COMMAND AND SUPERVISORY FUNCTIONS

a. Command Functions. Commanders should take the following actions to support physical readiness training:

(1) Lead by personal example.

(2) Instill command interest and indicate to subordinate personnel the importance of this training to the welfare of the organization.

(3) Allot sufficient time for the achievement of objectives and monitor the use of such allotted time. The substitution of other training or routine duties for scheduled physical readiness training is unsound and unwise.

(4) Assign and properly utilize qualified personnel to supervise and conduct physical readiness training. If leaders are not competent, take action to ensure they become competent quickly. Reassign those who do not meet this standard to other duties.

(5) Make necessary facilities and funds available to support a program to develop physical readiness within all personnel.

(6) Measure the physical fitness of individuals and units in order to evaluate progress and to determine if the program is successful.

b. Supervisory Functions. Leaders responsible for planning, conducting, and supervising physical readiness training should take the following actions:

(1) Prepare physical readiness training schedules which apply the principles of physical conditioning and which aim for a particular type of program plan.

(2) Provide for wide participation of as many Marines as possible. All Marines, regardless of position or age, will benefit from regular exercise. In some instances, special efforts are necessary to overcome obstacles to regular and frequent training. Special effort is also necessary to ensure remedial conditioning. Such conditioning should occur for those who are physically substandard and after extended absence due to leave, sickness, injury, and travel.

(3) Prevent waste or unwise use of time allotted for physical readiness training. Time-wasters include unprepared instructors; assignment of one instructor to a group larger than a platoon; progression which does not keep pace with the physical development of the Marines; extreme formality; inadequate equipment or facilities which require waiting turns to exercise; and lengthy rest periods between exercises which interfere with the application of overload.

(4) Ensure that the program contains vigorous physical activity. Such activity places progressively greater demands upon the body during each exercise session and also over the duration of the training program. To be of benefit, exercise must tire the muscles and cause the heart to increase its rate of beat.

(5) Set an overall objective for each physical fitness program.Observe the training as necessary to ensure that the established objectives are being achieved.

(6) Observe physical readiness training to insure the use of a positive approach. To implement a positive attitude, small-unit leaders and instructors should personally set the example; have an understanding, fair, and sympathetic attitude; recognize individual differences; and motivate Marines toward their best effort.

(7) Guide and inform small-unit leaders and instructors concerning approved techniques, directives, and literature. As necessary, arrange for local training of instructors to include clinics, conferences, schools, and demonstrations.

(8) Determine the effectiveness of physical readiness training by personal participation in and observation of training, analysis of field inspection reports, and analysis of individual physical fitness test scores. Scores may be combined to reflect the fitness of the unit.

1403. SMALL-UNIT LEADERS AND INSTRUCTORS

a. Responsibility. The instruction and conduct of physical readiness activities are the function of company/battery commanders, platoonleaders and persons assigned as instructors. Experience has proved the effectiveness of physical fitness development when conducted in company-and platoon-size units under direct control of the leader with overall supervision by the parent-unit commander. For example, all the platoons of a company may exercise at the same time under the general supervision of the company commander, with each platoon conducting the assignment separately and under its own leadership.

b. Leader's Assignment. A small-unit leader or an instructor in a school or training activity is assigned to a combat unit or to a support unit. In this assignment, the leader is responsible for all training to include physical readiness training. In a different situation, the noncommissioned officer (NCO) or officer may beassigned as a full-time physical readiness instructor. This contrasts to a unit leader assignment where only part of the time is devoted to such training. In either case, the leader will hold an important and vital position for the physical fitness of Marines.

c. Leader's Training. Leaders may come to the assignment either fully or partially trained. It may be their first responsibility for the development of physical fitness. If they have had previous training through experience, make certain that their information is supplemented with study of this manual. If they have had professional training in physical education during civilian life, but no military experience, they should also use this manual to learn the methods used by the Marine Corps. A new leader should take advantage of various ways to learn including attendance at leader training courses, self-study, practice, and discussion with more experienced leaders.

d.Leader's Objective. As a physical readiness training instructor, the leader has two general objectives. The first is to motivate Marines to want to be physically fit. The second is to conduct a program that will develop a high degree of physical fitnss. Motivated Marines will react enthusiastically to such a program. It aids greatly in achieving local program objectives.

e. Leader's Personal Fitnes. A unit leader who must instruct and demonstrate physical activities must be in physical condition to do the job without undue physical stress.The leader should be able to do those things that must be demonstrated. The leader's strength, endurance, posture, and skill should set the example. This does not mean that the leader must excel, as other Marines do not expect championship performance. However, they do expect, and deserve, a creditable showing of fitness for the job.

f. Leader's Knowledge. The leader must have three types of knowledge to properly administer physical readiness training. Theyare- -

(1) Knowledge of Marines. The leader must understand Marines, know how to lead and motivate them, understand how they learn, and apply this knowledge wisely in the day-to-day training situation.

(2) Understanding of Body Functioning. A more intelligent exercise program results from understanding and applying the principles which govern physical conditioning of the body. The leader with such knowledge can better prescribe, adjust, and regulate exercise types, amounts, and progression to attain fitness.

(3) Understanding Exercise Activities. The leader needs to understand the contribution each type of physical activity makes to physical fitness, and how to use each activity to develop fitness. Skill to demonstrate and lead the various activities is a necessary part of technique and is invaluable to the instructor or small-unit leader.

Chapter 2

PHYSICAL READINESS TRAINING PROGRAMS

Section I. Development of a Program

2101.GUIDANCE FOR PLANNERS

This chapter instructs planners on physical readiness training procedures. It contains program planning guidance including factors to consider when developing programs; steps in assembling a program; definitions of activity packages and systems of exercise; selection of activity packages; and selection of systems used in implementing those packages.

2102. MARINE CORPS MISSION

The mission of the Marine Corps is to seize and defend advanced naval bases and to perform other missions as the President of the United States may direct. Traditionally, to perform this mission, Marines have been projected into the area of operations by Navy vessels and have attacked hostile beaches over the shore in landing craft, assault amphibian vehicles, and helicopters. Recently, performance of this task has been somewhat complicated by the introduction of the concepts of maritime prepositioning and the air-landed Marine expeditionary-brigade. Under these concepts, Marines, landed at airfields in foreign countries, will join their equipment at commercial ports. The distance between these airfields and ports is often 10 or more miles. In time of international crisis, it can be expected that transportation assets will not be adequate. Marines from all elements of a Marine Air-Ground Task Force may have to march from the airfield to the site of their equipment. All Marines must be physically conditioned to rigors of conducting foot marches with individual weapons and equipment.

2103. EVALUATION OF FITNESS

The Marine Corps mission is such that all Marines must constantly and consistently achieve a high level of fitness that prepares them for the demands of combat. While Marines have many different MOS's, all Marines must be prepared for the demands of marching under load and performing basic infantry tasks, such as rear security and patrolling. While the physical fitness test remains our universal measure of individual fitness, it is the commander's responsibility to observe and evaluate the unit's ability to perform effectively in combat. Often the commander's most effective evaluation tool is weekly physical training sessions. The commander should participate in these sessions and see that they are fully integrated into the unit physical conditioning program.

2104. INTEGRATION OF TRAINING

Training time must be used efficiently and wisely. Every opportunity to integrate physical training into other training activities should be seized. For example, foot marches under load can be integrated into the normal daily activities as a way to move from one training or work site to another. This integration reflects realism in training and should be used to maintain overall proficiency. To accomplish this requires imagination and enthusiasm on the part of the commander. The area of operations in which the training is conducted determines what can be integrated and how. If possible, training should also consider the terrain and climate of the area in which the unit will subsequently conduct training or operations. Training should include familiarization with special equipment and the application of specialized techniques to tactical principles. Activities such as gun drills and command post emplacement drills also have physical conditioning value, particularly if supervised and done competitively or measured against time standards.

2105. ACTIVITIES AT THE UNIT LEVEL

Physical readiness training is a command responsibility and is generally conducted in Marine units at the company and platoon levels. Battalion/squadron commanders must constantly supervise, making sure that all Marines are physically prepared for combat.

a. Physical readiness training should usually be conducted weekly by each battalion/squadron-level unit, using events designed for all personnel (e.g., marches under load, battalion/squadron runs, military field days, etc.).These events enable the commanders to demonstrate personal leadership, and observe and evaluate the physical condition and combat physical readiness of their Marines. This exercise period should be designed to elevate the unit's morale and emphasize unit identity through the wearing of uniforms and the carrying of battalion and company guidons. The battalion/squadron commander should make every effort to be a visible participant in the unit's physical readiness program on a weekly basis. These activities could be conditioning marches under load or a weekly calisthenics/unit run.The suggested times for these activities are early on Monday morning or late on Friday afternoon after the training week has been completed. The uniform may be physical training gear orutilities and boots.

b. Commanders should be cautious of long runs in boots and utilities. This is not to suggest that units should not run in utilities, but to remind commanders that boots are designed for marching, not running. If commanders choose boots and utilities, they should select off-road routes which will provide a cushioning effect.

c. Units should be made well aware of the commander's intent before the unit falls out for exercise. If the exercise period is conducted early in the morning, it need not necessitate an extensive shower/clean-up period afterward, so long as all Marines from the commander on down remain dirty. Commanders should inform their Marines not to fall out for these sessions in clean, pressed uniforms.

d. The physical readiness training of headquarters units is often difficult to manage. Often physical training is left for individuals or sections to develop and manage on their own. This approach can be effective, but if not supervised, it can result in a "paper" program. Headquarters unit commanders should endeavor to conduct weekly integrated training which will enable them to evaluate the physical readiness of their personnel. Often this training has to be scheduled before or after normal working hours to maximize participation of the entire unit.

2106. STEPS IN PLANNING

To implement workable and effective programs (as directed by CMC ALMAR 261/87) the planner must-

a. Determine the Type of Program Needed. Marine units are inherently different in organization and mission. The physical readiness program must be tailored to the mission and to the current physical condition of most unit personnel. Programs to meet this need are of the following types:

(1) **Developmental Programs.** Marines in a beginning or poor state of physical readiness need a program which will develop strength, endurance, physical skills, and character traits which are beneficial to successful accomplishment of military missions. Such programs should be applied progressively to rise gradually to a peak of fitness and skill.

(2) Maintenance Programs. Once Marines reach the sustaining stage of conditioning, their goal is then to maintain this level by participation in a maintenance program.

(3) Remedial Programs. The term "remedial" is usually applied to those individuals or groups who possess substandard physical fitness. For example, a remedial physical conditioning program could be applied to persons who are overweight, who fail to reach physical fitness test standards, or who have missed extended periods of conditioning due to illness, injury, extended hospitalization, or other absence. Saturday training sessions are designed as remedial sessions for all individuals who either miss the daily training or have failed to meet adequate standards. Note that personnel in light duty or no-duty status should be expectedto attend all training sessions, observing, supervising, or assisting as needed.

b. Determine the Time Required. The amount of time for training operations varies considerably. However, every unit can find time to conduct physical readiness training. Frequent (i.e.,daily) physical training of short duration-30 to 60 minutes-is preferred over occasional longer periods. Other demands for training time are so urgent that every minute of time allotted for physical readinesstraining should be used. Determine the time required per week anddivide it into daily blocks.

c. Organize for Various Group Size. It is essential to stress exercise rather than formality. Marines must complete the program where they are- -on the training field, in the motor pool, on the range, next to the classroom, in the office area, in the shop, aboard ship, or elsewhere. Although desirable, it is not always possible to assemble company-size units for physical training. Platoon-size groups are appropriate for the proper conduct of physical conditioning activities. Certain situations may require exercise programs for section- or squad-size units. This manual outlines programs for all situations and types of organizations.

d. Allow for Weather and Exercise Area. In programing and scheduling, the climate and terrain often govern the selection of activities.

(1) Weather changes cause differences in temperature, rainfall, wind chill, and snow. These changes should be anticipated as they dictate the type of program. Alternate plans should be part of the schedule. Nevertheless, normal weather changes and rain should not drive training inside.

(2) Local terrain and available exercise areas may also influence the selection of activities and the type of program which it is possible to support. Some activities can be completed in nearly any area.

e. Plan for Seasonal Change. As most physical readiness training is conducted outdoors, it is necessary to recognize seasonal change. A program should be divided into fall, winter, spring, and summer parts. Seasonal change also causes change in light. For example, an early morning program started in the summer will have ideal lightconditions, yet in the fall or winter, darkness will occur at that same hour and interfere with the conduct of the program. Develop programs in seasonal blocks, and make provisions for anticipated changes.

f. Consider Needed Facilities. An excellent program can be conducted with practically no facilities since there are exercises which require no equipment. However, a better program can be developed when supported by certain facilities and items of equipment. Proper command support, plus ingenuity, will solve this problem. (Items of equipment, when necessary to support the recommended exercises, are included in the chapters on exercise activities.)

g. Specify Appropriate Uniform. The uniform worn for exercising depends upon the season of the year, the state of the weather, and local regulations. Whenever possible, Marines should be dressed alike. Undershirts are preferred as the upper garment when the weather permits. A uniform that restricts the free movement of the body should not be worn when exercising.

h. Consider Availability of Instructors. Leaders who can lead and direct the scheduled activity must be available. Organizational units should train junior officers and noncommissioned officers down to squad or section leaders to instruct and lead the various activities.

i. Select Activity and System. With the type of program needed and the objectives in mind, select an activity package or an integrated training package for each day's scheduled physical training. At this same time, the system to be used in employment of the selected activity or activities must be determined. This selection affects equipment, areas, instructors, transportation, and other support requirements.

j. Secure Command Participation and Support. Prepare and brief the commander to assure full understanding of the objectives and administration of the program. The full participation and support of the commander will greatly improve the level of success.

k. Supervise Execution. Determine needs, publish the program, and supervise its execution as a necessary part of developing the unit's training schedule.

2107. SELECTION OF ACTIVITIES AND SYSTEMS

a. Exercise Activities. The planner must determine exercise activities which will be appropriate to include in the program. An exercise activity is a single means of exercise usually identified by the name applied; for example, running, log exercises, and obstacle course. Many exercise activi ties can become part of activity packages.

b. Activity Packages. Many of the physical activities described in this manual are arranged in prescribed sequences and are known as activity packages. An activity package is a number of exercises of the same type, assembled as a group or a set, and arranged in a specific sequence. Exercise packages are organized in such manner that not more than 15 minutes will be required to complete the execution of any package. Each type of activity is explained in laterchapters. The number of available drills, tables, or circuits; the manner of organization; and the contribution each makes to the total program are covered. Full understanding of this information will greatly assist in developing effective programs. Various designations are used to identify exercise packages; for example, conditioning exercises when arranged in a set order are known as drills, and other packaged activities are designated as tables or circuits. The following are activity packages:

(1) Activity packages for groups:

o Conditioning Drill Two
o Conditioning Drill Three
o Rifle Drill
o Log Drill
o Grass Drill
o Running Tables
o Guerrilla Tables
o Circuit-Interval Table
o Combatives Tables
o Relay Tables

(2) Strength Circuits:

o Fixed Circuit
o Movable Circuit
o Simplified Circuit (Circuit Interval Table)

(3) Activity packages for individuals:

o The 6-12 Plan
o Weight Training
o Isometric Exercise

c. Advantages of Using Packages. These of exercise packages simplifies scheduling and conducting of exercise and results in the following benefits:

(1) Schedule development is simplified as the planner assembles packages which will satisfy the training objective. There is no need to deal with selection of individual activities or to be concerned about the amount of time to be expended on each.

(2) Any 15-minute period, and in some cases less time, can be scheduled or used to perform an activity.

(3) If longer periods of time are available or if the objective demands, several packages can be assembled to provide a more complete period of activity.

(4) Marines are assured a balanced set of exercises or activities as each package is carefully arranged to reach all muscle groups.

(5) The instructor can concentrate on the conduct of a vigorous workout as the type and duration of the activity have already been determined.

d. Nonpackaged Activities. Several types of activity are not packaged. Activities in this category are conditioning marches, unit runs, obstacle courses, combat water survival swimming, teamcontests, and team sports. These activities can be scheduled in combination with packaged activities, or they may be scheduled separately. Often they require a longer period of time. Most nonpackaged activities require a 50-minute period to satisfactorily complete their objective. For conditioning marches, a few hours are required. The benefits of these activities should not be overlooked as some desirable objectives cannot be attained without them.

e. Systems of Exercise. Several methods or systems of organizing exercise and activity packages can be used. Each system is based upon a specific organization as follows:

(1) Single Activity System. The unit leader immediately assumes command of the unit at the beginning of the exercise period. The leader moves the unit to a predesignated exercise site at double time, forms the unit in a circle around the leader, groundsclothing and equipment as appropriate, and quickly moves into the exercise routine. There is usually no time to teach; therefore,the Marines must know the activity to be used. At the conclusion of a 5- to 15-minute period, the unit leader returns the unit to the instructional area at double time and releases the unit for the next scheduled activity.

(2) Progressive Activity System. All Marines (company or platoon) complete activities in the same order during the period. For example, Drill One is followed by dual combatives, and finally a 1-mile run. This system is usually progressive from a warm up activity, such as Drill One, to an activity which contributes in a major way to one of the objectives, such as aggressiveness development through combatives or cardiovascular development, such as running.

(3) Rotating Activity System. Set up the same number of activities or stations as there are platoons in the company. Each platoon rotates through each station in turn. With four platoons in a 50-minute period, about 10 minutes can be devoted to each station. With three platoons, approximately 15 minutes may be spent at each station. Activities must be of a type that can be covered in the time allotted. For example, with three platoons, Station 1 could be Conditioning Drill One; Station 2, running; and Station 3, a team contest.

(4) Circuit System. Set up a number of stations to provide various types of exercise equipment and/or items of apparatus. The idea is to keep all Marines busy and exercising vigorously for a short period of time at each station. The fixed strength circuits and the movable strength circuits are examples of this system. Station changes must be rapid, and the exercise must be started quickly after each change. Since the objective is to exercise at top speed, the motivation comes from frequent change of activity by moving to another station. Rotation by station groups continues until all Marines have covered all stations.

(5) Interval System. This system stresses the development of strength and endurance. It involves heavy work for a given distance within a specified time, alternated with lighter work and recovery, but never stopping during the workout. This procedure is repeated, and the intensity is increased gradually as physical condition improves, but always with adequate recovery. The important concept is stress, recover, stress, recover, and so on. This system is often applied through running, but other activities of a continuous nature may also be used. An example is the Circuit-Interval Table.

f. Steps in Selecting Activities and Systems. The planner should follow these steps.

(1) Evaluate the needs of one unit with emphasis upon unit mission, objective, and time available.

(2) For each day, select an activity package or a combination of activities which will contribute to the objective.

(3) Then determine the system to be used in implementing the selected activities. There are several possible choices and much opportunity for flexibility in program development. For example, the weekly program may contain various exercise packages, systems, and time periods. Additional guidance is contained in following chapters.

2108. SAMPLE WEEKLY PROGRAMS

The variety of weekly programs given shows available alternatives. Programs and schedules other than those illustrated may be assembled to provide schedules to fit any situation. Many other scheduling combinations are possible through use of the packaged activities, varied time periods, and different types of assigned duty.

a. Sample A. Figure 2-1 illustrates a 60-minute time allotment in which the single activity system is used with activity packages that vary from day to day. In this program, 5 hours of training are scheduled for the week. This program needs some type of integrated unit training to supplement it each week. A bimonthly conditioning march of 10 miles in 3 hours is suggested to round out this program. The Saturday session is designed for remedial physical training.

NOTE: A physical training activity must be conducted at a minimum of once every 72 hours. After 72 hours, muscles begin to atrophy.

Figure 2-1 is available in the illustration section at the end of the book

b. Sample B. A more comprehensive weekly program (fig. 2-2) illustrates the scheduling of various-length periods, a variety of activities, and the use of three systems during the week. In this schedule, 3.5 hours are included for the week. The following should be noted concerning this schedule:

Figure 2-2 is available in the illustration section at the end of the book

(1) On Monday, a 60-minute period is available for physical readiness training. Three 20-minute packages are scheduled. Each platoon, under its own leadership, will progress through each scheduled activity in turn.

(2) On Tuesday, Wednesday, and Thursday, only 30-minute periods are available. Here each platoon leader supervises each platoon in a single activity system.

(3) On Friday, the company commander directs training to exhibit personal leadership example and supervision. In this example, only one strength circuit and one cross-country course are available, and each will accommodate only one platoon. For this reason, Conditioning Drill One station is included. A platoon rotates to each station within the 60-minute period.

(4) The Saturday session once again is designed for remedial physical training.
(5) Bimonthly, a 10-mile/3-hour march under load will be necessary to round out this program.
NOTE: Cardiovascular/respiratory exercise to improve endurance must be of long enough duration (20 minutes or more) to induce and maintain oxygen debt.

c. Sample C. A third sample schedule (fig. 2-3) illustrates the use of the single activity system for 5.25 hours of training. This schedule includes sustaining-type activities for a unit that has passed through both the toughening and slow improvement stages of conditioning. The use of competitive activities is featured to holdinterest and provide self-motivation. A bimonthly 10-mile/3-hour march under load will be necessary to round out this program. The Saturday session once again is designed forremedial physical training.

Figure 2-3 is available in the illustration section at the end of the book

NOTE:Thc body will respond to hard work or physical training without soreness and stiffness provided the work or physical training is conducted regularly.
NOTE:Excrciscs of short duration increase strength and bulk but do little to effect endurance.
NOTE:Thc times shown in the charts arc participation times.

Section II. Remedial Physical Training

2201. NEED FOR ATTENTION

Remedial physical training is a process by which physically substandard individuals are conditioned to meet standard levels of performance. To achieve a full degree of operational readiness, it is necessary to bring all Marines up to the prescribed standard. Experience has demonstrated that some Marines have difficulty due to a poor state of fitness, obesity, or lack of motivation. Attention to these individual deficiencies will help improve unit combat readiness.

2202. IDENTIFICATION OF INDIVIDUALS

The company or battery commander identifies persons who cannot achieve the prescribed level of proficiency through the physical fitness test or as a result of observation during training. The commander notes particular weakness of body parts as indicated by failure of certain test events. These Marines are then placed in a special remedial program either at platoon, company/battery, or battalion/squadron level. Extra time is devoted to overcoming the weakness. These extra periods of conditioning may be during or after duty hours, as determined by the commanders. Other persons need reconditioning as result of hospitalization. Restoring physical fitness to damaged areas of the body is accomplished through progressively graded physical activities under professional supervision, not by the physical fitness trainers.

a. Types of Deficiencies. Physical deficiencies which can be corrected by exercise fall into several categories.

(1) Lack of Strength in One or More Body Parts. The major muscle areas concerned are the arms and shoulder girdle, back, abdomen, and legs.

(2) Lack of Overall Endurance. Usually, there is deficient muscular and cardiovascular endurance.

(3) Deficiency in Coordination and Agility. In these cases, physical skill is not developed to a satisfactory degree in activities such as crawling, running, jumping, climbing, traversing, vaulting, pushing, pulling, lifting, and carrying.

(4) Overweight or Underweight. Either condition may interfere with physical fitness and mission accomplishment. Lack of exercise is not always the cause. The cause may be malfunction of normal physiological functions or it may be poor health habits such as over- or under-eating, lack of adequate rest, or overconsumption of alcohol.

(5) Lack of Motivation. Not all persons are motivated to attain or maintain **a** desirable state of fitness. Some Marines do not understand the importance of physical fitness, some find proper exercise too difficult, and others find it inconvenient.

b. Causes of Deficiencies. There are several causes for Marines being deficient in physical fitness:

(1) Absence of exercise.

(2) Exercise which fails to develop all muscle groups and components of fitness.

(3) Exercise which is not vigorous enough or which lacks progression.
(4) Injury or illness which depletes fitness.
(5) Inadequate amounts of sleep or rest.

c. Methods of Detection. The commander has several means by which to detect physical deficiencies:
(1)
from those who pass. Make an analysis to determine the cause of failure as revealed by test scores.
(2) Observe Marines as they perform physical tasks-
o Marines who have difficulty during training or in physical types of work,
o Marines who have become obese and therefore experience difficulty.
(3) Be alert to those Marines who are often on sick call, returnees from hospitalization, or newly assigned.

2203. ADMINISTRATION OF REMEDIAL ACTION

a. Group. Attitude Marines who are singled out as being physically deficient are self-conscious and are not always convinced they need extra help. Within a deficient group, motivation may be low. These Marines must be convinced that a remedial program which is tailored to their needs will help them and will eliminate their deficiencies. Encouragement is often needed and desirable.

b. Leadership. The leader of this group must study individuals and know their deficiencies. The unit leader should counsel Marines individually, observe them closely as they progress through the remedial program, maintain records, and adjust the program as required.

c. Measurement.. Whenever improvement in performance is noted, Marines should be measured by the physical fitness test either individually or by group. When testing reveals an individual to be satisfactory, that individual should be released from the remedial group. There may be exceptions to this policy in the case of Marines who are overweight or in the execution of an exercise program prescribed for some special purpose.

d. Organization of a Remedial Group
(1) A remedial group is usually a small group of Marines within a company or battalion. In some situations, it may be a larger group numbering 50 to 75 Marines. In the case of a larger group, divide the Marines initially into subgroups according to ability. Prescribe exercise loads commensurate with their ability. General conditioning is sufficient in the early part of the program to qualify Marines who are on the borderline. Qualification will reduce the size of the group, permitting more individual and specialized attention to those who remain.
(2) As the program progresses, regroup individuals who have like deficiencies in order that they can concentrate on their weakness. For example—

(a) One group may be weak in the arms and shoulders as revealed by failure in the chin up event. This group, with an assistant instruction charge, can work on pull-ups, rope climb, pushups, rifle or logexercises, horizontal ladder, and similar types of exercise.

(b) Another group may be weak in general endurance as revealed by thc3-milc run. This group could profit from participation in conditioning drills, running, grass drill, and strength circuit.

(3) It may be necessary to form some groups to overcome weaknesses in skills, such as an inability to throw, to quickly change direction while running, to crawl rapidly, or to carry a load. Lack of coordination or lack of practice may cause these deficiencies. In this instance, instructors must provide an opportunity to practice and correct poor form and other errors as they are noted.

Chapter 3
PHYSICAL CONDITIONING ACTIVITIES
Section I. Marching Under Load

3101. GENERAL

Few physical fitness activities are as directly related to readiness for combat as foot marches under load. In addition to the obvious physical conditioning and unit cohesiveness benefits, marches under load prepare most Marines for numerous foreseeable tasks in combat. The idea that only infantry and reconnaissance units actually require training marches to prepare them for combat is clearly a mistake. In addition to the very real possibility of Marines from all types of units being used to fill combat-depleted infantry units, it is likely in maritime prepositioning force operations that aviation, combat service support, and command element personnel will have to march with weapons and equipment, from the arrival airfield to the equipment marshalling site. In American coastal cities, the harbor is nearly always 10 or more miles from the airport. It is difficult to imagine any different condition in a lesser developed country .A further combat-related benefit is that foot marching under load is the most effective way to develop leg strength in the context of a unit training program. In his classic study of troop performance in World War II, Men Against Fire, S.L.A. Marshall noted that the greatest single weakness of replacements of all occupational fields was leg strength. Truck drivers must push trucks which are stuck in mud, and headquarters communicators must climb hills to emplace antennas. The demands on the lower body in combat apply to all Marines, and foot marching is one activity which can help build these muscles while being conveniently integrated with the rest of the training program.

3102.TRAINING GOALS

The standard for success of a foot march is very simple to measure: did the unit arrive at the destination at the prescribed time with Marines in condition and required equipment present to accomplish the mission?A progressive program can increase a unit's readiness. It can instill pride by increasing distances and rates of march and by selecting routes over increasingly challenging terrain as the Marines become better conditioned. However, the ability to execute the mission at the conclusion of the march must remain the standard of success.

3103.MOVEMENT PLANNING

The basic considerations in planning a foot march are the mission, tactical situation, terrain and weather, and the units to participate. The success of the march will depend largely upon the thoroughness with which it is planned. A successful march is characterized by adherence to prescribed routes and time schedules, the efficient employment of the means available, and the ability of the unit to accomplish its assigned mission upon arrival at the destination.

a. Movement Orders. Movement planning culminates in the preparation and issuance of an operation order prepared in the standard, five-paragraph format. Necessary annexes are attached to furnish detailed information required for the movement. Written movement orders are rarely prepared at company level. A discussion of the annexes is contained in this chapter.

b.March Planning. March planning, as discussed here, is the planning conducted at battalion level. March planning may be organized into the following steps:

(1) **Preparation and Issuance of the Warning Order.** In order to afford subordinate units the maximum possible time to prepare for a pending move, a warning order containing all available information about the march is issued. The amount of planning time available will determine the time of issuance and the content of the warning order.

(2) Estimate of the Situation. In his estimate, the commander considers the mission, terrain, weather, time, and space factors, available routes, available transportation for the movement of equipment and/or shuttling of Marines, enemy capabilities, disposition of own forces, physical condition and training of Marines, and courses of action available to the command.

(3) Organization and Dispatch of a Reconnaissance. Party very march plan is based on as thorough and complete a ground reconnaissance as time and the situation will permit. Map and aerial reconnaissance are valuable in formulating a plan, but are not a substitute for ground reconnaissance. Route reconnaissance is accomplished by a reconnaissance party which usually consists of a reconnaissance element, an engineer element from the attached or supporting engineer unit, and a traffic control element. Unit standing operating procedures generally establish the basic composition of the reconnaissance party. It is modified as necessary to meet the requirements of a particular march. A recommended method of reporting information obtained by the route reconnaissance party is shown in figure 3-1.The minimum information required from the reconnaissance party is—

(a) Available routes and conditions.(Routes may be specified by higher headquarters.)

(b) Recommended rate of march.

(c) Selection of start point and release point, or confirmation of the suitability of start point and release point previously selected by map reconnaissance.

(d) Confirmation of location of the assembly or bivouac area.

(e) Location of critical points on the route.

(f) Distance between critical points on the route and total distance from start point to release point.

Figure 3-1 is available in the illustration section at the end of the book

(g) Location of obstacles and estimation of necessary Marines and equip- ment needed to repair and maintain routes.

(h) Number of guides required and their location on the route.

(4) Development of Detailed Movement Plan. Some of the elements included in a detailed movement plan arc-

(a) Organizing the Column. To facilitate control and scheduling, units will be organized into serials and march units and given an order of march. In determining the order of march, the march planner must consider the enemy situation and the desirable order of arrival of the units at the destination. Where dispersion is required, a unit may be organized into two or more columns, each assigned a different route.

(b) Using Reconnaissance Information. Results of the route reconnaissance will be used to select the route(s); determine the start point, critical points along the route(s), and the release point; and select the rate of march.

(c) Determining March Computations. March unit pass time is based on the strength, formation, and rate of march. The pass time of the marching columns, plus necessary time distance computations, will be used to determine the completion time of the march.

(d) Drafting of Road Movement Table. Using the completed march computations, a draft road movement tabic is compiled.

(e) Checking the Plan. Using the draft road movement table and a road movement graph, the movement plan is checked to ensure that it conforms to the directive of the higher headquarters and the battalion commander's instructions.

(5) Preparation and Issuance of the Road Movement Order. After the plan has been checked and approved by the commander, an operation order is prepared and issued. The operation order may be in written form or issued orally, and is accompanied by a road movement table, overlay and/or strip map, and appropriate administrative details.

(6) Road Movement Table. A road movement table, prepared as an annex to the operation order, provides serial commanders with arrival and clearance times at critical points along the route of march. It also provides the column commander with information as to the proposed location of elements of the column at various times.

(7) Overlay/Strip Map. An overlay serves the normal purpose and should show, as a minimum, the present location of units, route of march, critical points, and the new location of units at the destination. A strip map is a schematic diagram of the route of march and shows landmarks and critical points with the distances between them. A strip map may be issued as an annex to the road movement order, in addition to or in lieu of an overlay.

c. March Computations. Prior to issuance of the operation order, the S-3 must verify time and space computations as they provide him with the necessary data for the preparation of a road movement table.

(1) Time-Distance. Time distance (TD) is determined by dividing the distance to be traveled (D) by the rate of inarch (R):

TD = D/RWhere TD = hours D = distance in kilometers R = kilometers per hour

(2) Length of ColumnThe length of column (LC) is used to determine the pass time of a column. The sum total of the following two parts determines the length of column. (See pars, (a) and (b)).The two parts are the space occupied by the Marines alone (including the distance between Marines) and the sum of the distances between the units of the column (column gap).

(a) The length of column of Marines alone is determined by multiplying the number of Marines by the appropriate factor selected from the table below. The length of column does not include distances between units. LC Marines = No. of Marines x factor.

(b) The total distance in meters between units is determined as follows:

Step One. Determine the number of gaps between serials (total serials minus one).

Step Two. Multiply the number of serial gaps from Step One by the length (in meters) between respective units.

Step Three. Determine the number of gaps between march units (total march units minus one, minus the number of serial distances).

Step Four. Multiply the number of march unit gaps from Step Three by the length (in meters) between respective units.

Step Five. Add the totals from Step Two and Step Four to get the total meters for the column gap.

EXAMPLE:A battalion foot column is organized into 12 platoon-sized march units and 3 company-sized serials. REQUIRED: total column gap distances when there are 100 meters between serials and 50 meters between march units.

(3)Pass-Time.Pass-time (PT) is the time a unit takes to pass a specified point. For foot columns, the pass-time is determined by applying the following formula:PT (minutes) = LC x FACTOR (for appropriate rate of march).

PASS-TIME FACTORS - FOOT TROOPS

.0150 for 4.0 km/h

.0187 for 3.2 km/h

.0250 for 2.4 km/h

.0375 for 1.6 km/h

EXAMPLE: Determine the pass-time of a unit whose length of column is 1,500meters and is marching at a rate of 4 km/h.PT (min) = 1,500 x .0150 (the factor for 4.0 km/h) = 22.5 mm.

(4) Completion Time. Completion time is the time of day that a march will be completed. Completion time is determined by using the following formula: Completion Time = SP time + TD + PT + Scheduled Halts (other than normal hourly halts).

EXAMPLE: A column's starting point (SP) time is 0700.The time-distance is6 hours and 40 minutes. Pass-time of the column is 30 minutes. A 35-minutelunch halt has been scheduled. What is the completion time of the march?Employing the 24-hour clock system, the formula can be applied as follows for simplified addition of the times:

HrMin SP time.................0700

TD......................0640

PT......................0030

Lunch halt..............0035

Completion time.........13 105or 1445

The march will be completed atl445.

(5) Experience Tables. Based on previous movements made by a unit, data is accumulated to facilitate march planning. Such data includes approximate pass-times lor various elements of the battalion. The S-3 can utilize these data rather than computing them each time a march is scheduled. Such experience tables serve to reduce the time required to complete the computation phase of march planning. Matter appropriate to the unit standing operating procedure should be integrated therein.

d. Road Movement Graph

(1) A road movement graph is a time-distance diagram used in planning, preparing, or checking road movement tables, and for controlling marches. The graph provides a visual representation of a march plan so that conflicts and discrepancies may be prevented in the planning stage, before congestion occurs on the route. It is not usually issued as a part of the order. Road movement graphs may be applied to small units, to movements of a single column, or to a large organization scheduling separate elements, moving by various means, with different rates of march, over one or more routes.

(2) To construct a road movement graph, use the following steps: (See fig. 3-2.)

Figure 3-2 is available in the illustration section at the end of the book

Step One. Determine the number of hours available for the march. Designate the lower left corner of a sheet of graph paper as the SP time or the earliest even hour before the march is to begin. Select a scale and plot the hours available in sequence from the left on the horizontal axis.

Step Two. Determine the distance to be moved in kilometers. Indicate the SP at the lower left corner of the graph sheet and, using an appropriate scale, plot the number of kilometers on the vertical scale from the SP to the release point. Indicate the location of critical points on the vertical scale.

Step Three. At the proper distance from the SP, draw a horizontal line indicating the location of the RP. Indicate by a vertical line the hour when the movement must be completed. Plot lines representing route restriction, if any, at the proper distances and times on the graph.

Step Four. Determine the pass-times of foot and motor elements of the column.

Step Five. Starting at the SP at the specified hour, plot the movement of the head of the leading element. If the rate of march is 3 kilometers per hour, the lead element will have moved 1.5 kilometers in 30 minutes, 3 kilometers in 1 hour, etc. Plot the trace of the lead man to the RP. The last Marine will cross the SP in pass-times after the lead man. Measure this time on the graph and plot the trace of the last Marine of the column. The lines describing the head and tail of a march serial are parallel. Indicate the time subsequent serials reach the SP and plot the trace of the head and tail. Check to see that the plan complies with all restrictions and orders. If there are any violations or conflicts, the plan may be made to conform by changing the order of march, the starting time of the serials, the planned rate of march, or the organization of the column.

e. Conversion to the Metric System. The adoption of the metric system for expressing linear distances requires a simple means of conversion.

(1) To accomplish this, use the following tables:

o Multiply inches x 2.54 to obtain centimeters. Multiply centimeters x .39 to obtain inches,

o Multiply yards x .91 to obtain meters. Multiply meters x 1.1 to obtain yards,

o Multiply miles x 1.6 (or 8/5) to obtain kilometers. Multiply kilometers x .62 (or 5/8) to obtain miles.

(2) Linear distances can be accurately measured on maps which have been converted to the metric system by using a straightedge graduated in centimeters (cm).The centimeter scale can be used in lieu of the graphic scale when the scale of the map is shown, as follows:

Map Scale:	One cm Equals:
1:500,000	5,000 meters
1:250,000	2,500 meters
1:100,000	1,000 meters
1:50,000	500 meters
1:25,000	250 meters

3104. INDIVIDUAL LOAD

Since the backbone of Marine ground combat forces is the infantry, tactical mobility is largely dependent on the marching ability of the individual Marine. The load that each Marine must carry is the greatest impediment to mobility. In general, the commander must make every effort to reduce the individual load of his Marines to the absolute minimum. While the load has the greatest effect on Marines marching on foot, the value of carrying the minimum essential supplies and equipment applies for all Marine forces.

a. Fighting Load. The fighting load consists of items of clothing, equipment, weapons, and ammunition that are carried by, and essential to, the effectiveness of the combat Marine and the accomplishment of the immediate mission. The fighting load should not weigh more than 40 pounds. The commander must ensure that his Marines' loads- -and his own- -be stripped to the bare minimum. The addition of small, light-weight items in excess of the limit should not be tolerated, since cumulatively, these items will take a toll in energy. Every extra pound a Marine must carry decreases his combat effectiveness.

b. Existence Load. The existence load consists of items other than those in the fighting load that are required to sustain or protect the combat Marine, that may be necessary for increased personal and environmental protection, and that are not normally carried by the individual. The existence load is normally carried by the Marine's unit. Marching, but not engaged in combat, a Marine may be expected to carry a maximum of 50 pounds of supplies and equipment.

c. Considerations

(1) The primary consideration is not how much a Marine can carry, but how much he can carry without impaired combat effectiveness-moral or physical. The combat strength of a unit is not counted simply in numbers of Marines, but in the numbers of willing and physically able Marines.

(2) It is generally better to risk temporary inconvenience from lack of health and comfort items than to exhaust Marines due to overloading.

(3) It is a fundamental truth that men become physically exhausted more quickly when under the stress of combat. Marines must be required to carry less into battle than they are conditioned to carry in training. Marines should be conditioned for carrying weight, but should be equipped in combat for fleetness of foot.

(4) A common mistake is to base the fighting load on the gear and supplies necessary to meet every contingency. The commander cannot reasonably expect the Marines to carry enough gear for every possible eventuality. The items to be carried must be based on reasonable expectations.
(5) It is the commander's responsibility to procure the transport to carry any additional gear. As a rule of thumb, a rifle company, or unit of similar size, requires one 5-ton truck and trailer in support to carry additional gear. In cold weather, or other conditions in which the necessary personal gear increases, this requirement will increase.
(6) The commander must ensure that the supply system provides, on a dependable and timely basis, the balance of essential supplies and equipment not carried by the unit. Marines must feel confident that they will be supported with the necessary supplies and equipment.
(7) In training, commanders must implant in their Marines the pride in operating under austere conditions. To effectively demonstrate the need for this spartan mentality, the commander must personally set the standard. Marines should be trained in field craft, foraging techniques, and the use of caches and field expedients. Maximum use should be made of captured stores.

3105. MARCH TECHNIQUES AND PROCEDURES

a. Organization. A command executing a march is organized into march units, march serials, and march columns, as necessary for control.
(1) March Units. A march unit is a unit of command which moves and halts at the command of a single commander. The march unit normally corresponds to one of the smaller Marine units such as a platoon or company.
(2) March Serials. A march serial consists of one or more march units organized under the senior officer and given a numerical or alphabetical designation to facilitate control. A serial is usually a battalion, but may be a company if the battalion is marching alone.
(3) March Columns. A march column is composed of elements of a command moving over the same route. It may be composed of one or more serials. To facilitate control, a column commander is designated. A column is normally a regiment or larger unit, but may be a battalion if marching alone.

b. Order of March. The order of march is determined by the mission, terrain, and the commander's desired order of commitment of units into action.

c. Control and Coordination Measures. The commander establishes initial control of the march by designating control measures in his march order. The most frequently used measures include:

- Start point and release point.
- Time at which head or tail of column passes the start point.
- Rate of march.
- Order of march.
- Route of march.
- Location of command post.
- March communications.

d. The Pacesetter. The pace-setter is an experienced individual carrying the same load as the majority of the Marines and marching from 4 to 10 meters at the head of the column. (See fig. 3-3.)The pace-setter's primary duty is to maintain the rate of march ordered by the column commander.

(l) Setting the Pace. The pacesetter does this by establishing the pace (length of step) and cadence (steps per minute) to obtain the prescribed rate of march. He should be of medium height so normal strides will be taken. Overstriding or understriding tends to tire leg muscles quickly and affects the combat efficiency of marching Marines. The officer marching at the head of the column supervises the pacesetter to ensure that he takes normal strides and maintains a uniform cadence.

Figure 3-3 is available in the illustration section at the end of the book

(2) Reducing Accordion Effect. Accordion effect occurs when the column alternately lengthens and shortens, causing the rear elements of the column to double time to maintain distance. It is caused by changes in the rate of march by the lead elements, usually after passing a slowing obstacle. To reduce this effect as much as possible, lead elements should slow their rate of march for a sufficient time after crossing obstacles to allow the rear elements to maintain the prescribed distance without running. The rear elements can signal the head when the last man has cleared the obstacle and the rate can be resumed. Since some accordion effect is nearly inevitable, the order of march should be periodically rotated to prevent units in the rear from becoming physically exhausted at the conclusion of the march.

e. Length, Rate and Pace

(1) **Length of March.** The length of march varies depending on terrain and weather, enemy situation, and physical and mental condition of Marines. The normal length of march for a 24-hour period is from 20 to 32 kilometers (12 to 20 miles) marching from 5 to 8 hours at a rate of 4 kilometers (2.5 miles) per hour (km/hmi/h).A march in excess of 32 kilometers in a 24-hour period is considered a forced march. Well-trained units that have been progressively trained will be able to achieve a unit goal of 60 miles in a 3-day period.

(2) Rate of March. The same factors of terrain, weather, load to be carried, and condition of Marines affect the rate of march. The commander considers these and selects a rate which will place his unit at the destination in the shortest possible time in condition to accomplish the mission. Unit standing operating procedure should state the standard rate of march for that unit over normal terrain for both road and cross-country movement and in both day and night. The commander then modifies this rate if the situation requires. Normally prescribed standards are as follows:

Figure is available in the illustration section at the end of the book

(3) Pace and Cadence. The normal pace is 30 inches. A pace of 30 inches and a cadence of 106 steps per minute result in a speed of 4.8 kilometers per hour or 3 miles per hour and a rate of 4 kilometers per hour or 2.5 miles per hour if a 10-minute rest halt per hour is taken. Since the pace of each man may vary, the cadence may have to be adjusted to maintain the prescribed rate. Slope of ground and surface footing affect pace. A normal cadence is relatively simple to maintain on level or gently rolling terrain on firm footing. Mud, sand, loose gravel, and hills all greatly affect pace.

f. Halts. Halts during a day's march are taken at regular intervals to rest personnel and adjust loads.Halts are specified by standing operating procedures or march order. Such factors as midday heat or enemy action may force the adoption of long midday halts or night marches. At long halts, each unit disperses to a previously selected location near the route of march.

(1) Normal Time and Frequency. Under normal conditions, a 15-minute halt is made after the first 45 minutes of marching. Following the first halt, a 10-minute halt is made after each 50 minutes of marching. Variations of this schedule are made when a scheduled halt time occurs when passing through a built-up area or when cover and concealment are required by the tactical situation and none is available. Observation posts may be established, if required for security of the unit during halts.

(2) Actions at Halts. All units in the column should be halted at the same time. At the halt signal, Marines should move to the side of the road, staying within the immediate vicinity of their unit. Marines should remove or loosen gear and sit or lie with feet elevated. Commanders inspect Marines and equipment, and corpsmen administer medical treatment, as required.

g. Communications. The four primary means of communications used in controlling foot marches are radio, visual, sound, and messenger.

(1) Radio. Radios are normally used for communications between platoons and higher headquarters in the march column. Radio transmissions should be held to the absolute minimum necessary for control and should be at minimum power required.

(2) Visual.. Visual communications may include flashlights, luminous markers, panel sets, pyrotechnics, and hand and arm signals. When contact with the enemy is not expected, the loss of light discipline from some of these signals may be a lesser security risk than the risk of disclosure from radio transmissions.

(3) Sound. Sound communications include voice and such signaling devices as horns and whistles. Some of their uses include assembling Marines at the conclusions of halts and providing warnings of air or chemical attacks.

(4) Messengers. Messengers are particularly effective during periods of radio silence and reduced visibility. Messages should be simple and brief to preclude mistakes by the messenger.

h. Security. Both active and passive measures are used against attack by enemy aircraft and indirect fire weapons during movement. Active measures include the use of air sentries and organic and attached weapons in accordance with the unit air defense standing operating procedure. Passive measures include the use of concealed routes and assembly areas, night marches, and extended interval between elements of the column. With imagination and planning, all of these measures can be practiced in training without detracting from the physical conditioning benefit of the march. In fact, such drills as rapidly taking cover during simulated attacks and then quickly resuming the march can prove physically challenging.

i. Reconnaissance. Reconnaissance determines in advance critical points along the march route such as bridges, fords, and obstacles in order that congestion or delay may be held to a minimum and local security provided. Paragraph 3103.C describes organization of the reconnaissance party.

3106.TACTICAL MARCHES

These marches are movements of Marines and equipment not in direct ground contact with the enemy but expecting early ground contact either en route or upon arrival at the objective. Some characteristics of this type of movement include security elements to the front, flanks, and, if necessary, to the rear of the march column. Additional considerations include the selection of concealed areas on defendable terrain for start and release points and rest stops. In training, infantry units of regimental size or smaller will normally derive the greatest benefit from tactical marches. The security requirements usually dictate a slower rate of march and, therefore, reduce the physical conditioning value for other types of units which would seldom move in combat with the expectation of early ground contact with the enemy. Two key points should be remembered regarding tactical marches in a training program.

a. Security. Although security requirements often slow the rate of march, it is a valid and important training objective for infantry units to work toward reducing march times while still practicing sound tactical security. March standing operating procedures, good land navigation skills, and simple but complete march orders contribute as much to this goal as physical conditioning. The idea that proceeding slowly enhances security is seldom true. Keeping the enemy off balance by rapid movement frequently is the best guarantee of security.

b. Training. In peacetime training, it is critical that support personnel who normally are located with infantry units in combat, participate completely with infantry units in forced march training. Not only must they not slow the infantry rate of march because of inferior conditioning, but they must be thoroughly familiar with the infantry unit's standing operating procedure for security on the march. Personnel from within the ground combat element such as artillery forward observer teams and combat engineer units, can be easily identified for this training and should be habitually associated with the same infantry units. Personnel from other elements, such as low-altitude air defense teams and radio battalion detachments, are more difficult to identify and earmark for support of the same infantry unit. Their normal mode of employment is in vehicles and their units frequently provide general support vice direct support. This does not, however, lessen their requirement to be prepared to provide support to dismounted infantry over terrain or in a tactical situation not conducive to vehicles. Such training needs to take place, and planning and cooperation among the elements of the MAGTF can overcome the obstacles.

3107.ADMINISTRATIVE MARCHES

These marches are movements of Marines and equipment made when no enemy interference is expected except from aircraft or indirect fire. During these marches, units are administratively grouped for ease of control and speed of movement. The term administrative foot marches should not be taken to imply that tactical considerations are disregarded. There are simply less stringent security measures used than when ground contact with the enemy is a distinct possibility. The normal formation is the route column with one file moving on either side of the road and with negligible security. Air sentries are used and dispersion between individuals and units is practiced. Although roads usually provide the fastest route, there may be reasons to conduct these marches cross-country (particularly as part of a training program). Administrative marches are the type of foot marches most practical for training programs for non-infantry units. These marches may also play a key role in the physical conditioning program of infantry units, because of the greater speeds possible due to lessened security requirements.

3108. FORCED MARCHES

A forced march is a march which covers a greater distance than 32 kilometer (20 miles) in a 24-hour period. Normally the extra required distance should be achieved by increasing the number of hours marched in a day beyond the norm (8 hours) rather than increasing the rate of march beyond the norm (4 kilometers per hour/2.5 miles per hour). However, there will be occasions when the situation or mission demands an increase in the rate of march.

a. Maximum Recommended Distances. The maximum recommended distances for forced marches are-56 kilometers (35 miles) in 24 hours; 96 kilometers (60 miles) in 48 hours; or 128 kilometers (80 miles) in 72 hours.

b. Sample Time Breakdown. A sample time breakdown for a forced march of 52 kilometers, beginning at first light is as follows:

Figure is available in the illustration section at the end of the book

3109. NIGHT MARCHES

Night marches are characterized by closed formations, more difficult control and reconnaissance, and a slower rate of march than day marches.

a. Control. Control is increased by reducing the distance between individuals and units, and by using connecting files to maintain contact between platoons and companies. Connecting files are normally constituted from personnel from the rear most march unit.

b. Safety on Roads. Because of reduced visibility, night marches on roads used by vehicles require attention to safety procedures to prevent accidents. Commanders should not think these safety measures apply only to peacetime training. In combat, nearly all vehicles operate with blackout lights at night, making visibility even more difficult than in garrison driving. If consistent with the tactical situation, the following measures can reduce the possibility of marchers being struck by vehicles:

(1) Use off-road trails and routes as much as possible.

(2) Guards to the front and rear of march units should be marked with strips of luminous tape and carry red-filtered flashlights. These measures are generally consistent with light discipline, while still affording reasonable warning to approaching drivers.

Section II. Conditioning Drills (One, Two, and Three)

3201. GENERAL

a. Description and Objective. Conditioning drills are calisthenic exercises. Each drill contains seven exercises organized and numbered in a set pattern. Each drill takes 15 minutes to complete. The objective of conditioning drills is to exercise all major muscle areas in order to develop strength, endurance, coordination, and flexibility.

b. Area and Equipment. Any level area is satisfactory for conduct of the drills. Drills One and Two contain ground exercises. If ground conditions are unsatisfactory, Drill Three should be used as it contains no ground positions. Usually, no equipment is required;however, if the group exceeds a platoon in size, an instructor's stand is necessary.

c. Formation. The extended rectangular formation is used in this drill. (See app. A.)

d. Starting Level and Progression. The starting level is six repetitions of each exercise. An increase of one repetition for each three periods of exercise in which the drill is performed is an acceptable rate of progression. This rate is continued until 12 repetitions can be completed. To maintain, continue the drill at 12 repetitions. To progress, move to 6 repetitions at a more difficult drill which exercises the same muscle group. Progression can also be gained by moving from Drill One to Drill Two, as Drill Two is more demanding.

e. Starting Positions. Starting positions vary with the exercise and are explained in each exercise.

f. Leadership. A principal instructor demonstrates and leads the drills. The instructor must be familiar with leadership techniques peculiar to conditioning drills to include the exercises, commands, counting cadence, cumulative count, formation, method of teaching the exercises, and utilization of assistant leaders.

g. Use With Other Programs. Conditioning Drills One, Two, and Three reach all major muscles of the body. They are easy to learn and to perform, and they are simple to administer and supervise. These features, coupled with the short time required for completion, the fact that no equipment is necessary, and adaptability to most areas of execution, make these drills possible in any programs.

3202. CONDITIONING DRILL ONE

This conditioning drill is similar to the calisthenics drill commonly known throughout the Marine Corps as the "Daily Seven."The "Daily Seven" may be used as a substitute for Conditioning Drill One.

a. Exercise l:High Jumper

(1) Starting Position. Feet separated shoulder width, knees flexed, body bent forward at the waist, arms aligned with the trunk and hips, elbows locked, palms facing, fingers extended and joined, head and eyes to the front.(See fig. 3-4, A.)(Elbows remain locked throughout the exercise.)

Figure 3-4 is available in the illustration section at the end of the book.

(2) Cadence. Moderate.
(3) Movement. A four-count exercise: at the count of—

(a) ONE -- Take a slight jump into the air, swinging the arms forward and up to shoulder level.

(b) TWO -- Take a slight jump into the air and swing the arms downward and back, returning to the starting position.

(c) THREE -- Take a vigorous leap into the air, swinging the arms forward and up to an overhead position, momentarily looking skyward, on returning to the ground the knees are flexed, head and eyes return to the front.

(d) FOUR--Repeat the action of count two.

b. Exercise 2: Bend and Reach
(1) Starting Position. Feet spread more than shoulder width, arms overhead, elbows locked, palms facing, fingers extended and joined, head and eyes to the front.(See fig. 3-4, B.)
(2) Cadence. Moderate.
(3) Movement. A four-count exercise: at the count of—

(a) ONE-Bend at the knees and waist, swing the arms straight downward and reach between the legs. Touch the ground as far to the rear as possible and look to the rear. (Elbows remain locked throughout the exercise).

(b) TWO-- Recover sharply to the starting position.

(c) THREE--Repeat the action of count ONE.

(d) FOUR-- Repeat the action of count TWO.

c. Exercise 3: Pushup
(1) Starting Position. Front leaning rest position. To assume this position there is a silent one-two count:on the silent count of one, assume the squatting position, heels together, elbows locked inside the knees, hands flat on the ground directly beneath the shoulders; on the silent count of two, thrust the legs to the rear, toes and heels together, body straight from head to heels.(Seefig. 3-4, C.)
(2) Cadence. Moderate.
(3) Movement. A four-count exercise: at the count of—

(a) ONE -- Flex the elbows lowering the body until the thick portion of chest touches the ground.

(b) TWO -- Raise the body until elbows are straight and locked.

(c) THREE--Repeat the action of count ONE.

(d) FOUR -- Repeat the action of count TWO. (On returning to position of attention, the silent one-two count is used in reverse).

d. Exercise 4: Trunk Twister

(1) Starting Position. Feet are spread more than shoulder width apart, fingers placed behind neck, thumbs pointing downward, elbows back. (See fig. 3-4, D.)(Elbows remain well back throughout the exercise).

(2) Cadence. Slow.

(3) Movement. A four-count exercise: at the count of-

(a) ONE-- Keeping the knees locked and the back straight, bend forward at the waist sharply, with a slight recovery.

(b) TWO-- Twist the trunk to the left vigorously at the waist, keeping the elbow back. The left elbow is higher than the right.

(c) THREE-- Twist vigorously to the right, so the left elbow comes under the right.

(d) FOUR-- Straighten sharply to the starting position. NOTE: Do not attempt to touch the elbows to the knees on counts two and three.

e. Exercise 5: Squat Bender

(1) Starting Position. Feet are spread less than shoulder width apart, hands on hips, thumbs in small of back, elbows back. (See fig. 3-4, E.)

(2) Cadence. Moderate.

(3) Movement. A four-count exercise: at the count of—

(a) ONE-- Assume the squatting position, maintain balance on the balls of the feet, with trunk erect thrust arms forward to shoulder level, elbows locked, palms down.

(b) TWO-- Recover to starting position. Elbows are well back.

(c) THREE-- Keeping the knees locked, bend forward at the waist, touching the ground in front of the toes.

(d) FOUR-- Vigorously recover to the starting position.

f. Exercise 6: Body Twist

(1) Starting Position. On the back, arms extended sideward on the ground, palms down. The legs are raised to a near vertical position, feet together, knees locked.(See fig. 3-4, F.)

(2) Cadence. Slow-fast.

(3) Movement. A four-count exercise: at the count of-

(a) ONE-- Lower legs slowly to the left until they touch the ground near the left hand, keeping the knees straight and shoulders on the ground.

(b) TWO-- Recover the starting position by quickly raising the legs, keep knees straight and feet together.

(c) THREE-- Repeat movement of count ONE, except the movement is to the right side.

(d) FOUR-- Recover sharply to the starting position.

g. Exercise 7: Stationary Run

(1) Starting Position. Position of attention. (See fig, 3-4, G.)

(2) Cadence. Fast.

(3) Movement.

(a) At the command of execution, start running in place at double-time cadence, lifting the left foot first time cadence. Follow the instructor as he counts two repetitions of cadence;

e.g., 1, 2, 3, 4--1, 2, 3, 4. The instructor then gives informal commands such as FOLLOW ME, running on the toes and balls of the feet, keeping the back straight, speeding up the cadence to a sprint, raising the knees high, leaning forward at the waist, and pumping the arms vigorously.

(b) To halt the exercise, the instructor will count two repetitions of cadence as the left foot strikes the ground: 1, 2, 3, 4 - 1, 2, 3, HALT.

NOTE: When counting cadence, the instructor counts only as the left foot strikes the ground. The duration of the exercise is approximately 1 1/2 minutes.

3203. CONDITIONING DRILL TWO

a. Exercise 1: Jumping Jack

(1) Starting Position. Feet separated more than shoulder width, arms overhead. (See fig. 3-5, A.)

(2) Cadence. Moderate.

(3) Movement. A four-count exercise: at the count of~

(a) ONE-- Jump to position with the feet together and assume the squatting position, swinging the arms sideward and downward, placing the hands palms down on the ground, elbows locked inside the knees.

(b) TWO-- Recover to the starting position by jumping to the side straddle and swinging the arms sideward overhead.

(c) THREE-- Repeat the action of count ONE.

(d) FOUR-- Recover to the starting position.

b. Exercise 2: Turn and Bend

(1) Starting Position. Side straddle, arms overhead. (See fig. 3-5, B.)

(2) Cadence. Moderate.

(3) Movement. A four-count exercise: at the count of-

(a) ONE-- Turn the trunk to the left and bend forward over the left thigh, attempting to touch the fingertips to the ground outside the left foot. Keep the left knee straight. On successive repetitions attempt to touch farther and farther to the side.

(b) TWO-- Recover to the starting position.

(c) THREE-- Turn the trunk to the right and bend forward over the right thigh, trying to touch the hands to the ground outside the right foot. Keep the right knee straight.

Figure 3-5 is available in the illustration section at the end of the book.

(d) FOUR-- Recover to the starting position.

c. Exercise 3: Eight-Count Pushup

(1) Starting Position. Position of attention. (See fig. 3-5, C.)

(2) Cadence. Moderate.

(3) Movement. An eight-count exercise: at the count of—

(a) ONE-- Assume the squatting position, palms on the ground directly beneath the shoulders, elbows locked inside the knees.

(b) TWO-- Thrust the legs to the rear assuming the front leaning rest position.

(c) **THREE**-- Flex the elbows until the thick portion of the chest touches the ground.

(d) **FOUR**-- Raise the body on a straight plane until the elbows are locked.

(e) **FIVE**-- Repeat the action of count THREE.

(f) **SIX**-- Repeat the action of count FOUR.

(g) **SEVEN**-- Recover to the squatting position as in count ONE (elbows locked inside the knees).

(h) **EIGHT**—Return sharply to the position of attention.

d. Exercise 4: Turn and Bounce

(1) Starting Position. Feet spread more than shoulder width apart, arms sideward at shoulder level, palms up. (See fig. 3-5, D.)

(2) Cadence. Slow.

(3) Movement. An eight-count exercise: at the count of—

(a) **ONE**-- Turn sharply to the left as far as possible, then recover slightly.

(b) **TWO**-- Again turn to the left as far as possible and recover as in ONE.

(c) **THREE**-- Repeat the action of count TWO.

(d) **FOUR**-- Recover sharply to the starting position.

(e) **FIVE**—Turn sharply to the right as far as possible, then recover slightly.

(f) **SIX**—Again turn to the right as far as possible and recover as in FIVE.

(g) **SEVEN**-Repeat the action of count SIX.

(h) **EIGHT**-Return to the starting position.

NOTE: The head and hips remain to the front throughout the exercise and the knees and elbows are locked at all times.

e. Exercise 5: Squat Stretch

(1) Starting Position. Attention. (See fig. 3-5, E.)

(2) Cadence. Moderate.

(3) Movement. A four-count exercise: at the count of—

(a) **ONE**-- Squat, placing the hands on the ground about 12 inches in front of the feet.

(b) **TWO**— Keeping the fingertips on the ground, straighten the knees completely and raise the hips.

(c) **THREE**-- Recover to position ONE.

(d) **FOUR**-- Recover to the starting position.

f. Exercise 6: Leg Circular

(1) Starting Position. On the back, arms stretched sideward, palms down, feet raised foot from ground, knees straight. (See fig 3-5,F.)

(2) Cadence. Slow.

(3) Movement A four-count exercise: at the count of—

(a) **ONE**-- Swing the legs as far as possible to the left, keeping the knees straight and the legs together.

(b) **TWO**-- Swing the extended legs over head with the thighs as close as possible to the trunk.

(c) **THREE**-- Swing the legs as far as possible to the right.

(d) **FOUR**-- Recover to the starting position.

g. Exercise 7: Stationary Run

(1) Starting Position. Position of attention. (See fig. 3-5, G.)

(2) Cadence. Fast.

(3) Movement

(a) At the command of execution, start running in place at double-time cadence, lifting the left foot first. Follow the instructor as he counts two repetitions of cadence; for example: 1, 2, 3, 4--1, 2, 3, 4. The instructor then gives informal commands such as FOLLOW ME. Running on the toes and balls of the feet, keeping the back straight, speeding up the cadence to a sprint, raising the knees high, leaning forward at the waist, and pumping the arm vigorously.

(b) To halt the exercise, the instructor will count two repetitions of cadence as the foot strikes the ground: 1, 2, 3, 4-1,2, 3, HALT.

NOTE: When counting cadence, the instructor counts only as the left foot strikes the ground. The duration of the exercise is approximately 1 1/2 minutes.

3204. CONDITIONING DRILL THREE

a. Exercise 1: Side Straddle Hop

(1) Starting Position. Position of attention. (See fig. 3-6, A.)

(2) Cadence. Moderate.

(3) Movement. A four-count exercise: at count of-

(a) ONE-- Take a slight jump into the air, moving the legs sideward (more than shoulder width apart); at the same time, swing the arms overhead (to an overhead position) clapping the palms together.

Figure 3-6 is available in the illustration section at the end of the book.

(b) TWO-- Take a slight jump into the air, swing the arms sideward and downward returning to the starting position.

(c) THREE-- Repeat the action of count ONE.

(d) FOUR-- Repeat the action of count TWO.

b. Exercise 2: Back Bender

(1) Starting Position. Standing, feet 12 inches apart, fingers placed behind the head. (See fig. 3-6, B.)

(2) Cadence. Slow.

(3) Movement. A four-count exercise: at the count of-

(a) ONE-- Bend the upper trunk backward, raising the chest high, pulling the elbows back, and looking upward. Keep the knees straight.

(b) TWO-- Recover to the starting position.

(c) THREE-- Repeat the action of count ONE.

(d) FOUR-- Recover to the starting position.

c. Exercise 3: Squat Thrust

(1) Starting Position. Position of attention. (See fig. 3-6, C.)

(2) Cadence Moderate.

(3) Movement. A four-count exercise: at the count of—

(a) ONE-- Assume the squatting position; heels together, placing the hands flat on the ground, shoulder width apart, elbows locked and inside the knees.

(b) TWO-- Thrust the legs to the rear, assuming the front leaning rest position, body in line from head to toe, heels and toes together.

(c) THREE-- Return to the squatting position as in ONE.

(d) FOUR-- Return to position of attention.

d. Exercise 4: Side Bender

(1) Starting Position. Feet are spread more than shoulder width apart, arms are raised sideward and overhead, thumbs interlocked palms to front, fingers extended and joined, elbows locked. (See fig. 3-6, D.)

(2) Cadence. Slow.

(3) Movement. An eight-count exercise: at the count of—

(a) ONE-- Bend to left as far as possible, then recover slightly.

(b) TWO-- Again bend to the left as far possible, then recover slightly.

(c) THREE-Repeat the action of count TWO.

(d) FOUR-- Recover sharply to the starting position.

(e) FIVE-- Bend to the right as far as possible, then recover slightly.

(f) SIX-- Again bend to the right as far as possible, then recover slightly.

(g) SEVEN-- Repeat the action of count SIX.

(h) EIGHT-- Recover sharply to the starting position.

NOTE:Keep the elbows and knees locked throughout the exercise. The bend should occur to the side and not the front.

e. Exercise 5: Knee Bender

(1) Starting Position. Feet are spread less than shoulder-width apart, hands on hips, thumbs in small of back, elbows back. (See fig 3-6, E.)

(2) Cadence. Moderate.

(3) Movement. A four-count exercise: on the count of—

(a) ONE-Do a knee bend, lean trunk forward at the waist, thrust arms between legs until the extended fingers touch the ground palms to the ground, hands 6 inches apart.

(b) TWO~Recover sharply to the starting position.

(c) THREE-Repeat the action of count ONE.

(d) FOUR-Repeat the action of count TWO.

f. Exercise 6: Bottoms Up

(1) Starting Position. Front leaning rest position. A silent one-two count is used as in the pushups. (See fig. 3-6, F.)

(2) Cadence. Moderate.

(3) Movement A four-count exercise: at the count of-

(a) ONE-- With the weight on the hands, and knees locked, jump forward bringing the feet as close to the hands as possible; look to the rear.

(b) TWO--Keeping the knees locked, thrust the legs backward assuming the front leaning rest position.

(c) THREE-- Repeat the action of count ONE.

(d) FOUR-- Repeat the action of count TWO.

g. Exercise 7: Stationary Run

(1) Starting Position. Position of attention. (Sec fig. 3-6, G.)

(2) Cadence. Fast.

(3) Movement

(a) At the command of execution, start running in place at double time, lifting the left foot first. Follow the instructor as he counts two repetitions of cadence; for example: 1, 2, 3, 4--1, 2,3, 4. The instructor then gives informal commands such as FOLLOW ME, running on the toes and balls of the feet, keeping the back straight, speeding up the cadence to a sprint, raising the knees high, leaning forward at the waist, and pumping the arms vigorously.

(b) To halt the exercise, the instructor will count two repetitions of cadence as the left foot strikes the ground: 1, 2, 3, 4 - 1, 2, 3, HALT.

NOTE: When counting cadence, the instructor counts only as the left foot strikes the ground. The duration of the exercise is approximately 1 1/2 minutes.

Section III. Rifle and Log Drills

3301. RIFLE DRILL

a. Description and Objective. Rifle exercises are conditioning exercises performed with a rifle. Each drill contains six exercises (fig. 3-7) and they are numbered in a set pattern. The drill takes 15 minutes to complete. The objective of rifle drills is to exercise the arms, shoulders, and back muscles in order to develop strength and endurance, particularly in the upper body. In units without rifles, log drills may be substituted.

b. Area and Equipment. Any level area is satisfactory for conducting this drill. Each Marine completes these exercises from a standing position and no ground contact is required. Each Marine will need a rifle and, if the group exceeds a platoon in size, then the instructor will need an instructor's stand.

c. Formation. The extended rectangular formation is used in this drill. (See app. A.)

d. Starting Positions. Starting positions vary with the exercises and are explained in each exercise. As in all set conditioning drills, the command used to start the exercise is STARTING POSITION, MOVE. The following directions apply to rifle drill.

(1) In those exercises which start from the rifle downward position, on the command MOVE, execute port arms as prescribed in NAVMC 2691, Drill and Ceremonies Manual, and then assume the starting position. The command to return to the men to the position of attention at the conclusion of the exercise is POSITION OF ATTENTION, MOVE.

(2) In exercises which terminate in the rifle downward position, on the command MOVE, execute the position followed by order arms as prescribed in NAVMC 2691.

(3) In the exercises which terminate in a position other than the rifle downward position, Marines first assume the rifle downward position before executing port arms and order arms.

(4) These movements are executed without command. This procedure promotes uniformity, but precision is not expected. To be effective, rifle exercises must be strenuous enough to tire the arms, but not to the point where the arms cannot move with precision.

e. Leadership. A principal instructor demonstrates and leads the drill. He must be familiar with leadership techniques for conditioning exercises and the peculiar techniques for rifle drill.

3302. EXERCISES PERFORMED WITH RIFLES

The exercises of rifle drill are outlined in the following paragraphs,

a. Exercise 1: Foreup, Behind Back

(1) Starting Position. Rifle downward, feet together.(See fig. 3-7, A.)

Figure 3-7 is available in the illustration section at the end of the book.

(3) Movement. A four-count exercise: at the count of—

(a) ONE-- Swing the arms forward and upward to the overhead position. Inhale.

(b) **TWO**-- Lower the rifle to the back of the shoulders. Exhale.

(c) **THREE**-- Recover to position ONE and inhale.

(d) **FOUR**-- Recover to the starting position and exhale.

b. Exercise 2: Lunge Side, Turn and Bend

(1) **Starting Position.** Rifle downward, feet together.(See fig. 3-7, B.)

(2) **Cadence.** Moderate.

(3) **Movement.** An eight-count exercise:at the count of—

(a) **ONE**-- Lunge sidewards to the left, swing the rifle forward and upward to the overhead position.

(b) **TWO**-- Turn the trunk to the left and bend forward over the left hip. At the same time, swing the rifle to a low horizontal in front of the left ankle.

(c) **THREE**-- Recover to position ONE.

(d) **FOUR**-- Recover to the starting position.

(e) **FIVE, SIX, SEVEN, and EIGHT**-- Repeat on the right side.

c. Exercise 3: Foreup, Back Bend

(1) **Starting Position.** Rifle downward, feet together.(See fig. 3-7, C.)

(2) **Cadence.** Moderate.

(3) **Movement.** A four-count exercise:at the count of—

(a) **ONE**-- Swing the arms forward and upward to the overhead position.

(b) **TWO**-- Bend backward, emphasizing the bend in the upper back. The face is up.Keep the knees straight.

(c) **THREE**-- Recover to position ONE.

(d) **FOUR**-- Recover to the starting position.

d. Exercise 4:Up and Forward

(1) **Starting PositionRifle** downward, feet together.(See fig. 3-7, D.)

(2) **Cadence.** Fast.

(3) **Movement.** A four-count exercise:at the count of—

(a) **ONE**-Swing the arms forward and upward to the overhead position.

(b) **TWO**- Swing the arms forward to shoulder level.

(c) **THREE**-Recover to position ONE.

(d) **FOUR**-Recover to the starting position.

e. Exercise 5: Foreup, Full Squat

(1) **Starting Position.** Rifle downward, feet in narrow stance. (See fig. 3-7, E.)

(2) **Cadence.** Moderate.

(3) **Movement.** A four-count exercise: at the count of—

(a) **ONE**-- Swing the arms forward and upward to the overhead position.

(b) **TWO**-- Swing the arms down to shoulder level and assume the squatting position.

(c) **THREE**-- Recover to position ONE.

(d) **FOUR**-- Recover to the starting position.

f. Exercise 6: Arms Forward, Side Bend

(1) Starting Position. Sidestraddle, regular stance, rifle forward.(See fig. 3-7, F.)
(2) Cadence. Moderate.
(3) Movement. A four-count exercise:at the count of-

(a) ONE-- Bend the trunk to the left. Keep the knees straight.
(b) TWO-- Recover to the starting position.
(c) THREE-- Bend the trunk to the right. Keep the knees straight.
(d) FOUR-- Recover to the starting position.

NOTE:Keep the rifle on the same level as the shoulders throughout the exercise.

3303. LOG DRILL

a. Description and Objective. Log exercises are conditioning exercises performed with a log. Each drill contains six exercises and they are numbered in a set pattern. The drill takes 15 minutes to complete. The objective of log drills is to develop strength and muscular endurance and, in this instance, under maximum loads. Log exercises also develop teamwork. Log may be used in lieu of conditioning drills after the Marines have become somewhat conditioned.

b. Area and Equipment. Any level area is satisfactory for conducting this drill. Each Marine completes these exercises from a standing position and no ground contact is required. If the group exceeds a platoon in size, then the instructor will need instructor's stand. Each six-person group or eight-person group will need a log. The logs should be from 6 to 8 inches in diameter. They may vary in length from 14 feet (for 6 people) to 18 feet (for 8 people).The logs should be skinned, smoother, and dried. The 14-foot logs should weigh approximately 300 pounds and the 18-foot logs, approximately 400 pounds. Rings should be painted on the logs to indicate each person's position. When not in use, the logs should be stored on a rack to keep them off the ground.

c. Formation. All the Marines assigned to the same log team should be about the same height at the shoulders. The recommended method of dividing the platoon is to have the Marines form a single file or column with short people to the front and tall people to the rear. Have the Marine assume their positions in the column according to shoulder height, not head height. When they are in position, they are given the command COUNT OFF BY SIXES (OR EIGHTS), COUNT OFF, to divide them into six- or eight-person log teams. Each team in turn, can then proceed to the log rack, shoulder a log, and carry it to the designated exercise area. The log teams form in columns in front of the instructor. With the Marine holding the log in the chest position, have them face the instructor and ground the log at least 10 yards between columns and 10 yards between log teams within the columns.

d. Starting Positions. The Marines fall in, facing the log, with their toes about 4 inches from it. The basic starting positions and commands are as follows:(See fig. 3-8.)
(1) **RIGHT-HAND STARTING POSITION, MOVE.** It the command MOVE, move the left foot 12 inches to the left, and lower the body into a flat foot squat. Keep the back straight, head up, and arms between the legs. Encircle the far side of the log with the left hand. Place the right hand underneath the log.(See fig. 3-8, A.)

(2) LEFT-HANDED STARTING POSITION, MOVE. These commands are executed in the same manner as in paragraph a. except that the left hand is underneath the log and the right hand encircles its far side. (See fig. 3-8, B.)

Figure 3-8 is available in the illustration section at the end of the book.

(3) RIGHT SHOULDER POSITION, MOVE. At the command MOVE, pull the log upward in one continuous motion to the right shoulder. At the same time, move the left foot to the rear and stand up, facing left. Balance the log on the right shoulder with both hands.(See fig. 3-8, C.)This movement cannot be performed from the left-hand starting position because of the position of the hands.

(4) LEFT SHOULDER POSITION, MOVE. These commands should be given from the left hand starting position. At the command MOVE, pull the log upward in one continuous motion, to the left shoulder. At the same time, move the right foot to the rear and stand facing right. Balance the log on the left shoulder with both hands.(See fig. 3-8, D.)This movement cannot be performed from the right-hand starting position.

(5) WAIST POSITION, MOVE. From the right hand position pull the log waist high. Keep the arms straight and fingers laced underneath the log. The body is inclined slightly to the rear and the chest is lifted and arched. (See fig. 3-8, E.)

(6) CHEST POSITION, MOVE. This command should be given after the waist position has been assumed .On the command MOVE, shift the log to a position high on the chest, bring the left arm under the log and hold the log in the bend of the arms.(See fig. 3-8, F.)Keep the upper arms parallel to the ground.

(7) To move the log from the right shoulder to the left shoulder, the command is: LEFT SHOULDER POSITION, MOVE. On the command MOVE, push the log overhead and lower it to the opposite shoulder.

(8) To return the log to the ground from any of the above positions, the command is: STARTING POSITION, MOVE. At the command MOVE, slowly lower the log to the ground. The hands and fingers must be kept from under the log.

e. Leadership. A principal instructor demonstrates and leads the drill. He must be familiar with the leadership techniques for conditioning exercises and the peculiar techniques for log drill.

3304. EXERCISE PERFORMED WITH LOGS

The exercises of log drill are outlined in the following paragraphs. Figure 3-9 graphically explains log drill.

a. Exercise l:Two-Arm Pushup

(1) Starting Position. Right or left shoulder position. Regular stance. (See fig. 3-9, A.)

(2) Cadence. Moderate.

(3) Movement. A four-count exercise:at the count of—

- **(a) ONE--** Push the log overhead until the elbows lock.
- **(b) TWO--** Lower the log to the opposite shoulder.
- **(c) THREE--** Repeat the action of count ONE.
- **(d) FOUR--** Recover to the starting position.

b. Exercise 2: Forward Bender

(1) Starting Position. Chest position. Regular stance.(See fig. 3-9, B.)

(2) Cadence. Moderate.

(3) Movement. A four-count exercise:at the count of-

(a) ONE-- Bend forward at the waist, keeping the back and legs straight.

(b) TWO-- Recover to the starting position.

(c) THREE-- Repeat the action of count ONE.

(d) FOUR-- Recover to the starting position.

c. Exercise 3: Straddle Jump

(1) Starting Position. Right or left shoulder position, feet together, fingers interlaced on top of the log.(See fig. 3-9, **C.)**

(2) Cadence. Moderate.

(3) Movement. A four-count exercise:at the count of—

(a) ONE-- Jump to a side-straddle. Pull down on the log with both hands to keep it from bouncing on the shoulder.

(b) TWO-- Recover to the starting position.

(c) THREE-- Repeat the action of count ONE.

(d) FOUR-- Recover to the starting position.

Figure 3-9 is available in the illustration section at the end of the book.

d. Exercise 4: Side Bender

(1) Starting Position. Right shoulder position, feet regular stance.(See fig. 3-9, D.)

(2) Cadence. Moderate.

(3) Movement. A four-count exercise:at the count of—

(a) ONE-- Bend sideward to the left as far as possible, bending the left knee.

(b) TWO-- Recover to the starting position.

(c) THREE-- Repeat the action of count ONE.

(d) FOUR-- Recover to the starting position.

(4) After completing the required number of repetitions, change shoulders and execute an equal number of repetitions to the other side.

e. Exercise 5: Knee Bend

(1) Starting Position. Right or left shoulder position. Narrow stance. Fingers interlocked on top of the log.(See fig. 3-9, E.)

(2) Cadence. Slow.

(3) Movement. A four-count exercise:at the count of—

(a) ONE-- Flex the knees to a quarter-squat position.

(b) TWO-- Flex the knees to a half-squat position.

(c) THREE-- Lower the body to a three-quarter squat position.(Lean slightly forward.)

(d) FOUR-- Recover to the starting position.

NOTE:Pull forward and downward on the log throughout the exercise.

f. Exercise 6: Overhead Toss

(1) Starting Position. Right or left shoulder position, regular stance. The knees are bent to a quarter-squat.(See fig. 3-9, F.)

(2) Cadence. Moderate.

(3) Movement. A four-count exercisc:at the count of-

(a) ONE-Straighten the knees and toss the log into the air approximately 12 inches overhead. Catch the log with both hands and lower it toward the opposite shoulder. As the log is caught, lower the body into a quarter-squat.

(b) TWO-- Again toss the log into the air and when caught return it to the original shoulder.

(c) THREE-- Repeat the action of count ONE.

Section IV. Grass Drills

3401. GENERAL

a. Description and Objective. Grass drills are extremely strenuous exercises and are performed at top speed for only short periods of time. No cadence is counted but the Marines continue to execute the multiple repetitions of the command until the next command is given. The grass drill consists of two drills:Drill One and Drill Two. Each drill contains six exercises. The objective of the drills is to decrease reaction time, to develop cardiovascular endurance, and to provide a vigorous workout for all major muscles.

b. Area and Equipment. Any level area suitable for ground contact and of a size to accommodate the group is adequate. No equipment is needed.

c. Formation. All movements are executed in place. The extended, rectangular formation is recommended for a platoon- or company-size unit. The circle formation is suitable for groups of squad or section size. At the beginning of an exercise program, 2 to 3 minutes of grass drills will insure a good workout.

d. Progression. Progression is gained by gradually increasing the length of time devoted to the drills. As the physical condition of the Marines improves, the periods should be gradually lengthened to 5 minutes. As the second drill is more difficult than the first, some progression can be attained by initially executing Drill One; then as the program and the Marines progress, introduce Drill Two. To extend the duration of the drill, it may be necessary to repeat the drill.

e. Starting Position

(1) The drills are started from the GO position. Other basic positions are FRONT, BACK, and STOP. (See fig. 3-10, A.)

(a) GO. Running in place (top speed): on the toes and balls of feet, knees raised high, arms pumping, body bent forward at waist.

(b) FRONT. Prone position: elbows bent (along body), palms flat on ground directly under the shoulders, legs together and straight.

(c) BACK. Supine position (flat on back): arms extended near side on ground with palms down, legs together and straight, feet toward the stand or instructor.

(d) STOP. Football lineman stance: feet spread and staggered, left arm across left thigh, right arm straight, knuckles on ground, head up, back parallel with ground.

(2) To assume the FRONT or BACK position from the STANDING, GO, or STOP position, vigorously get into the prescribed position as quickly as possible. (See fig. 3-10, B.)

(3) To change from the FRONT to the BACK position, quickly do a pushup, move the feet several short steps to the right or left, lift the arm on the side toward which the feet move, and thrust the legs vigorously to the front. (See fig. 3-10, C.)

Figure 3-10 is available in the illustration section at the end of the book.

(4) To move from the BACK to the FRONT position, sit up quickly, place both hands on the ground to the right or the left of the legs. Move the feet several short steps to the rear on the side opposite the hands. When the feet are opposite the hands, thrust the legs vigorously to the rear and lower the body to the ground.(See fig. 3-10, D.)

f. Leadership. A warm up activity of lesser intensity should proceed grass drill. During the instructional phase and conduct of these drills, the following points should be applied.

(1) The instructor executes only GO and STOP with the Marines. This allows the instructor to supervise the drill.

(2) The commands for grass drills are given in rapid succession without the usual preparatory command.

(3) To prevent confusion, the instructor should give the commands sharply to distinguish them from comments or encouragement.

(4) As soon as the Marines know the drill, they should respond to the instructor's commands and perform all exercises vigorously and as rapidly as possible. All exercises are executed continuously until the next command is given. Insist on top speed performance; anything less is not effective.

(5) The commands peculiar to each exercise are identical to the name of the exercise.

(6) Marines are not to be required to assume the position of attention once the drills are started. To halt the drill for instructions or for rest, the command UP is used. At this command, the Marines assume a relaxed standing position. Do not demand formality. At the conclusion of a fast and vigorous 5-minute grassdrill, it is physically impossible for people to stand at attention.

(7) The sequence of commands for the execution of grass drills should occur as follows:Drill One. GO, FRONT, Bouncing Ball; GO, BACK, Bicycle; GO, Full Squatter; GO, BACK, Situps; GO, FRONT, Mountain Climber; GO, FRONT, Roll Left; GO, STOP, UP.

g. Use With Other Programs. Since grass drills can be executed in a short period of time, they may be executed where only a few minutes are available for exercise or in conjunction with another type of activity. Grass drills are an excellent substitute for running when time is a factor.

3402. GRASS DRILL ONE AND TWO

a.Grass Drill One

(1) Bouncing Ball. From the FRONT position, push up, supporting the body on the hands (shoulder-width apart) and feet. (See fig. 3-11.)Keep the back and legs in line and the knees straight. Bounce up and down by a series of short, upward springs from the hands, hips, and feet simultaneously,

(2) Bicycle. From the BACK position, raise the legs and hips. Keep the elbows on the ground and support the hips with the hands. Move the legs vigorously as if pedaling a bicycle.

(3) Full Squatter. From the STOP position, assume a full knee bend, the feet on line, hands on hips. Bounce up and down in place by short, bouncing jumps.

Figure 3-11 is available in the illustration section at the end of the book.

(4) Situps. From the BACK position and with arms stretched overhead, sit up, reach forward, and touch toes. Return to the supine position.

(5) Mountain Climber. From the STOP position, place both hands on the ground directly under the shoulders. Thrust the right leg to the rear, knee straight. The left foot should be close to the left hand, the left knee outside the left arm. Shift the weight to the hands, thrust off with the rear (right) foot and bring that foot up close to the right hand, the right knee outside the right arm. At the same time, thrust the left leg vigorously to the rear, knee straight. Continue at a fast cadence, alternating the legs.

(6) Roll Left. From the BACK or FRONT position, make one complete roll in the direction commanded. On completing the roll, return to the FRONT or BACK position.

b. Grass Drill Two

(1) Legs Over. .From the BACK position and with arms stretched overhead, palms up, raise the legs upward and then swing them backward over the head until the toes touch the ground behind the head. Return legs to the starting position. (See fig. 3-12.)

(2) V-Up and Touch Toes. .From the BACK position, raise the legs with the knees straight, sit up until the trunk and legs form a V, and touch the toes with the hands. Return to the BACK position.

(3) Rocker. .In the FRONT position, clasp the hands behind the back, arch the body, holding the head back. Start rocking, using the front part of the trunk as a rocker.

(4) Bounce and **Clap Hands**. The procedure is the same as for bouncing ball, but while in the air, clap the hands. This requires a more vigorous bounce or spring. (See fig. 3-12.)

(5) Leg Spreader. .From the BACK position, raise the legs so that the heels are 10 to 12 inches from the ground, spread them apart as far as possible, then close them together. Open and close legs as rapidly as possible.

(6) Forward Roll. .For forward roll from the STOP position, place both hands on the ground, tuck the head, and do one complete forward roll, keeping the legs tucked as you roll, and come back to the STOP position.

Figure 3-12 is available in the illustration section at the end of the book.

Section V.Guerrilla Exercise

3501.GENERAL

a. Description and Objective. Guerrilla exercises are individual exercises of various basic skills that are performed rapidly while moving forward in a circle formation. There are two tables of guerrilla exercises. Each table takes 15 minutes to complete. The objective of this exercise is to increase strength and endurance, aid flexibility, and develop coordination.

b. Area and Equipment. Any level area is suitable for conducting guerrilla exercises. No ground contact is required other than the hands. There is no equipment requirement.

c. Formation
(1) The circle formation (app. A) is used for guerrilla exercises. Each platoon forms its own circle and engages in guerrilla exercises under a platoon instructor. If the platoon exceeds 30 Marines, double or concentric circles may be used.
(2) When the circle is formed, the instructor steps into the center of the circle and moves clockwise in a small circle. He commands:QUICK TIME, MARCH, 1-2-3-4. (Rapid cadence of approximately 130 counts per minute. Cadence and step are maintained between exercises.)
(3) To reform the platoon after completing guerrilla exercises, the instructor halts the Marines and places the base man or platoon guide where he wishes and commands:
(a) BASE MAN (or platoon guide),POST.
(b) FALL OUT AND FALL IN ON THE BASE MAN (or platoon guide).

d. Progression. Progression may be attained by moving from table I to table II or by shortening the quick time marching periods between exercises and performing all exercises a second time.

e. Leadership
(1) To execute the exercises, the Marines continue at quick time while the instructor simultaneously explains and demonstrates the exercise to be performed, and then commands the Marines accordingly. In each instance, the preparatory command will be the name of the exercise and, in all instances, the command of execution will ***MARCH**.. To* terminate each exercise, the command *is* ***QUICK TIME, MARCH.*** *The* Marines immediately pick up the step as the instructor counts cadence.
(2) Unless specified differently, each exercise should be continued for 20 to 40 seconds depending upon the vigor of the exercise. The leader can determine the duration of each exercise by observing its effect upon the Marines.
(3) To form for double guerrillas, the commands for pairing the Marines (who are in circle formation) are-
(a) PLATOON, HALT.
(b) FROM (designate an individual), BY TWO'S, COUNT OFF(.Examplel-2; 1-2; 1-2; etc.)
(c) EVEN NUMBERS MOVE UP BEHIND ODD NUMBERS(At this time, adjust pairs according to height and weight.)

(d) YOU ARE NOW PAIRED UP FOR DOUBLE GUERRILLAS. To change the Marines position, merely command CHANGE.)
(e) FORWARD, MARCH.

f. Place in the Program. Many Marines have not had the opportunity to perform the simple skills involved in guerrilla exercises. The conduct of these exercises is a simple matter since they can be performed easily and quickly in almost any situation. The tables of exercise are applicable to all personnel. The tables can constitute a station within a 1-hour period or be completed within a separate 15-minute period.

3502.GUERRILLA TABLES

a. Table I

(1) Double Time.(See fig. 3-13, A.)Hold arms in the thrust position. Execute a double-time run, maintaining the circle formation and the prescribed distance between your fellow Marines. Duration-1 minute.
(2) All Fours.(See fig. 3-13, B.) Face downward. Support the body with the hands and feet. Walk forward hands first.
(3) Crab Walk.(See fig. 3-13, C.) Get in the sitting position, face upward, and lift the hips. Support the body with the hands and feet. Walk forward feet first.
(4) Squat Walk.(See fig. 3-13, D.) Assume a full knee bend position. Grasp the ankles (left ankle with the left hand, right ankle with the right hand). Walk forward.
(5)Broad Jump.(See fig. 3-13, E.) Jump forward on both feet in a series of broad jumps. Swing the arms vigorously to assist the jumps.
(6)Toe-Touch Walk.(See fig. 3-13. F.) Walk forward, bending at the waist and touching one hand to the toe of the opposite foot while it is on the ground. Raise the trunk to the vertical position between steps. Keep the knees straight.
(7)Bottoms-Up Walk.(See fig. 3-13, G.) Assume the front leaning rest position and move the feet toward the hands in short steps, keeping the knees locked. When the feet are as close to the hands as possible, walk forward on the hands to the front leaning rest position.
(8) Straddle Run.(See fig. 3-13, H.) Run forward, leaping to the right from the left foot and to the left from the right foot.
(9) Fireman's Carry .(See fig. 3-13,1.)
(lO)Single-Shoulder Carry(See fig. 3-13, J.) Two men execute the carries as indicated by the diagram. No. 1 man executes one type; No. 2 man executes the other.

Figure 3-13 is available in the illustration section at the end of the book.

b. Table II

(1) Double Time.(See fig. 3-14, A.)Hold arms at the thrust position. Execute a double-time run, maintaining the circle formation and the prescribed distance between your fellow Marine.Duration-1 minute.
(2) Toe-Grasp Walk.(See fig. 3-14, B.) Bend forward and grasp toes. With knees slightly bent, walk forward.

(3) Hand-Kick Walk..(See fig. 3-14, C.) Walk forward, kicking the moving foot upward on every step. At the same time, lean forward and touch the elevated toe with the hand of the opposite arm.

(4) Pike Juinping.(See fig. 3-14, D.) Jump forward and upward from both feet, keeping the knees straight, and at the same time, swing the legs forward and touch the toes with the hands at the top of each jump.

(5) Squat Jump.(See fig. 3-14, E.) Leap forward from the squatting position, with the hands on the ground and the arms between the legs. Land on the ground with hands and legs extended. Bring up the legs to the squatting position.

(6) Steam Engine..(See fig. 3-14, F.) Lace the fingers behind the neck and walk forward in the following manner: as the left leg moves forward, raise the knee high, bend the trunk forward, and touch the outside of the right elbow to the outside of the knee. Then lower the left leg and step forward on the left foot and raise the right leg. Repeat with the right leg and left elbow.

(7) Knee-Touch Walk.(Scc fig. 3-14, G.) Walk forward, bending the knees and touching the ground on each step. The knees are bent and straightened on each step.

(8)Hobble Hopping.(See fig. 3-14, H.) Hold foot behind back with opposite hand and hop forward. On the command CHARGE, grasp the opposite foot with opposite hand and hop **forward.**

(9) Cross Carry.(See fig. 3-14,1.)(10)Saddle Back Carry. (See fig. 3-14, J.) Two Marines execute the carries as indicated in the diagram. No. 1 Marine executes onetype;No. 2 Marine executes the other.

Figure 3-14 is available in the illustration section at the end of the book.

Section VI. Running and Orienteering

3601. RUNNING

a. General

(l)Description. The general form and technique for all types of running are fairly constant. (See fig. 3-15.)The head is erect, body slightly forward without bending at the waist, and the arms are at a loose thrust position alternating from front to rear in straight planes. A cross-body arm movement wastes energy .The movement of the legs and feet will be discussed in subsequent paragraphs dealing with the different types of running. Of primary importance is the fact that in all types of running, the toes should be pointed straight ahead. Toeing out is a common error in both running and walking and should be an item of individual correction.

(2) Objective. The objective of running is to develop cardiovascular endurance. Despite the fact that Marines have developed their muscle structure and the strength of their muscle tissue, unless they have developed cardiovascular endurance to a satisfactory degree, they are not entirely physically fit or combat ready. Cardiovascular endurance (wind) depends on the efficiency of the lungs and heart. The efficiency of the lungs and heart depend on the amount of oxygen the lungs can absorb with each breath inhaled and the amount of carbon dioxide the lungs can expel. The process of absorbing oxygen and expelling carbon dioxide (cardiopulmonary process) is performed by the blood that circulates through the lungs. The condition of this process will determine the amount of effort a person can exert over a period of time. Running is one of the best activities to develop this vitally important endurance.

b. Running Skills. In the development of running skills, individuals may require instruction to improve their proficiency. Some important skills to consider are-

Figure 3-15 is available in the illustration section at the end of the book.

(1) Action of the Arms. Arm action is important. Check to see that arms are held loosely and that the action is relaxed. The faster the run, the more rapid the arm action.

(2) Breathing. Allow the individual to breathe through the mouth as the body demands a large supply of oxygen. Oxygen can be inhaled in greater quantities through the mouth.

3602. TYPES OF RUNNING

a. Double Time

(1) Description and Objective. Double timing is marching at the rate of 180 steps per minute, each step being 36 inches in length. It takes practice to double time with precision in formation. The Marines should keep in step, placing their feet flat on the ground. This, however, should not be a stamping motion, but should be done with as slight a jolt as possible. Double timing is like a jog, the difference being that in a jog the feet are lifted well off the ground and the running motion is bouncy. In double timing, the feet skim the ground and there is no bounce to the ran. Double timing is a vehicle for teaching proper running form and for the development of the cardiovascular system.

(2) Area and Equipment. This type of running can be completed over a variety of surfaces. Usually a Marine uses a field or road. There is no equipment requirement.

(3)Progression. There is no set standard for alternating quick time and double time in the early conditioning of Marines. A general rule is to begin with enough quick-time marching to ensure a thorough warming up, then double time about 100 paces. Change again to quick time until the individuals have made a reasonable recovery from the running, then double time another 100 paces. The amount of double time can be increased and the quick time decreased from week to week, until the individuals are double timing about 1800 yards. This type of training should be given at least twice a week, but by no means is it adequate as the sole means of conditioning.

(4) Leadership

(a) The instructor should be to one side of the column or group and toward the rear so there is a full view of all Marines. Inexperienced instructors have a tendency to supervise from a position too far forward.

(b) Select an individual who can maintain the proper cadence to act as the guide during double-time running.

(c) There are several ways for the instructor and group to count cadence while double timing. If not contrary to local policy, learn several methods and use them for variety.

b. Wind Sprints

(1) Description and Objective. This type of running involves a series of 30- or 40-yard dashes, usually conducted in successive waves of squads. Each squad is in line and the squad leader on the right flank. Wind sprints assist indeveloping speed and cardiovascular endurance. Any flat and level area may be used which will permit the squad to form a line and run the required distance.

(2) Progression. One or two 30-yard sprints will be adequate at the beginning. As time passes, sprints can be lengthened and up to six or seven sprints may be used.

(3) Leadership

(a) At the command READY (given by squad leader), each runner assumes the sprinter's starting position. At the command GO, the squad sprints approximately 30 yards, takes 10 yards to stop, and lines up immediately with the squad leader who repeats, READY, GO, and again the squad sprints. At the conclusion of the third sprint, the squad waits until all the squads of the platoon have made three sprints. Then they all line up and the squad leaders conduct three more wind sprints in the opposite direction.

(b) Valuable time is gained by having each squad ready to go when the preceding squad has moved off its second sprint mark.

c. Cross-Country Running

(1) Description and Objective. Cross-country running is a distance run conducted on a course laid out along roads, across fields, over hills, through woods, and on any irregular ground. The cross-country ran may be utilized as a conditioner or as a competitive event; the objective is to cover the distance in the shortest possible time. The course should be 2 to 2.5 miles in length and be laid out to avoid heavy vehicular traffic. The course should be marked by directional arrows until the runners know the course. These runs build leg muscles, increase lung capacity, and develop endurance. Any local area of varied terrain is suitable.

(2) Progression. In the mass training of a large group, leaders should be stationed at the head and the rear of the column and should make every effort to keep the individuals together .After determining the abilities of the individuals in cross-country running, it is advisable to divide the unit into three groups. The poorest conditioned group starts first, and the best conditioned group, last. The starting time of the groups should be staggered so that all of them finish about the same time. In preliminary training, the running is similar to ordinary road work in that it begins with rather slow jogging, alternating with walking. The speed and distance of the run is gradually increased. As the condition of the individual improves, occasional sprints may be introduced. At first the distance run is from 1/2 to 1 mile. It is gradually increased to 2 or 2.5 miles. Well-conditioned personnel can run 2 to 2.5 miles within a 15-minute period.

(3) Leadership. Marines should not be required to take part in distance running until they have been through a progressively scheduled training program which requires a considerable amount of running. Cross-country runs should be scheduled occasionally to provide variety in the program. Cross-country running has the advantage of allowing mass participation. Interest can be stimulated by putting the runs on a competitive basis. As a single activity, short cross-country runs can be scheduled once a week, gradually increasing the distance as the physical conditioning improves.

(4)Use With Other Programs. Cross-country running can be combined with other activities such as conditioning exercises.

d. Fartlek Training

(l)Description and Function. Fartlek training is various running exercises conducted along an intense combat-like course. The running is conducted to overwork the lungs, allowing only partial recovery which is followed by another intense period of overload. This sequence is repeated for the duration of the workout. Fartlek training is a useful combat training method and general cardiovascular conditioner. The entire unit must be in uniformly good to excellent condition to qualify for this training.

(2) Progression. The following is an example of one fartlek session:

(a) Warm up by stretching 3-5 minutes.

(b) Warm up running easily 5-10 minutes.

(c) Run at a fast, steady pace for 3/4 to 1 3/4 miles (dependent on terrain).

(d) Walk/jog at a moderate pace for 5 minutes (recovery).

(e) Run easily sprinting 15 to 20 meters occasionally.

(f) Run full speed uphill for 175 to 200 meters.

(g) Warm down by running easily for 1/2 to 1 mile.

(h) Warm down by stretching 3 to 5 minutes.

(3)Leadership. To add variety, try incorporating combat movements, terrain association, a series of exercises or carrying table of equipment weapons (hit and roll). This will take some initiative in setting the course, but it allows for variety. In this manner, the fartlek course can be used as a training session or part of one. It works extremely well with highly motivated, competitive Marines.

3603.ORIENTEERING

Orienteering is land navigation over a prescribed course as a timed, competitive event. It is an excellent way to integrate land navigation training with physical conditioning and requires Marines to be skilled in both areas. Commanders adjust the training as needed to change the degree of difficulty or to emphasize either navigation or conditioning. Taken to the extreme, an orienteering exercise is conducted in full combat gear over rugged terrain against aggressors. Orienteering requires the following physical skills: endurance, speed, strength, and all-around conditioning.

Section VII. Circuit Training

3701. GENERAL

a. Description and Objective. A strength circuit consists of a series of stations where individuals in small groups exercise vigorously for a short period of time and then move (on signal) to the next station where a different form of exercise is conducted. This rotation continues until all groups move through all stations. Strength circuits contain no set or specific types of exercise stations within the circuit. The objective of circuit training is to develop strength. There are three general types of circuits.

(1) Fixed Circuit. This is a circuit in which apparatus of an immovable type (fixed into the ground) is used.

(2) Movable Circuit. This circuit consists of individual exercise apparatus which is portable and can be moved to and from the training area.

(3) Simplified Circuit. This circuit requires no equipment or apparatus.

b. Formation. The exercises are done at will, but rapid, steady, and continuous work is required of all. Each Marine's nervous and muscular system reacts differently to timed vigorous exercises. His performance should be measured on how many movements per exercise he can complete as an individual. For example, one Marine may be able to complete 5 movements, while another may be able to complete 20, and yet each is receiving the maximum benefit. All three circuits contained in this chapter are designed for platoon-sized groups. Expansion beyond this capacity requires a large amount of equipment, as each Marine in the fixed and movable types of circuits must have an item of equipment available for exercise at each station. A group larger than a platoon could be exercised through use of the simplified type of circuit; however, the group would be unwieldy and control could be a problem.

c. Place in the Program. All circuits illustrated can be completed in a 15-minute period. This feature allows the exercise of a platoonor smaller group on the circuit for a single 15-minute period, or the scheduling of the circuit as a 15-minute period within a longer period. A circuit can thus be utilized within a rotating activity system of scheduling. Choice of a circuit by the unit depends upon area, facilities, and other local factors; however, there is a circuit for every need.

3702. FIXED STRENGTH CIRCUIT

a. Description and Objective. The strength circuit is an arrangement of various types of exercise apparatus which are fixed in position. (See fig. 3-16.)Seven basic exercises are used and each exercise requires an apparatus. All apparatus of one type are positioned together to constitute a station. Each station will accommodate 10 Marines. The circuit is designed to be accomplished in 15 minutes when conducted on a time-rotation basis, as normally executed for unit training. For individual use, a Marine may complete the circuit by executing a specific number of repetitions for each exercise. In this case, the time required for completion of the circuit would vary slightly depending on the number of repetitions the Marine completed. The objective of this circuit is to provide a series of exercises which will improve and maintain the strength of the body's major muscle groups.

Figure 3-16 is available in the illustration section at the end of the book.

b. Starting Level. Marines must be thoroughly warmed up prior to participating in the circuit system. If Marines have not engaged in vigorous exercise immediately prior to starting the circuit, then an instructor should execute the following warm up exercises. These exercises should be conducted in the normal formation for set drills. Seven repetitions of each exercise will normally provide sufficient warm up. The exercises are-

(1) High jumper, Conditioning Drill 1.
(2) Bend and reach, Conditioning Drill 1.
(3) Squat bender, Conditioning Drill 1.

c. Progression. The instructor can adjust the circuit system through several methods to accommodate participants of varying physical ability. One method of adjusting an exercise is to change the method of executing the leg lift or to select a heavier weight for the barbell curls. When it becomes apparent during a unit program that the overall fitness of the group has Improved, then the exercise may be made more strenuous in two additional ways. First, the instructor can have one or more of the supplementary stations added. Second, the time spent exercising at each station can be increased in 5-second increments to a maximum of 60 seconds. Also, as Marines learn the circuit, the 45-second period for movement and instruction between stations can be eliminated, thus allowing only enough time to change stations.

d. Starting Positions. The Marine starts on any station, exercises steadily for a certain period (45 seconds initially), then moves on command to the next type of apparatus where he again exercises steadily for an equal period. The Marine continues until the required exercise is completed at each station. If it is desired to expand the number of stations in the circuit to accommodate more participants at one time, the instructor can provide four additional supplementary exercises, requiring no equipment. This will increase the amount of time required to complete the circuit.

e. Leadership. Close control of all Marines is necessary to ensure that a minimum amount of time is spent in moving them to their initial stations and in moving them between stations. One instructor can control the activity on the strength circuit. A stop watch or wrist watch with a second hand is required. When Marines arrive at the strength circuit, the instructor will form them for exercise and conduct the warm up drill. The group is then reassembled and formed into a number of files equal to the number of stations being used in the circuit. Each file is then directed to a station. As soon as all participants have reached an exercise position at a station, the instructor gives the command READY, GO. After 45 seconds of exercise, the instructor gives the command STOP, CHANGE OVER. The instructor allows Marines 45 seconds for moving to the next station and for preparing for the next exercise before the command EXERCISE is again given. In lieu of verbal commands, a whistle may be used to stop and start the exercises. For large groups, a megaphone or loudspeaker is useful.

f. Individual Conditioning Program. For the Marine working alone on the strength circuit, it would be impractical to time the exercise periods. For individual exercise, the Marine should select a number of repetitions of each exercise to complete, then rotate to the next stations after completing these repetitions. The number of repetitions selected should be at or near the maximum that the Marine is capable of doing without halting for rest.

g. Fixed Circuit Stations

(1) Primary Stations (See fig. 3-16.)

(a) Pullups/Chinups. A horizontal bar placed 8 feet above the ground is necessary for each Marine at this station. The Marine will also need a space on the bar that is 45 inches wide. On the command**EXERCISE**,the Marine grasps the bar with both palms facing either forward or to the rear, arms fully extended, and feet free of the ground and executes the exercise as described in MCO 6100.3_.The exercise is repeated as many times as possible until the command **STOP** is given. Then the Marine moves to the next exercise. If a Marine has done his maximum number of pullups/chinups prior to the command **STOP**, he will remain in the "dead" handing position until the command **STOP** and move to the next station.

(b) Twist Grip. The apparatus is a horizontal bar, free to turn, held between uprights placed 30 inches apart. The bar is 52 inches above the ground. A weight of 20 pounds is attached to the center of the bar by a light rope long enough to permit the weight to rest on the ground. The Marine stands at arm's length from the bar and grasps it with his hands on either side of the rope, palms down, thumbs under the bar. On the command **EXERCISE,** the hands are rotated so that the backs of the hands are rotated away from the body, thus winding the rope on the bar. The elbows are kept straight to ensure that the exercise is performed by the hand and forearm. When the weight is drawn up to the bar, the bar is then rotated in the opposite direction to lower the weight to the ground. This exercise is continued until the command **STOP** is given. The Marine then moves to the next station.

(c) Pulley Weights. The apparatus is a T frame with a system of pulleys that suspends a weight of about 90 pounds. The weight is attached to a light steel cable which has a drawbar attached to the other end. The Marine grasps the drawbar and sits down directly under the bar, legs extended to the front and arms extended overhead. The exercise is executed by pulling the drawbar down behind the head, then extending the arms slowly again until they are fully extended overhead. The exercise is repeated as many times as possible until the command STOP is given. The Marine then will move to the next exercise. Upon completion of the exercise, the weight is lowered slowly to the ground.

(d) Barbell Curls. A barbell is necessary for each Marine at this station. The barbell is constructed of 1 1/4-inch pipe 5 feet long, and two concrete-filled No. 10 cans. Each barbell should weigh about 40 pounds. Variance in the weight of the barbells, up to about55 pounds, will allow appropriate overload to be applied to Marines who are above average in strength or weight. The Marine grasps the bar with the palms forward and assumes a standing position with the barbell held in front of the hips, hands approximately shoulder's width apart. On the command **EXERCISE,** the Marine flexes the elbows and draws the barbell up until it touches the upper chest. The elbows remain at the sides. Breath is inhaled with the upward movement and exhaled as the barbell is lowered to the starting position. The exercise is repeated as many times as possible until the command **STOP** is given. The Marine moves to the next apparatus.

(e) Step-Up. The apparatus is a platform or ledge 18 inches high and of such size. At accommodate 10 Marines. The Marine faces the platform and on the command **EXERCISE,** steps up onto the platform, bringing his trailing foot up beside the leading foot. He then steps back down to the original position, stepping down first with the same foot he initially used in stepping up. After 10 repetitions of the exercise, he changes the order of moving the feet to use the opposite leg for stepping up. Repeat this exercise until the command **STOP** is given. The Marine then moves to the next station.

(f) Leg Lift. The apparatus is a horizontal bar constructed as described in (a). To prevent the body from swaying, a horizontal back support is added 40 inches below the horizontal bar. The arms are kept fully extended. On the command **EXERCISE,** the Marine jumps up, grasps the bar with the palms forward and the back support behind him. The exercise is executed by raising the legs to a horizontal position then slowly lowering them to the vertical position. The Marine does not flex his knees. He does not swing the legs to the rear of a vertical position to gain momentum for raising them in the next repetition of the exercise. The movement is repeated until the command **STOP** is given. The Marine then moves to the next exercise. If unable to raise his legs to a horizontal position without flexing his knees, the Marine flexes his knees and draws them up to the chest, then lowers his legs to the vertical position.

(g) Rope Climb. The rope climb is 20 to 30 feet high with five ropes suspended from a horizontal bar which forms the uppermost part of the framework. To prevent the horizontal bar from sagging and to provide safety, only five ropes are attached to it. There are two frameworks per station. The ropes are 6 feet apart. Any method may be used to climb the rope, and the Marines climb as high as possible. Marines who are proficient should climb therope several times during the time allotted. Inexperienced Marines should be cautioned to take care during descent to avoid rope burns on their hands.

(2) Supplementary Stations. The following exercises are designed to expand the basic circuit by being inserted in specific places within the system. For each supplementary station used, there should be adequate room for 10 Marines to exercise.

(a) Bent Knee Situp or Bottoms-Up. These calisthenics are designed to strengthen the abdominal muscles. These exercises will be inserted between the pullup/chinup and twist grip stations. The primary stomach exercise is the situp. In case of inclement weather or other conditions that make ground contact undesirable, the bottoms up exercise is used. In situps on the command **EXERCISE,** the Marine lies on his back (supine position) with knees flexed and both feet flat on the ground and executes the exercise (minus an assistant) as described in MCO 6100.3_.The Marine then returns to the starting position, repeating the exercise until the command **STOP** is given. The Marine then moves to the next station. In bottoms up on the commandEXERCISE,the Marine assumes the front leaning rest and executes the bottoms-up exercise as described in Conditioning Drill 3. He continues this exercise at a moderate cadence until the command **STOP** is given. The Marine then moves to the next station.

(b) Pushup. This exercise is designed to strengthen the arm- and shoulder-girdle muscles. It should be included between the twist grip and pulley weight stations. Upon the command **EXERCISE,** the Marine executes the pushup as described in Conditioning Drill 1. He continues this exercise at a moderate cadence until the command **STOP** is given. The Marine then moves to the next station.

(c) Knee Bender. This exercise is designed to build leg muscles and is included between the pulley weight and barbell curl stations. On the command **EXERCISE,** the Marine executes the knee bender as described in Conditioning Drill 3.He continues this exercise at a moderate cadence until the command **STOP** is given. The Marine then moves to the next station.

(d) Trunk Twister. This exercise strengthens the major muscles of the trunk and is included between the step-up and pullup/chinup stations. On the command EXERCISE, the Marine executes the trunk twister as described in Conditioning Drill 1. He continues this exercise at a moderate cadence until the command **STOP** is given. The Marine then moves to the next station.

3703. MOVEABLE STRENGTH CIRCUIT

a. Description and Objective. The exercises in this circuit are progressive and the course is planned to gain and hold the interest of the participating groups. The circuit consists of a series of stations, with each station designed to develop a particular group of muscles. Along with muscular development, correct posture and deep rhythmic breathing should be stressed on this circuit at all times.(See figs. 3-17 and 3-18.)

Figure 3-17 is available in the illustration section at the end of the book.

b. Equipment. The equipment is set up in files. Six files of 8 stations will accommodate a platoon

c. Formation. The platoon marches to the area where the equipment is positioned and forms a file within each lane of stations, covering down on a piece of equipment. Movements are made on the double, the important factor being that no time is wasted in getting to work.

d. Progression. Initially 40 to 45 seconds per station is adequate. As individuals become stronger the time should be increased in 5-second increments until a minute to a minute and a half is reached.

e. Leadership

(1) The leader stands in front of the barbell station and controls the rotation from this position. The leader supervises the entire group, with the assistance of several instructors who move about in the platoon correcting and encouraging the Marines.

(2) The leader starts each group but does not count cadence nor lead them through the exercises. Each individual exercises rapidly but individually.

(3) As each Marine finishes his repetitions with the barbell, he places the barbells on the ground. The leader calls, READY, followed by the command, FALL OUT ONE. All will double-time to the station directly in front of them, while the individuals on the barbell stations will do an about face to the rear station in their lane.

f. Movable Circuit Stations. The Marine can obtain best results on the movable circuit if the exercises on the various stations are given in the following manner:

(1) Station 1—The Barbell. (See fig. 3-17, 1.)The exercises at this station stress proper posture; deep, rhythmic breathing; and development of the muscles of the arms, shoulders, and upper body. An instructor teaches the proper methods of lifting before the exercises begin. The methods are to lift with the legs, to keep the back straight, and to merely grip with the hands. Two recommended exercises are given below. The instructor will use only one exercise per period .He may use either exercise. At successive periods, the other exercise is used.

(a) Exercise 1--Two-Hands Military Press. (See fig. 3-18, 1.)Grasp the barbell with both hands, knuckles up at shoulder width, and lift to the chest. Steadily press to arm's length overhead; lower to the chest resisting weight all the way. Inhale as the weight is pressed up and exhale as the weight is brought down.

(b) Exercise 2--Two-Hands Regular Curl. (See fig. 3-18, 1.) Lift the weight to the waist, with the palms of the hands out, heels together, stomach in, chest lifted and arched, shoulders back, elbows in close to the sides; inhale deeply and curl the weight to the shoulders, using the arms only, at the same time keeping the elbows close to the sides; exhale Emphasize posture and the use of the arms only. This is a very valuable exercise for the development of the biceps and the grip and should be repeated from 8 to 16 times, depending on the ability of the Marine.

(2) Station 2--The Jump Rope. (See fig. 3-18, 2.) This exercise develops strength and agility in the legs and stamina of the whole body. It makes the Marine agile on his feet and increases his footwork efficiency and timing. Each Marine should progress until able to jump rope at least 3 minutes at top speed.

(3) Station 3--The Twist Grip. (See fig. 3-18, 3.) The twist grip is an excellent exercise for the hands and forearms, and adds greatly to the Marine's ability in hand-to-hand combat. The handle is gripped and twisted, winding the rope until the weighted can is level with the height of the hands, which are held horizontal. The weight is lowered in the same manner; the Marine resists the weight all the way, occasionally stopping the twisting motion and alternately removing first one hand, then the other, from the handle. A variation of the above exercise is to wind the handle with the palms up and the arms bent and with the elbows held close in to the sides. Each Marine should maintain a good posture and keep the stomach muscles taut throughout this exercise.

(4) Station 4--The Incline Plane. (See fig. 3-18, 4,)The use of the incline plane is a very strenuous exercise and well designed for the development of the abdomen. Although 6 to 10 repetitions are sufficient for the beginners, more can be added as ability increases. Marines with hernias or recent operations will be excused from participation at this station.

(5) Station 5--The War Club. (See fig. 3-18, 5.) The war club is a simple and effective means of exercising the principal muscle groups of the body, especially those of the trunk, back, and shoulders. To gain the maximum benefit from this exercise, the Marine must keep both feet flat on the ground at all times. Throughout the exercise period, the weight is swung from arm's length as follows:

(a) As in chopping wood, first on one side, then on the other.

(b) As a batter warming up with a number of bats.

(c) In large circles, first with one hand and then with the other.

(6) Station 6--The Bicycle Ride. (See fig. 3-18, 6.) The bicycle ride is well suited to exercising many of the muscle groups of the body, particularly those of the abdomen. Vary the speed of the exercise, but keep everyone "riding" the entire period. A variation exercise may be performed by placing the legs together, raising them slowly to a height about 2 feet from the ground, and then lowering them slowly to the ground.

(7) Station 7--The Step-up. (See fig. 3-18, 7.) The step-up exercises the legs. The step-up is performed by initially stepping up with the left foot, followed by the right, then stepping down with the left foot followed by the right. Continue for 20 seconds, then change to the right foot as the lead foot for 20 seconds.

(8) Station 8--The Isometric Pull. (See fig. 3-18, 8.) Two trainees work at this station with a cable pull and perform the following exercises:

(a) Initially start with one Marine in the supine position and one Marine sitting. The sitting Marine lowers the upper body to the ground and pulls the partner up to the sitting position. The partner then performs this same action and this is continued for 20 seconds at a rapid rate.

(b) During the last 20 seconds, the same action takes place but, in this case, the Marine in the supine position resists the pull of the partner for approximately 5 seconds before allowing to be pulled up into the sitting position.

3704. CIRCUIT-INTERVAL TABLE

a. Description and Objective. The circuit-interval table is designed to develop strength and endurance within a short period of time, with no equipment requirement, through a rapid and vigorous routine of exercise. Fifteen minutes is an adequate period to execute all exercises and to secure a vigorous workout with the circuit-interval principle.

Figure 3-18 is available in the illustration section at the end of the book.

b. Formation and Starting Positions

(1) A leader forms platoons or smaller groups in an oval or circular formation with 3- to 5-yard intervals between Marines. The Marines face to the right and move forward at quick time and then double time. (See fig. 3-19, A.)After running several platoon circle laps, the leader calls out the name of an arm and shoulder exercise from the list below, orders quick time and commands, for example, PUSHUPS. On this command, all Marines immediately hit the ground and individually and rapidly begin doing pushups. No cadence is counted. (See fig. 3-19, B.) After 30 seconds of exercise, the leader commands. **QN YOUR FEET, FORWARD, MARCH** The platoon resumes the quick time cadence and the leader, when ready, gives the necessary commands for double time. The double time is continued for one or more laps and the leader calls out the name of the next exercise and the process is repeated. This continues, with running between each exercise, until every body part has been exercised.

(2) The instructor controls the running and quick time to observe the effects of the exercise upon the Marines. Cadence, step, and precision are not important to the objective and the instructor should not use them. What is important is speed and the instructor should stress this. After the exercise period is started, the Marines do not stop. This circuit method emphasizes stress and recovery, the recovery occurring during the quick time periods.

c. Activities. The leader can use the following exercises. These exercises can be repeated if necessary, during a second round. The leader can use other calisthenic exercises in the circuit-interval table.

(1) Arms and shoulders-push-ups.

(2) Stomach situps.

(3) Back-squat thrusts.

(4) Legs-bicycle (on back).

d. Progression. lhe progress is controlled by the leader, who must pace the running, quick-time movement, and exercise in such a way that Marines will receive a vigorous workout yet be able to participate throughout the 15-minute period. Marines who are in the initial stages of physical condition will not be able to double time or exercise as long as those who are better conditioned. The idea is to set a pace which can be increased during each workout, thus progressing gradually to a higher level of physical fitness.

e. Leadership. The platoon leader, platoon sergeant, or section leader can lead the group. The leader must execute the exercise with the unit to feel the effects and thereby adjust the pace.

f. Place in the Program. This activity may be scheduled whenever a short period of time is available. The only requirement is that enough space, indoors or out, be available to form the circle.

Figure 3-19 is available in the illustration section at the end of the book.

Section VIII. Basic Physical Skills and Obstacle Courses

3801. GENERAL

The purpose of this section is to list the basic military skills and the methods for their development. Many of these skills are best developed by obstacle courses but other drills are also discussed. The section explains types of obstacle courses, details of construction, and methods of negotiating the various obstacles.

3802. BASIC PHYSICAL SKILLS

a. Objective. The objective is to develop proficiency in the various military physical skills which are essential to personal safety and effective combat operations. In travel by foot over rugged terrain and in the execution of combat duties, Marines must be trained to perform basic skills, such as running, jumping, climbing, and carrying. During training, Marines will develop agility and coordination in these skills .Fast and skillful execution of these skills may mean the difference between success and failure on the battlefield.

b. Place in the Program. These skills are practiced throughout the entire program in many different activities. Many of these skills are best practiced on obstacle courses.(See par. 3803.)

c. Description. These basic skills are the minimum skills required by all Marines. The basic skills are as follows:

(1) Running. Running is used to strengthen the legs and develop cardiovascular endurance. Marines should be exposed to running in various situations: on roads, over rough ground, up and down hills, cross-country, and running over low obstacles.

(2) Jumping. In broad jumping, the take-off foot is planted firmly and the spring comes from the extension of this leg as the other leg reaches for the far side of a ditch or similar obstacle. (See fig. 3-20.) The arms are forcibly raised forward and upward to assist in propelling the body up and forward. Landing may be on one or both feet depending upon the length of the jump. In vertical jumping downward from a height, the jumper should aim his feet at the desired landing spot and he should also jump with the knees slightly bent and feet together, with the trunk inclined slightly forward. As the feet touch the ground, the shock is absorbed by bending the knees into a full squatting position. If the height is too great or the ground too hard to absorb the shock, then the jumper should forward roll or side roll thus eliminating some of the momentum.

Figure 3-20 is available in the illustration section at the end of the book.

(3) Dodging In combat situations, it is often times necessary to change directions quickly. To execute this movement while running, a lead foot is firmly planted, left foot if the direction is to the right and right foot if the direction is to the left. The opposite foot is moved toward the new direction. The knees are slightly flexed during the movement and the center of gravity is low and balanced. (See fig. 3-21.) At the time of the change of direction, the head and trunk are turned quickly in the new direction.

Figure 3-21 is available in the illustration section at the end of the book.

(4) Climbing and Surmounting. All Marines should know how to climb and surmount various types of obstacles. (See fig. 3-22.)

Figure 3-22 is available in the illustration section at the end of the book.

(a) Vertical Climbing, as in Climbing a Rope or Pole. The hands grasp the rope or pole overhead with the palms toward the face. Gripping the object, the body is pulled upward with the arms and shoulders, assisted by the feet which grip the object and assist by pushing downward. If shoulder girdle strength and body coordination are not adequate to permit alternating the hands, the arms act together in pulling upward.

(b) Climbing as in Surmounting a Wall. Ingoing over a wall, the body should be kept as close to the top as possible, since in combat operations it is important to offer as small a target as possible to the enemy. If an individual climbs a wall while carrying a rifle, both hands should be freed by slinging the rifle over the back. There are two methods commonly used for surmounting a wall of moderate height, but only one for dropping from it.The methods are as follows:

1 Run, Jump, and Vault. Approach the wall at a run, jump forward and upward at the wall and place one foot against it as high up as possible. Use the foot in contact with the wall to help push the body upward while grasping the top of the wall with the hands. Pull the body up with the arms, assisted by pressure of the foot against the wall and swing the legs over, propelling the body weight over the wall.

2 Hook and Swing. Approach the wall at a run and jump forward and upward. Hook one elbow over the wall, locking the arm in place by pulling up until the top of the wall is underneath the armpit. Grasp the top of the wall with the other hand. Draw the leg which is closer to the wall up as far toward the top as possible. Then swing the other leg over the top of the wall.The body is then carried over with a rolling motion. A variation of this leg action can be used by Marines who are unable to draw up the leg as described. While hanging with both legs fully extended, start a swinging motion with the legs together. When the legs have enough momentum, swing the outside leg over the top of the wall with a vigorous kick, then follow with the body.

3 **Dropping.** All drops from a height are executed in the same manner regardless of the method used to gain the top. One hand is placed against the far side of the wall while the other hand grasps the top. From this position, the body is rolled over the wall and vaulted away from it with the legs swinging clear. As the body passes over the wall and drops, it should at all times face the wall. This will keep the rifle and other equipment clear. Break the fall by retaining a grasp on the top of the wall as long as possible.

(c) Climbing Ladders and Cargo Nets. Rope ladders, stationary vertical ladders, and cargo nets employ the same general technique. The important element is to grasp the side supports firmly in the hands about shoulder height and place the feet on a rung which would cause the body to be fully extended. In movement upward, one hand is moved upward and a new grasp is secured and, at the same time, the opposite leg moves up a rung. As the knee straightens, the body is elevated. This process is repeated using the opposite arm and leg. Alternation continues in this manner until the climber reaches the objective.

(5) Traversing Horizontal Objects. The traversing of horizontal objects puts heavy stress on the arms and shoulder girdle area as the feet are usually suspended in the air with all of the body weight on the arms and shoulders.(See fig. 3-23.)

(a) Traversing Horizontal Ropes or Pipes. The hands grasp the horizontal support overhead with the palms facing. To propel the body forward, one hand is released and moved forward to secure a new grasp. At the same time, the opposite side of the body is swung forward (some people are able to "walk" in the air, keeping the body to the front and moving the legs in time with the arms as in walking on the ground). The other hand is then released and moved forward; this alternation is continued until the objective is reached.

Figure 3-23 is available in the illustration section at the end of the book.

(b) Traversing Horizontal Ladders. In this situation, the movement is the same as used in traversing a rope or pipe. The hands, however, are placed on the rung with the palms away from the face. Other than this difference the technique is the same.

(6) Crawling. Crawling in combat situations is an often used skill. Crawling may be high or low.(See fig. 3-24.)

Figure 3-24 is available in the illustration section at the end of the book.

(a) High Crawl.

In the high crawl, the Marine moves on hands and knees, moving one hand and the opposite knee and then continuing to move the hands in alternation with the opposite knee following the companion hand.

(b) Low Crawl. The Marine is in the prone position usually with the forearms and palms of the hand on the ground. He propels forward by bending the knee of one leg and pushing with the inside edge of the shoe. At the same time, the opposite arm moves forward and pulls to the rear. The body remains low and movement is continued by bending the opposite knee and pushing, and at the same time sliding the opposite arm forward and pulling. Alternation of hands and legs continues until the objective is reached.

(7) Throwing. Throwing may be executed from the kneeling or standing positions. The object to be thrown is held in the hand, and the throwing arm is bent at the elbow; the hand is then moved to the rear until the hand is behind the ear. The body is turned so that the lead foot and balance arm on the side toward the target are pointing at the target. (See fig. 3-25.) The balance arm is used to sight over and align the throwing hand and the target. When properly aligned, the elbow is moved rapidly forward until it is at a pointjust in front of the body where the arm is straightened and the wrist snapped. This whip motion propels the object to the target. Underhand throws secure momentum by the thrower bending his knees and swinging the throwing arm to the rear. As the knees are straightened, the arm is forcefully swung forward from the shoulder and the object released.

Figure 3-25 is available in the illustration section at the end of the book.

(8) Vaulting. Vaulting is employed to overcome low barriers or fences. (See fig. 3-26.)The object to be surmounted is approached at an angle. The hand on the side next to the obstacle is placed on the top of the obstacle and, with a straight arm, the body weight is pushed upward. At the same time, the leg on the side next to the obstacle is thrown upward and over the top followed by the other leg (side approach). In landing, the weight comes down on the leading leg first followed by regaining the balance on both legs. The free arm serves as a balance. A direct (front) approach can be used at which time both legs go over the object together.

(9) Man Carrying. There are three basic individual means of carrying personnel in combat situations and one of these methods may be used in carrying objects.

Figure 3-26 is available in the illustration section at the end of the book.

(a) Fireman's Carry."A" stands sideways in front of "B", "A" bends his knees and leans forward, placing one arm through "B's" crotch, grasps the wrist of "B's" arm, which is hanging over the shoulder, and then "A" runs forward.

(b) Saddle-Back Carry. With his back toward "B", "A" stands in front of "B". "B" mounts "A's" hips and clasps his arms in front of "A's" chest. "A" grasps "B's" thighs.

(c) Single-Shoulder Carry. "A" stands facing "B"."A" assumes a semi-squatting position. "B" leans forward until "B" lies across "A's" right shoulder. "A" clasps his arms around "B's" legs and straightens up, lifting "B" from the ground. "A" then runs forward. This method may also be used to carry heavy objects.

(10) Balancing. Balancing the body while walking or running on a narrow object when crossing obstacles is a skill which requires practice and confidence. Balance is required in negotiating a log placed across a stream, in crossing a narrow beam or rail, and in similar situations.(See fig. 3-27.) To perform this skill, place the feet on the object to be crossed, hold the arms to the sides at shoulder level, and fix the eyes on the object approximately 5 yards in front of the feet. Generally, it is not a good practice to look down at the feet. Walk the beam by placing first one foot and then the other in the center of the beam, thereby moving forward, using the arms to aid in maintaining balance.

Figure 3-27 is available in the illustration section at the end of the book.

(11) Falling. Injury can be avoided if Marines are taught to fall properly. They should know how to use the body momentum to their advantage during a fall rather than to try resisting that force. (See fig. 3-28.) If enough force is present, such as occurs during a fall while running or in jumping downward from a height, individuals can extend their hands to catch the weight. At the same time, duck the head and roll forward onto the feet. The key to falling without injury from the standing position is relaxation and rolling on the outside of the leg, hip, and buttocks to take the brunt of the fall.

Figure 3-28 is available in the illustration section at the end of the book.

3803. OBSTACLE COURSES

a. General. Obstacle courses are a valuable part of physical readiness training. The challenge presented by the obstacles assists in developing and testing the basic physical skills. In many combat situations, success will depend upon the Marine's ability to perform one or more of these skills, often while carrying field equipment and when fatigued. In this section, two types of obstacle courses will be discussed-Conditioning Obstacle Course and Confidence Obstacle Course.

b. Course Safety Precautions. Commanders and course instructors should take certain precautions to prevent injury on obstacle courses. A few of the precautions are:

(1) Inspect the course for faulty construction of obstacles, protruding nails, rotten logs, condition of the landing pits, and other hazards to safety.

(2) Conduct warm up exercises before the unit runs the course.

(3) Explain and demonstrate the correct techniques for negotiating all the obstacles before allowing the Marines to try them.

(4) Give Marines at least two weeks of conditioning exercises before scheduling the obstacle and/or confidence courses.

(5) Ensure that negotiation of the higher and more difficult obstacles are under the supervision of an instructor.

(6) Do not permit individuals who have neither practiced the basic skills nor run the conditioning obstacle course to participate in the confidence obstacle course.

(7) Weather conditions may cause footing or handhold surfaces to be slippery. If such is the case, postpone training on the course.

3804.CONDITIONING OBSTACLE COURSES

a. Description and Objective. The Conditioning Obstacle Course is commonly known as the Obstacle Course. This course consists of fairly low obstacles which are designed to be negotiated quickly. The obstacles serve to test various basic skills, and running the course is a test of the Marine's physical condition. After receiving instruction and an opportunity to practice the skills, Marines run the course against time.

b. Area and Equipment

(1) Complete standardization of obstacle courses should not be attempted since topographical conditions always vary. Commanders should use ingenuity in constructing a course, making good use of streams, hills, trees, rocks, and other natural obstacles. Since the course is eventually run at high speed, it should not be dangerous.

(2) The course should be wide enough for six or eight men to run simultaneously, encouraging competition. The lanes for the first several obstacles should be wider and the obstacles themselves easier than those that follow. This avoids congestion until the contestants scatter out over the course. The last two or three obstacles should not be too difficult and should not involve high climbing. This prevents injuries and falls resulting from fatigue.

(3) The total distance of the course should range from 300 to 450 yards and include from 15 to 25 obstacles. Normally the obstacles should be 20 to 30 yards apart and arranged so that those which exercise the same groups of muscles are separated.

(4) The obstacles should be substantially built. Peeled logs, 6 to 8 inches in diameter, are ideal for many of the obstacles. Sharp points and corners should be eliminated. Landing pits for jumps or vaults should be filled with sand or sawdust to prevent injuries.

(5) The course should be constructed and marked so that it is not possible to sidestep or detour obstacles. However, it is desirable to provide alternate obstacles of varying degrees of difficulty.

(6) The course should be in the shape of a horseshoe or figure eight so that the finish is close to the start and signs should be placed to indicate the course route.

c. Leadership. Before Marines run an obstacle course, they should be instructed in the proper technique of negotiating each obstacle. In each case this technique should be explained and demonstrated in detail, with emphasis on avoiding injury. Every individual should be given an opportunity to practice on each obstacle until he becomes reasonably proficient at negotiating it. Before the course is run against time, it is advisable for individuals to make several runs at a slower pace. During such practice or trial runs, the instructor should observe the performances and make appropriate corrections. Marines should never be permitted to run the course for time until they have practiced on all obstacles. The best method of timing the runners is to have the timer stand at the finish and call out the minutes and seconds as each individual finishes. If several watches are available, each wave may be timed separately. If only one watch is available, the different waves should be started at regular intervals, such as every 30 seconds. If an individual fails to negotiate an obstacle, a previously determined penalty should be exacted.

d.Types of Obstacles

(1) Jumping-Type Obstacles. These obstacles may be ditches which are cleared with one leap, trenches which the individuals can jump into, heights which require jumping downward, or hurdles which an individual can jump over. (See fig. 3-29.)

Figure 3-29 is available in the illustration section at the end of the book.

(2) Dodging-Type Obstacles. Obstacles of this type are usually mazes consisting of posts set in the ground at irregular intervals. (See fig. 3-30.)The intervals between posts should be rather narrow so that the Marines must pick their way carefully through and around them. Lane guides may be constructed to guide the Marines to dodge and change direction. Obstacles may be put into a maze pattern to cause the Marines to change direction.

Figure 3-30 is available in the illustration section at the end of the book.

(3) Vertical Climbing and Surmounting Type Obstacles. These obstacles may be climbing ropes, either plain or knotted and 1-5 inches in diameter; cargo nets or walls 7 or 8 feet high; or vertical poles 6 to 8 inches in diameter and 38 feet high.(Seefig. 3-31.)

Figure 3-31 is available in the illustration section at the end of the book.

(4) Horizontal Traversing-Type Obstacles. Horizontal obstacles may be pipes, beams, ladders, or ropes. (See fig. 3-32.)

(5) Crawling Type Obstacles. Obstacles which require crawling may be constructed of large pipe sections, low rails, and wire. (See fig. 3-33.)

Figure 3-32 is available in the illustration section at the end of the book.

(6) Vaulting-Type Obstacles. Obstacles of 3 to 3.5 feet in height such as low walls or fences may be used as a vaulting obstacle.(See fig. 3-34.)

(7)Balancing-Type Obstacles. Beams, logs, and planks may be used as balancing-type obstacles.(See fig. 3-35.)These items may be used to span water obstacles and dry ditches, or raised off the ground somewhat to simulate natural depressions.

Figure 3-33 is available in the illustration section at the end of the book.

3805.CONFIDENCE OBSTACLE COURSES

a. Description and Objective. The Confidence Obstacle Course is commonly known as the Confidence Course. This course is composed of higher and more difficult obstacles than those used in the conditioning course. The confidence obstacle course is designed to give Marines confidence in their mental and physical capacities and to cultivate their spirit of daring. They are encouraged but not compelled to negotiate this course and the course is not run against time. The negotiation of a confidence course, however, is strenuous enough to be an excellent physical conditioner. Marines should NEVER attempt to take the obstacles at high speed and should not compete for speed. The obstacles vary from fairly easy to extremely difficult. Some are of considerable height to accustom Marines to climbing such heights without fear. Considerable emphasis is placed on obstacles that train and test an individual's balance.

Figure 3-34 is available in the illustration section at the end of the book.

b. Area and Equipment

(1) The confidence course accommodates four platoons, one platoon at each group of six obstacles. The course should be made up of about 24 obstacles, numbered and marked as follows:! to 6, white numbers on red background; 7 to 12, black numbers on a white background;

13 to 18, white numbers on a blue background; and 19 to 24, white numbers on a black background.

(2)A few simple pieces of equipment should be provided for individuals who do not have the strength, courage, or ability to negotiate the obstacles.

Figure 3-35 is available in the illustration section at the end of the book.

c. Formation. The obstacles should be divided into groups of six, and each group is designated by a different color. Each platoon starts at a different color. Individuals are separated into groups of 8 to 12 at each obstacle. At the starting signal from the company commander, they proceed numerically through. Anyone may skip an obstacle who is afraid to try. Individuals proceed from obstacle to obstacle until time is called, then assemble as ordered.

d. Leadership. If the Marines are new to the confidence course, an instructor will demonstrate or will give a brief orientation at each obstacle, including an explanation and demonstration of a method of negotiating it. Marines are encouraged to try the various obstacles, but they are not compelled to do so. No compulsion is to be used. The manner of negotiating any obstacle is left to the discretion of the Marine. However, the instructor assists anyone who experiences difficulty. Instructors must supervise closely at all times to prevent injuries, as some of the obstacles are quite high. Also, some of the obstacles should not be used when slippery or wet. The example of instructors and especially selected demonstrators will serve to inspire the individuals to greater effort.

e. Negotiating the Obstacles. Although personnel need not conform to any one method of negotiating the obstacles, there should be some uniformity in the approach to them. A general method of negotiating the obstacles is indicated below.

(1) Red Group This group contains the first six obstacles. (See fig. 3-36.)

(a) The Belly Buster Individuals may vault, jump, or climb over. Warn them that the log is not stationary.

(b) Reverse Climb. Climb the reverse incline and go down the other side to the ground.

(c) The Weaver. Move from one end of the obstacle to the other by weaving the body under one bar and over the next.

(d) Hip-Hip. Step over each bar, either alternating legs or using same lead leg each time.

(e) Balancing Logs. Step up on log, retaining the balance, walk or run along log.

(f) Island Hopper. Jump from one log to another until the obstacle is negotiated.

(2)White Group. This group is composed of the second six obstacles. (See fig. 3-37.)

(a) Tough Nut. Step over each X in the lane.

(b) Slide for Life. Climb the tower, grasp the rope firmly, and swing the legs upward. Hold the rope with the legs to distribute the weight between them and the arms. Braking the slide with the feet and legs, proceed down the rope. Warn Marines that there is danger of getting rope burns on their hands. When the rope is slippery or wet, this can be a dangerous obstacle.

(c) Low Belly Over. .Mount the low log and jump onto the high log, both arms grasping over the top of the log, the stomach area in contact with it. Swing the legs over the log and lower the body to the ground.

(d) Belly Crawl Move forward under the wire, belly down, to the end of the obstacle.

(e) The Dirty Name Mount the low log and jump to or reach the higher logs in succession, then jump or drop to the ground. Warn the Marines about the height of the final log.

(f) The Tarzan Mount the lower log and walk the length of it and each successive, higher log until reaching the horizontal ladder. Grasp two rungs of the ladder and swing the body into the air. Negotiate the length of the ladder by releasing one hand at a time and swing forward, grasping a more distant rung.

(3) Blue Group. This group is formed by the third group of six obstacles. (See fig. 3-38.)

(a) High Step over. Step over each log, alternating the lead foot or using the same lead foot.

(b) Swinger. Climb onto the swinging log and over to the ground on the opposite side.

(c) Low Wire. Move under the wire on the back, using the hands to raise the wire to clear the body.

(d) Swing, Stop and Jump. Gain momentum with a short run, grasp the rope, and swing the body forward to the top of the wall. Release the rope while standing on the wall and jump to the ground.

(e) Six Vaults. Vault over the logs, using one or both hands.

(f) Easy Balancer. Walk up one inclined log and down the one on the other side to the ground.

(4) Black Group. The last group is formed by the final six obstacles. (See fig. 3- 39.)

(a) Incline Wall. Approach the underside of the wall, jump up and grasp the top and pull the body up and over. Slide or jump down the incline to the ground.

(b) Skyscraper. Jump or climb to the first floor, climb up the corner posts or assist each other to any desired floor. Descend to the ground in any desired manner.

(c) Jump and Land. Climb up the ladder to the platform and jump to the ground.

(d) Confidence Climb. Climb the inclined ladder to the vertical ladder. Go to the top of the vertical ladder, then down the other side to the ground.

(e) Belly Robber. Step on the lower log and assume the prone position on the horizontal logs. Crawl over the logs to the opposite end of the obstacle.

(f) The Tough One Climb the rope or pole on the higher end of the obstacle, then go down the ladder and across the log platform. Climb over or between the logs at the end and go down the rope or pole to the ground. Vault over the final log.

Figure 3-36 is available in the illustration section at the end of the book.

Figure 3-37 is available in the illustration section at the end of the book.

Figure 3-38 is available in the illustration section at the end of the book.

Figure 3-39 is available in the illustration section at the end of the book.

Section IX Individual Exercise Programs

3901.GENERAL

Marines at times will **be** stationed on independent duty and consequently will be responsible for their own physical fitness program. This section will assist them in understanding the need for exercise and will aid in the planning and execution of an individual exercise program. Exercise activities included are the Bench Conditioner, 6-12 Plan, Weight Training, and Isometric Contraction. Keeping physically fit is a problem that faces every Marine. Even though we are frequently engaged in training that requires some physical effort, in many cases, it is not enough to prepare us to meet the intense physical demands of combat. Attaining a satisfactory level of physical readiness is not an insurmountable objective. Available time appears to be the most difficult obstacle to the development of physical readiness. In most cases, regular physical training programs are centralized, requiring the individual to temporarily leave the work area. The problems involved in setting an hour aside two or three times each are numerous. However, most of us can devote a few minutes each day to physical fitness with little, if any, impact on our daily work schedule, especially if it does not require us to leave our work area.

a. Type of Program. There are many good physical fitness programs available to the individual or group. Regardless of the type or duration, to be effective, the program must contain exercises that are strenuous and are challenging to the individual. Space will not permit the inclusion of all available means of individual exercise. The programs selected for this chapter have met the requirement of minimum space and minimum time.

b. Need to Augment Program. These programs are quite strenuous and will develop a satisfactory level of physical readiness. However, if the Marine desires additional development of endurance, it is recommended that he supplement these programs with a 15- to 30-minute period of wind sprints and double timing on an alternating daily basis.

c. Progressive Training. If Marines are performing duties which require little or no physical activity, they must plan a physical conditioning program that assures a moderate beginning, moderate but steady progression, and sufficient warmup before starting the vigorous exercise. To avoid organic of bodily harm, a Marine activity without adequate warmup. He should conduct conditioning programs on a daily basis over an extended period of time, never on an unduly accelerated or crash basis.

3902.THE BENCH CONDITIONER

a. Description and Objective. The bench conditioning program uses a modified bench to employ both isotonic (moving) and isometric (stationary) exercises are as the nucleus of the program. The exercises are designed to develop strength and endurance in all the major muscle groups of the body. The principles of progression, overload, and balance are employed when the exercises are performed properly.

b. Area and Equipment. The conditioning apparatus can be constructed in any unit motor pool with welding equipment found in most salvage yards.(See figs. 3-40, 3-41.)Additionally, there are available any number of commercially produced apparatus available. The important thing is that in utilizing this equipment the following exercise routine be adhered to.

c. Starting Level and Progression.. The program consists of two tables, each with 10 exercises. The Marine controls progression by required repetitions or, in some cases, by application of maximum effort. Each table can be completed within 15 minutes.

d. Starting Position. To start the program, the Marine begins with the Table 1 and executes each exercise for the required number of repetitions as indicated. The Marine controls the starting level and progression. When the Marine executes the maximum repetitions for Table I within a 15-minute exercise period, he progresses to Table II. To maintain this level of development, the Marine should also execute the maximum repetitions for Table II within a 15-minute exercise period. The Marine should keep substitution of exercises to a minimum. However, it he completes a full 15 minutes of strenuous exercise and exercises all muscles, then there should be no appreciable difference in the overall development.

e. Bench Conditioning, Table I

(1) Exercise 1: Side-Straddle Hop. This is a two-count warmup exercise done at moderate cadence. The starting position is the position of attention. On count ONE jump slightly into the air, swinging the arms out to the sides and up to a vertical position, hands touching.(See fig. 3-40, A.)At the same time, spread the feet wider than shoulder-width apart. On count TWO, using a slight flexing of the knees and ankles, jump slightly into the air and return to the starting position by swinging the arms back down to the sides. Twenty repetitions of this exercise is the standard dosage throughout the program.

(2) Exercise 2: Hand Walk. Remove the lower horizontal bar. Adjust the upper horizontal bar so that it is high enough to permit a "dead hanging" position, with the feet off the ground.(See fig. 3-40, B.j From the "dead hanging" position, release one hand and drop the arm to the side of the body. Then raise that arm and regrasp the horizontal bar. Release the bar with the other hand and drop that arm to the side. Repeat this as many times as possible.

(3) Exercise 3: Situps. Lie down with the fingers interlocked and placed behind the head. Hook the toes under the foot braces. Raise the trunk and upper body to an upright sitting position, twisting it to the left and then forward and downward until the right elbow touches the left knee.(See fig. 3-40, C.)Lower the body to the starting position .Sit up again but twist the body to the opposite direction as before, touching the left elbow to the right knee. Again lower the body to the starting position. The starting dosage is 20 situps. The Marine should continue the progression until he has attained 40 situps.

Figure 3-40 is available in the illustration section at the end of the book.

(4) Exercise 4: Double Step-upStarting at one end of the bench, step up onto the bench, and walk across it. Step down from the other end; turn around and repeat the process to return to the starting point.(See fig. 3-40, D.)Each return to the starting point constitutes a repetition. The starting dosage is 20 repetitions. Maximum dosage is 35 repetitions. This exercise should be done at a rapid cadence.

(5) Exercise 5: Isometric Bar Lift. Adjust the lower bar so that it is slightly higher than the beltline. Placing the feet on the footplates at the base of the frame, grasp the lower bar so that the hands are spread shoulder-width apart. Assume a crouched position and lift with maximum effort using the arms, back and legs.(See fig. 3-40, E.)Starting dosage is 4 repetitions of a stress time of 5 seconds followed by a 5-second rest prior to the next repetition. The Marine obtains progression by lengthening stress periods to 6 and later 7 seconds. Do not increase the number of repetitions.

(6) Exercise 6: Knee Lift. Adjust the upper bar to the same height used in Exercise 2.Adjust the lower bar so that is stops rearward movement of the hips when the "dead hanging" position is assumed.(See fig. 3-40, F.) Keeping the arms extended, flex the legs and raise the knees as high as possible. Hold this position for 5 seconds, then return to the starting position. After 2 seconds in the starting position, raise the knees again. Each return to the starting position constitutes one repetition. The dosage is five repetitions. Progression is the same as in Exercise 5.

(7) Exercise 7: Isometric Pull. Adjust the lower horizontal bar to a position where it is slightly higher than the beltline. Grasp the handles and pull outward.(See fig. 3-40, G.)Apply maximum effort and hold for approximately 5 seconds. Relax for 5 seconds between repetitions; perform four repetitions. Moving the body closer to or farther away from the bar will change the stress from the upper arms to the forearms. Progression is the same as in Exercise 5.

(8) Exercise 8: Isometric Compression. Maintain the position as in exercise 7.(See fig. 3-40, H.)Grasping the handles in the same manner, press in **with** maximum effort and hold for approximately 5 seconds. Relax for 5 seconds between repetitions. Progression is the same as in Exercise 5.

(9) Exercise 9: Isometric Press. Remove the lower horizontal bar. Adjust the upper horizontal bar until it is about 6 inches lower than the extended arms can reach. Stepping on the footplates at the bottom of the frame, grasp the bar with both hands and push up.(See fig. 3-40,1.)Keep both the legs and arms slightly flexed and the back straight. Apply maximum effort for 5 seconds then relax for 5 seconds. Complete four repetitions. Progression is the same as in Exercise 5.

(10) Exercise 10: Pushups. Grasping the foot braces with both hands, assume the front leaning rest position.(See fig. 3-40, J.)Keeping the back and legs straight, lower the body until the chest is lower than the **hands,** then return to the starting position. The Marine should complete the maximum possible number of repetitions.

f. Bench Conditioning, Table II. There is no limit on the maximum number of repetitions attainable in Exercises 3, 4, and 6 of Table II. The only limit imposed is that the entire program of 10 exercises outlinted in either table should not exceed 15 minutes.

(1) Exercise 1: Side-Straddle Hop. This is a two-count warm up exercise done at a moderate cadence. The starting position is the position of attention. On count ONE, jump slightly into the air, swinging the arms out to the sides and up to a vertical position, hands touching.(See fig. 3-41, A.)At the same time, spread the feet wider than shoulder-width apart. On count TWO, using a slight flexing of the knees and ankles, jump slightly into the air and return to the starting position by swinging the arms back down to the sides. Twenty repetitions of this exercise is the standard dosage throughout the program.

(2) Exercise 2: Pull-up. Adjust the horizontal bar so that is is high enough to permit a "dead hanging" position with the feet off the ground. Grasp the bar with both hands, palms facing forward. By flexing the arms, raise the body to a position where the chin is higher than the bar.(See fig. 3-41, B.)Then lower the body to the "dead hanging" position. Repeat as many times as possible.

(3) Exercise 3: Bench situps. Sit on the bench and hook the feet under the foot braces. With the fingers interlocked behind the head, lean back until the head touches the floor.(See fig. 3-41, C.)Return to the starting position. The starting dosage is 15 situps.

(4) Exercise 4: Step-Up. Face the bench and step up on it with one foot, bringing the trailing foot up next to the leading foot. Step back down again, leading with the same foot used first in stepping up.(See fig. 3-41, D.)Perlorm half of the total repetitions, then change the sequence of moving the feet to use the other leg in stepping up, and repeat the same amount of exercise. The starting level is a total of 40 step-up. **This exercise should be done at a rapid cadence.**

(5) Exercise 5: Isometric Bar Lift\djust the lower bar so that it is slightly higher than the beltline. Placing the feet on the footplates at the base of the frame, grasp the lower bar so that the hands are spread shoulder-width apart. Assume a crouched position and lift with maximum effort using the arms, back, and legs.(See fig. 3-41, E.)Starting dosage is four repetitions of a stress time of 8 seconds followed by a 5-second rest prior to the next repetition. Progression is obtained by lengthening stress periods to 10 seconds. Do not increase the number of repetitions.

Figure 3-41 is available in the illustration section at the end of the book.

(6)Exercise 6: Leg Lift. Adjust the bars and assume the starting position as shown in figure 3-41, F. Keeping arms and legs extended, raise the legs to a horizontal position and hold in that position for 2 seconds. Then lower the legs slowly to the starting position. Five repetitions is the starting level.

(7) Exercise 7: Isometric Pull. Adjust the lower horizontal bar so that it is slightly higher than the beltline. Grasp the handles and pull outward.(See fig. 3-41, G.)Apply maximum effort and hold for approximately 8 seconds. Relax for 5 seconds between repetitions; perform four repetitions. Moving the body closer to or farther away from the bar will change the stress from the upper arms to the forearms. The Marine obtains progression by lengthening the stress period to 10 seconds.

(8) Exercise 8: Isometric Compression. (See fig. 3-41, H.)Maintain the position as in exercise 7.Grasping the handles in the same manner, press in with maximum effort and hold for approximately 8 seconds. Relax for 5 seconds between repetitions; perform four repetitions. The Marine obtains progression by lengthening the stress period to 10 seconds.

(9) Exercise 9: Isometric Pres. Remove the lower horizontal bar. Adjust the upper horizontal bar until it is about 6 inches lower than the extended arms can reach. Stepping on the footplates at the bottom of the frame, grasp the bar with both hands and push up.(See fig. 3-41,1.)Keep both the legs and arms slightly flexed and the back straight. Apply maximum effort for 8 seconds, then relax for 5 seconds. Complete four repetitions. The Marine obtains progression by lengthening the stress period to 10 seconds.

(10) Exercise 10: Inclined Pushup. Assume the front leaning rest position with the feet on the bench.(See fig. 3-41, J.)Keeping the back and legs straight, lower the body until the nose touches the ground. By extending the arms, raise the body to the starting position. Repeat as many times as possible.

3903. THE 6-12 PLAN

a. Description and objective. The 6-12 plan of physical fitness has been developed to assist in regulating quantity and progression and to provide a convenient set of exercises. This is a basic program and will take 18 weeks to complete if you follow the moderate progression as to the time prescribed for each level of achievement. The time can be shortened as explained below. This plan consists of six basic exercises a day which can be completed in 12 minutes. There are six tables of six exercises each, thus allowing you to progress from table to table. The plan is progressive, fits any age group, contains balance and variety, and applies the principle of overload in a safe and gradual manner. Begin at Table I, Progression Guide, with the number of repetitions as indicated by age.

b. Formation. If just starting an exercise program, do not rush through the first table. Remember, individuals should remain at each level for about a week before moving upward. The time allotment stated for each exercise at the bottom of the tables is a guide; some people may take more and some less time on the individual exercises. At the end of a one-week period, when the individual can comfortable perform the six exercises in 12 minutes, move on to the next level. To a certain degree, the individual must be the judge of his ability to progress form level to level and table to table. If starting with a certain degree of fitness, some of the beginning tables may present little challenge.

c. Starting Level **and Progression.** There are three levels of achievement for each age group, indicated as A, B, and C. Start at the C level for the appropriate age group. At the end of a one-week period, or when the individual can do all exercises at that level within 12 minutes, progress to the B level. At the end of the second week, or when the individual can accomplish that level within 12 minutes, progress to the A level. At the conclusion of the third week or when the individual can achieve the A level within the time limitation, move on to table II.(See Tables I through VI, Progression Guide.)

d. Maintenance Level. Attempt to work through all six tables. If this proves to be too difficult, then maintain exercise at the—

(1) A-level on Table IV, Progression Guide it in the 45 to 49, 50 to 59, or over 60 age group.

(2) A-level on Table V, Progression Guide if in the 17 to 29, 30 to 39, or 40 to 44 age group.

e. Precautions. To achieve the maximum benefit, perform each exercise exactly as specified. Read the descriptions and study the illustrations. Do not slight the movements. Use a sensible approach and follow these four points as they apply before starting or during your exercise program.

(1) If you have the slightest doubt about your ability to participate in this exercise program, consult a physician.

(2) Stop immediately if you notice unusual breathlessness or chest pain while taking part in these exercises. If these conditions persist, consult a physician.

The following Tables of Progression Guides and Exercises are the progression of the 6-12 Plan program.

Table I. Progression Guide is available in the illustration section at the end of the book.

Table I. Exercises.is available in the illustration section at the end of the book.

Table II. Progression Guide.is available in the illustration section at the end of the book.

Table II. Exercises.is available in the illustration section at the end of the book.

Table III. Progression Guide.is available in the illustration section at the end of the book.

Table III. Exercises.is available in the illustration section at the end of the book.

Table IV. Progression Guide.is available in the illustration section at the end of the book.

Table IV. Exercises.is available in the illustration section at the end of the book.

Table V. Progression Guide.is available in the illustration section at the end of the book.

Table V. Exercises.is available in the illustration section at the end of the book.

Table VI. Progression Guide.is available in the illustration section at the end of the book.

Table VI. Exercises.is available in the illustration section at the end of the book.

(3) Unless you have exercised regularly and know yourself to be in good physical condition, start at table I with the C level appropriate to your age.

(4) If you are out of shape, admit that fact to yourself. Hide your pride; after all, you are in the privacy of your own quarters. Set your goal for the longer, steadier pull toward fitness. Resist the urge to pass over the lower numbered tables to find a table that will test your fitness. You are not trying to test, but rather to develop.

3904.WEIGHT TRAINING

Weight or barbell training should not be confused with the more common types of weight lifting used as a competitive sport. Weight lifting is designed to develop specific muscle groups so that the individual is capable of lifting a large amount during a single lift. In contrast, weight training is the systematic development of all the major muscle groups by the use of calisthenics reinforced with weight to provide resistance.

3905.WEIGHT LIFTING

a. Description and Objective. The weight lifting program is progressive and applies the principle of overload in a safe, gradual manner. The exercises of the table can be completed in 15 minutes. The objective of the exercises is to develop strength and muscular endurance, and muscle tone of the five major muscle groups:legs, arms, back, trunk, and shoulder girdle.(See fig. 3-42.)

b. Warmup. A warm up exercise is important to prepare the body for the more vigorous exercises that are to follow. Ten repetitions of the high jumper exercise are excellent for a warm up period. Also, an exercise which will require fast body movement is needed.To provide such exercise, 3 to 5 minutes of rope skipping is recommended to increase the individual's development potential.

c. Formation. Marines should take care in the completion of these weight lifting exercises. They should ensure that their back is straight during the lifting phase of all exercises. When exercises require assuming the standing position with the weight, they should always grasp the weight while in a squatting position and then rise to a standing position.

d. Progression. Each exercise has a starting number of repetitions and specified pounds of weight. After each 4th or 5th day of exercise, the Marine should increase the repetitions by one until he has reached the maximum of 10 repetitions. At this time, the Marine should increase the weight by 5 pounds. Then he should repeat the process again with the initial number or repetitions.

e. Weight Training

(1) Exercise 1:Squat(See fig. 3-42, A.)Starting level—six repetitions, 50 pounds (commonly called the flatfoot deep knee bend). Place the bar upon the shoulders. Stand with feet about 18 inches apart. Keeping the feet flat, lower the body into the low squat position. Stand erect and repeat. Exhale while lowering into the squat position and inhale while standing. This constitutes one repetition.

(2) Exercise 2:Waist Bender(.See fig. 3-42, B.)Starting level-6 repetitions, 40 pounds. Assume the standing position with the bar across the shoulders and the feet a shoulder-width apart .Bend forward at the waist until the upper body is parallel to the ground; return to the starting position. Each return to the upright position constitutes one repetition.

(3) Exercise 3:Curl (See fig. 3-42, C.)Starting level—6 repetitions, 40 pounds. Grasp the barbell with the palms facing to the rear and assume the standing position with the feet a should-width apart. With the barbell held in front of the hips, flex the elbows and lift the weight until the bar touches the upper chest. Lower the barbell back to the hip level position. Inhale deeply with the upward movement and exhale on the downward movement. Each time that the bar touches the chest will constitute one repetition.

(4) Exercise 4:Side Bender(.See fig. 3-42, D.)Starting level—6 repetitions per side, 40 pounds. Assume the standing position, with the bar across the shoulders, with feet a shoulder-width apart. Bend to the left as far as possible and return to the starting position. Repeat six times and then execute the same procedure to the right for six repetitions.

(5) Exercise 5:Standing Press(See fig. 3-42, E.)Starting level—6 repetitions, 45 pounds. Grasp the bar with the palms facing forward and assume the starting position. Curl the weight to the upper chest position. Inhale deeply and press the bar upward to an overhead position. Exhale while lowering the bar to the chest position. Each time that the bar is pressed upward constitutes one repetition.

(6) Exercise 6:Upward Row. .see fig. 3-42, F.)Starting level—6 repetitions, 40 pounds. Grasp the bar, hands close together, palms to the rear, and assume the standing position. Starting with the bar held in front of the hips, flexing the elbows and the shoulder girdle muscles, lift the bar straight up to an overhead position. Inhale deeply while lifting the bar. Exhale while lowering the bar to the hip position. Each time that the bar returns to the hips constitutes one repetition.

(7) Exercise 7:Shoulder Curl. See fig. 3-42, G.)Starting level—6 repetitions, 25 pounds. Grasp the bar, palms down, and assume the standing position. Keeping the elbows locked, curl the bar, pivoting the arms at the shoulders until the bar is in an overhead position and as far to the rear as possible. Return the bar in the same manner to the hip position. Each time that the bar returns to the hip position constitutes one repetition.

Figure 3-42 is available in the illustration section at the end of the book.

3906.ISOMETRIC EXERCISING

a. Description and Objective. Isometric exercising is the application of maximum effort during an exercise period. The isometric principle is to apply force gradually over a 5- to 10- second period until the maximum effort is applied. Relaxation follows for approximately 5 seconds and then force is applied again. The Marine continues this process at the prescribed level for each exercise for a period of 15 minutes or less. The objective of isometric exercises is to create muscle growth. Isometric exercises are the fastest means of creating muscle growth. They are founded on the fact that a muscle will grow only so fast regardless of the type or duration of the activity. The principle of overload—that the muscles develop commensurate with demand—reinforces that fact. However, isometrics will not develop cardiovascular or muscular endurance. Consequently, Marines who choose to use isometrics in their exercise routine must also include running or some other type of aerobic activity.

b. **Area and Equipment** isometric exercises may be designed to be performed with or without equipment.

c. Use With Other Programs. The isometric exercises presented in this section will provide a variety from which to choose. In addition to the exercises contained in this section, the application of isometric force is used in some of the exercises in section II.

d.Door Frame Exercises. The following exercises are designed for use with a standard door frame found in all offices or barracks.(See fig. 3-43.)

(1) Exercise l:Arm Press. Stand in the doorway with the legs straight, knees locked. Using the arm muscles, press hard upward against the top of the door frame. Repeat for three repetitions applying gradual effort to maximum contraction.

(2) Exercise 2:Leg Press. Stand in the doorway with the hands on the top of the door frame, elbows locked .With knees bent, press hard with the leg muscles .Repeat for three repetitions beginning with a gradual effort and increasing to maximum contraction. A low platform may be necessary to reach the top of the door frame and still maintain a bent knee position.

(3) Exercise 3:Side Press. Extend both arms to the side of the doorway. Palms are shoulder high, facing outward. With both arms, press hard against the sides of the door frame. Repeat for three repetitions. Begin gradually and increase to maximum contraction.

(4) Exercise 4:Lateral Raise. Extend both arms to the sides of the doorway, arms down, palms facing inward. With the back of the hands, press hard against the sides of the door frame .Repeat for three repetitions. Begin with a gradual effort and increase to maximum contraction.

Figure 3-43 is available in the illustration section at the end of the book.

(5) Exercise 5:Neck Press?lace the forehead against the door frame, hands clasped behind the back Using the neck muscles, press hard against the door frame. Repeat for three repetitions, then reverse position so that the back of the head is resting on the door frame. Again do three repetitions .Begin gradually with both exercises and increase to maximum contraction.

(6) Exercise 6:Door Pull Stand facing the edge of the open door and grasp the doorknobs. Pull outward with both arms (if doorknobs are not available grasp the edge of the door).Applying outward pressure, move the body toward and away from the door .Repeat the three repetitions. Begin with gradual effort and increase to maximum contraction.

Chapter 4
COMBAT WATER SURVIVAL
Section I. Marine Corps Water Survival Program

4101.MILITARY SWIMMING

Military swimming emphasizes strokes that result in staying power rather than speed. Marines need to swim easily, quietly, and with adequate vision. Strokes should be used that allow the Marine to carry basic combat equipment and to tow or push a wounded buddy while keeping the face out of the water to allow breathing. The best strokes to use are the side stroke and the breast stroke. Marines must be proficient in floating, drown-proofing, and in using their basic combat equipment to make flotation devices. They must be drilled on using the Kapok and Mae West buoyancy compensators and on emergency egress from helicopters, landing craft, and amphibious assault vehicles.

4102.PROGRAM DEVELOPMENT

Commanders of the Landing Force Training Commands (Atlantic/Pacific) are tasked with developing and maintaining the individual water survival and swimming training program. They are responsible for training all Marine Corps water safety and survival instructors and integrating this training in to their combat readiness training program. All Marines are deployable and as soldiers of the sea must constantly practice the skills necessary to achieve mission accomplishment. Commanders must be imaginative in program development, integrating basic skills development which enhances the total mission accomplishment. An example is teaching waterproofing of the Marine's backpack so that it can be used as a flotation device in streams or rivers. Once learned, this skill would be practiced first in a pool and then in a stream or a river. Routine inspections during deployments would ensure that backpacks could be used as flotation devices.

4103.QUAL STDS AND TRAINING GUIDELINES

The inherent nature of Marine Corps operations and training requires that Marines achieve an ability to survive in water. Water survival and swimming training is designed to reduce fear of water, instill self-confidence, and develop a Marine's ability to survive in water. As a minimum, every Marine should be qualified as swimmer, third-class (S3).This qualification should be met during recruit training or Officer Candidate School. Once qualified, a Marine need not be requalified unless requalification at a specific level (S3 or higher) is required by unit mission, military occupational specialty or duty assignment, or for other reasons determined by the commander. The following classifications and standards apply to Marine Corps water survival and swimming training. Abbreviations will be used for service, medical, and training record entries citing MCO 1510.29A, Individual Water Survival and Swimming Training, as authority.

a. Qualification standards and test procedures are described in paragraph 4104.

b. Swimmer classification and abbreviations are:

(1) Unqualified - UQ.

(2) Swimmer, third-class (S3).Minimum aquatic skill level for all Marines.

(3) Swimmer, second-class (S2)-

(a) Minimum standard for all naval aviators, naval flight officers, and aircrewmen with written waiver from the commanding officer.

(b) Requirements for military occupational specialty 1803 (assault amphibian vehicle officer), 1833 (assault amphibian vehicle crewman), 1302 (engineer officer), 1371 (combat engineer), 1381 (shore party specialist), and 1379 (engineer operations chief).

(4) Swimmer, first-class (S1)-

(a) Naval aviator, naval flight officer, and aircrewman qualification requirement.

(b) Requirements for military occupational specialty 0321 (reconnaissance man), 8652 (reconnaissance man, parachute jump qualified), 8653 (reconnaissance man, self-contained underwater breathing apparatus qualified), 8654 (reconnaissance man, parachute/self-contained underwater breathing apparatus qualified), 9952 (self-contained underwater breathing apparatus Marine (officer/enlisted)), and 9953 (parachutist/self-contained underwater breathing apparatus Marine (officer/enlisted)).

(5) Water Survival Qualified (WSQ).Ultimate water survival/swimming goal for Marines.

c.Instructor and assistant instructor classifications are-

(1) **American Red Cross Advanced Lifesaver.** Individuals who are currently certified as American Red Cross advanced lifesavers may assist qualified instructors during Marine Corps water survival/swimming training and testing. Certification is valid for 3 years.

(2) American Red Cross Water Safety Instruction. Individuals who are currently certified as American Red Cross water safety instructors may conduct swimming training and qualify individuals as 3d, 2d, and 1st class swimmers. Certification is valid for 2 years.

(3) United States Marine Corps Water Safety and Survival Instruction. Those personnel certified as water safety and survival instructors by one of the landing force training commands may conduct Marine Corps water survival and swimming training and American Red Cross swimming and water safety training. These instructors may also participate as assistant instructors during the conduct of landing force training command-sponsored water safety and survival instructor courses. Certification is valid for2 years.

(4) American Red Cross Water Safety Instructor Traine. Individuals qualified who have been certified as instructor trainers by the American Red Cross may train American Red Cross water safety and basic swimming instructors. Certification is valid for 2 years.

(5) United States Marine Corps Water Safety and Survival Instructor Traine. Individuals qualified who are certified as instructor trainers through special training courses conducted jointly by the American Red Cross and the Marine Corps, under the direction of the Commandant of the Marine Corps, may train water safety and survival instructors and American Red Cross water safety and basic swimming instructors. Certification is valid for 2 years.

d. Training Guidelines Instructors engaged in training Marines as swimmers will use the current edition of the American Red Cross swimming and water safety courses instructors' manuals as the authoritative textbooks for swimming procedures, strokes, breaks, and holds.

4104.WATER SURV/SWIM QUAL STD AND TEST PROCS

a. The following qualification standards and test procedures are applicable to Marine Corps water survival and swimming training. The utility uniform without cover, boots/oxfords or socks will be worn for all swimming tests through swimmer, first class. The utility uniform with boots/oxfords will be worn as prescribed for the water survival qualification.

(1) Swimmer, Third-Class (S3) To qualify as a swimmer, third-class, a Marine must enter the water feet first from a minimum height of 5 feet and remain afloat for 5 minutes. During this time the Marine must swim 50 yards using any stroke or combination of strokes. This test should be taken by all Marines as early as possible in their initial training. Those unable to pass this test will be classified as UQ and should be given instruction in fundamental swimming skills. Those who barely meet requirements are swimmers who need help and should receive additional instruction in fundamental swimming skills, whenever possible.

(2) Swimmer, Second-Class (S2.) To qualify, a Marine must enter the water feet first from a minimum height of 10 feet and remain afloat for 10 minutes. The Marine must swim 100 yards and use the three basic survival swimming strokes (side stroke, backstroke, and breast stroke) for a minimum distance of 25 yards each.

(3) Swimmer, First-Class (S1)A prerequisite to qualification as swimmer, first-class, is successful completion of the test for swimmer, second-class. To qualify as a swimmer, first-class, a Marine must be able to do each of the following:

(a) Approach a person of approximately the same size while in the water, demonstrate a release, get that person in a carry position, and tow the "distressed swimmer" 25 yards.

(b) Enter water feet first and immediately swim underwater for 25 yards. Swimmer is to break the surface twice for breathing during this distance at intervals of approximately 25 feet.

(c) Remove trousers in water, inflate for support, and remain motionless for a minimum of 1 minute.

(d) Swim 220 yards using any survival swimming stroke or any combination of survival strokes.

(4) Water Survival Qualification (WSQ)A prerequisite to water survival qualification is successful completion of the test for swimmer, first-class. To be water survival-qualified, a Mannemust:

(a) Enter water by jumping from a height of 10 feet.

(b) Stay afloat in full utility uniform (less cover) for 1 hour. Boots/oxfords will be removed after 5 minutes in the water but will be retained by the swimmer.

(c) Traverse 75 yards in deep water in full utility uniform (less cover) with boots/oxfords.

b. All water survival and swimming training will include instruction in the following:

(1) Employment of safety lines or other comparable expedients to assist in swift stream/river crossings.

(2) Adverse physiological effects caused by cold water (hypothermia) and the precautionary measures to be taken prior to exposure to such an environment.

(3) Employment of a standard and expedient flotation devices in a water survival situation.

(4) Removal of individual field equipment after unscheduled water entry. Entry will be made from a minimum height of 10 feet with a complete set of field equipment, properly packed as an expedient flotation device and waterproofed. The ALICE pack will then be removed, the rifle laid across the top of the pack, the helmet remaining or put back on the individual's head, and the individual will traverse 25 yards to the side of the pool or water's edge.

c. The additional water survival qualification (flight physiology training) and requalification required of personnel assigned to flight status will be conducted in accordance with OPNAVINST 3710.7_,Promulgation of NATOPS General Flight and Operating Instructions. Swimming qualification (first-class, SI) is a prerequisite to water survival (flight physiology) training and is a one-time requirement.

Section II. Water Sports and Competitive Activities

4201.GENERAL

The commander, when developing the combat readiness program, needs to look carefully at the mission. If the mission requires Marines who must perform mission-essential water-related activities of long duration, he may want to look to water sports/competitive activities as a method of developing endurance. Endurance and confidence can only be developed by extended time in the water. While defined training can accomplish this goal, the goal can be enhanced through training which is integrated with the physical conditioning program. This approach will provide variety and enthusiasm. Two activities which will accomplish this goal are water relays and water polo. The important consideration is that the unit achieve total participation by all members of the unit.

4202.WATER RELAYS

Development of water relays will generally conform to the guidelines in chapter 5, section IV. When contemplating relays, the commander should review chapter 5 and adapt the principles to the size of the unit and the facility to be used.

4203.WATER POLO

The official rules provided in this paragraph will need to be adapted to the unit's unique situation, its size, and the size of the facility. Commanders should make every effort to have all members of the unit participate.

a. NCAA Rules This style of play is based on two popular sports: swimming and basketball. Playing is done on the surface of the water by teams of seven players each. The size of the playing area is 25 yards by 14 yards, which is the size of the typical high school or college indoor pool. The deeper the water, the better. If the entire playing area is deep-6 feet or more—this is ideal, although most indoor pools have a shallow end. At each end of the pool is a goal. In deep water, the goal is 10 feet across and 3 feet high. In shallow water, where the goalie can stand on and jump from the bottom, the goal is 10 feet across and 5 feet high. One of the seven players on each team is the goalie. He is the only player on the team who can stand on or jump from the bottom (if it is shallow enough to permit this), catch and pass the ball with both hands at the same time, or hit the ball with a clenched fist. However, the goalie may not swim across or pass the ball across the middle of the pool.

b.Equipment. The goalie on the visiting team must wear a white cap with the number 1 clearly marked on it; his teammates, who are designated as guards and forwards, must wear white caps numbered 2 through 7, with the substitutes wearing white caps numbered 8 through 21.The goalie on the home team must wear a dark-colored cap with the number 22 clearly marked on it; his teammates designated as guards and forwards, must wear similar dark-colored caps numbered 29 through 42.The guards and forwards on each team may swim freely up and down the pool as they see fit, interchanging positions as often as they wish. They may not stand on or jump from the bottom, touch the ball with more than one hand at a time when catching or passing or shooting it, or enter inside the opposing team's 2-yard line unless preceded by the ball. The ball is similar to a soccer ball, except that it is yellow and covered with a waterproofing substance that makes it easier to handle with one hand.

c. Playing Time. In intercollegiate competition the game consists of four 7-minule quarters, the teams changing ends after every period of play. In high school competition the game consists of four-,five-, or six-minute quarters, depending on the ages of the participants. Here again the teams change ends after every period. There is a very brief interval between each quarter. At the start of each period the teams line up at their respective ends, and with ablast of the whistle, the head referee throws the ball into the middle of the pool. The fastest swimmers on each team then race to gain possession.

d. Officials. The head referee walks along one side of the pool watching for infractions. He is aided by the assistant referee, who patrols the opposite side of the pool. The referees are both equipped with whistles and two official's flags. These flags should be 12 inches square and should correspond in color to the caps worn by the respective teams. Whenever a referee sees an infraction, he blows his whistle and signals with his flag indicating that the player nearest to the ball with the appropriate colored flag will be given possession of the ball. Whenever the whistle is blown, all players should react by looking immediately to see which color flag is being waived which will indicate which team should be moving into the defense. Other officials are a timer and a scorer. As in all sports, it is important that the officials be skilled and competent.

e. Technical Fouls. The following are some of the common technical fouls: starting before the referee blows his whistle to open play; holding onto or pushing off from the side of the pool during play; taking hold of the ball underwater when tackled by an opposing player; swimming inside an opponent's 2-yard line unless preceded by the ball; touching the ball with both hands at the same time (goalie excepted); standing, walking on or jumping from the bottom when taking an active part in the game. When a technical foul occurs, thereferee blows the whistle and with his flag awards possession. The player on the team awarded possession nearest the point of infraction then has 5 seconds to put the ball into play by passing to a teammate or by dropping the ball into the water and swimming after it.

f. Personal Fouls. The following are some of the more common personal fouls: committing any of the technical fouls mentioned above for the purpose of scoring or preventing a goal; holding, ducking, pulling, pushing off from, swimming over, or impeding the arm or leg movement of any opponent who is not touching the ball; and splashing water in the face of an opponent. When a personal foul occurs, the referee will follow the same procedure as with a technical foul. While doing this, the referee will also call out clearly and loudly the number of the player who was guilty of the infraction and a personal foul will be marked against that player by the official scorer. When a player accumulates five personal fouls, he fouls out of the game and must be replaced by a substitute.

g. Penalty Shot. A penalty shot can be awarded by either of the officials or the scorer when:

o An offensive player inside the opponents' 4-yard line not touching the ball is held, ducked, pulled-back, kicked, or struck.
o A team has accumulated a total of ten personal fouls. When the former occurs, the head or assistant referee should immediately blow the whistle and by holding a two-flagged stick in a vertical position above his head, signal that a penalty shot has been awarded. When the latter occurs, the scorer should use a buzzer located at the scorer's tabic to signal that one team has accumulated ten personal fouls, thereby entitling the other team to a penalty shot. A penalty shot is taken from the 4-yard line in front of the goal. All players except the defending goalie must leave the 4-yard line until the shot is taken. No player can be within 1 yard of the shooter. After ascertaining that the shooting player is on the 4-yard line and the goalie is onThegoal line, the referee will ask the shooter to lift up the ball. When he docs, the referee will give a sharp blast from the whistle. At the whistle, the shooter must shoot without delay and without any faking at the goal. The goalie may try to block the shot. If he does, or the shot is otherwise missed, it is immediately in play and action continues.

h. When a Goal is Scored. When a goal is scored, either from a shot taken by a player out in the field or by penalty shot, the defending goalie must then pass the ball to the nearest referee. The referee will then pass the ball back to the goalie, who puts it into play by a pass to one of his teammates as soon as possible.

Section III. The Battle Swimming Test

4301.GENERAL

The battle fitness test is the evaluation contained in the Guidance for Basic Warrior Training (BWT) as part of the Basic Warrior Training Concept Plan.

4302.THE BATTLE SWIMMING TEST

The battle swim test consist of the following:

a. Be able to climb a three meter high board, while wearing a field uniform, carrying a M-16, web gear, and four canteens. Jump off blindfolded, recover and swim 100 meters with the M-16.

b. Repeat the same process, without the blindfold, and upon entering the water a second time, recover and retrieve all submerged gear.

c. Remain afloat and tread water for five minutes.

d. Demonstrate a knowledge of how to cross a stream using field expedient techniques and a single strand rope bridge.

e. Successfully complete helicopter egress training.

Chapter 5
COMPETITIVE CONDITIONING ACTIVITIES
Section I. Organization of Competitive Activities

5101.LEADERSHIP OF COMPETITIVE ACTIVITIES

a. Place in the Program. Competitive conditioning activities consist of dual combatives, relays, team contests, and team sports in which individuals or teams compete against an opponent to win. Competition is one of the best ways of maintaining interest in the physical training program during the sustaining stage. Organized competition provides enjoyable, vigorous physical activity that has proved to be one of the best supplements to conditioning drill activities. The benefits of competition are the development of aggressiveness, unit pride and identity, teamwork, and the will to win. These activities help develop cardiovascular and muscular endurance, strength, and coordination. Scheduling competitive activities in an orderly and progressive manner is desirable. The progression should be from relays to dual combatives, to team contests, and finally to team athletics. These activities should be conducted as part of the program after a basic level of conditioning has been developed, usually during the slow-improvement stage of conditioning. Muscles and joints should be strengthened by preconditioning to withstand the strain placed upon them by sudden stops and turns, body contact, bearing of weight, and falls. Competitive activities, however, should not be allowed to dominate the physical training program.

b. Area and Equipment. Some of the competitive activities included in this chapter require specific types of areas and equipment; others do not. The area requirement can usually be satisfied on available training fields. When items of equipment are required or specific courts or field layouts are to be marked off, such information is included.

c. Leadership. The principal factor for success as a leader of competitive activity is an energetic, dynamic, enthusiastic approach. The leader's attitude is reflected by the group, so the commander must carry on the activity in a snappy and vigorous manner. Confidence on the part of the commander will create an impression of decisiveness and certainty. Confidence grows out of experience and a thorough knowledge of the activity. Mastery of subject is the first step in developing confidence, assurance, and poise.

(1) The following suggestions are offered for leaders of competitive combatives, contests, and sports:

(a) Get the activity underway quickly by selecting and teaching only the essentials.

(b) Use rules to add to the enjoyment of the activity and do not allow them to interfere with the spirit of competition.

(c) Stop the activity before interest begins to lag.

(d) In team contests, clearly distinguish sides by attempting to maintain unit identity.

(e) Always insist on fair play, enforcing the rules impartially.

(2) The following procedure is recommended for presenting a competitive activity.

(a) Name the activity.

(b) Briefly explain the objective of the activity and give only the pertinent rules.

(c) Have a demonstration at slow speed and answer questions.

(d) Organize groups into teams and appoint captains.

(e) Arrange teams in the proper starting positions.

(f) Conduct the activity.

d. Competitive Units. Units for competition should be the same organizations in which Marines train: battalions, companies, platoons. In most situations, the unit is the squad.

e. Provisions for Instruction. One of the most effective methods of maintaining interest and participation in competitive activities is to provide instruction in activities with which most Marines are unacquainted. Such instruction can be conducted during regular physical training periods. Careful planning is required to keep all individuals continuously engaged in vigorous activity. The materials in the following sections should be used as a guide for instructional purposes.

f. Officiating and Control. Every effort should be made to provide good officials for all competitive activities. Poor officials quickly cause dissatisfaction among participants in team activities and create a situation which can turn an organized activity into a brawl. Each company should have or develop several competent and qualified officials available for games on company and platoon levels.

5102.MILITARY FIELD MEETS

When units reach the latter part of the slow-improvement and sustaining stage of training, interest in the program may lag. A change in course content can arouse the desire to participate. An event which does not require a high degree of skill yet demands strenuous activity is ideal for the military field meet. This is a series of team contests conducted on a station-to-station basis during a given period of time. Team contests carried on simultaneously provide essential training, vigorous exercise, and stimulating competition in an atmosphere that is enjoyable for all. Because of the healthy rivalry that a field meet arouses, it is an excellent form of inter-unit competition.

a. Objective and Advantages. The objective of the military field meet is to provide activity for everyone in the participating units. Activities are chosen that will develop aggressiveness, teamwork, a will to win, and competitive spirit, and that will stimulate interest and build esprit de corps. The military field meet can **be** included as part of the physical training program or as part of the off-duty recreational program. It is a form of contest that can be conducted in nearly all circumstances because it can be easily modified, requires a minimum amount of equipment, and can be readily organized.

b. Level of Competition. The military field meet is flexible. It can be adjusted to large or small groups. A company-size unit is the most desirable, but it may also be administered within a larger unit. If it is conducted within a company, the participating units will be squads. If it is held within a larger unit, platoons compose the teams. A larger-size unit requires more extensive organization and administration.

c. Selection of Events. In organizing a military field meet, select events that are simple and easy to administer. All rules and regulations should be clearly understood by everyone. No event should require previous practice. In selecting the events, consider the interest and capabilities of the Marines and available equipment and facilities. Select events which will require that all members of the units participate. Events should not be dominated by the athletes; rather, MOS-related events which have significance in producing skills and combat readiness should be used.

d.Equipment and Facilities. The site must be large enough to permit events to be grouped about a central control point. Use available facilities such as parade decks, Softball fields, or tracks. A public address system is desirable at the control point for the initial orientation of teams and for subsequent announcements of time lapses, cumulative scores, and final standings. A tally board is necessary so that all units can constantly monitor how the events are progressing.

e.Administrators. Efficient administration of the military field meet depends on the referees, judges, and scorers. For these positions, choose individuals within the units who have had athletic or officiating experience. Prior to the day of the event, all administrators should be briefed and assigned a specific task in order that they may become familiar with the rules and organization of the contests they will conduct. The following officials should be available:

(1) A primary instructor or supervisor who is in charge of the control center and who is responsible for the successful operation of the athletic carnival. The supervisor must have an assistant to act as a timer and scorer.

(2) One assistant instructor in charge of each event. This person should be-

(a) Familiar with the rules of the event.

(b) Effective as a leader to ensure proper supervision and control over the participating teams.

(c) Enthusiastic to provide proper motivation.

(d) Self-confident of the ability to judge infractions of the rules. The assistant instructor must be fair in judgment and penalize without hesitation when infractions occur.

(3) Runners between event stations and the control point. These individuals collect and deliver scores.

f. Team Organization for Competition. The size of the teams is determined by the level on which the field meet is organized. Maintaining the integrity of the unit promotes esprit de corps, but this does not preclude grouping two squads into one team.

g.Conduct of Events

(1) The assistant instructor at each station takes charge of the group and gives a brief explanation of the major rules of the event. The assistant instructor speaks clearly and distinctly from a position to be seen and heard by all. A short demonstration is desirable if it will help clarify the event.

(2) The assistant instructor should make certain that teams can be clearly distinguished by the use of, for example, T-shirts and fatigue jackets, or caps and no cap, or colored jerseys. The assistant instructor should teach a whistle response (teams stop play immediately upon hearing whistle), get the event started as quickly as possible, and make any necessary corrections as the contest progresses. The rale of good officiating is to use a minimum of calls, yet maintain control of the contest. Penalize when necessary, but refrain from disqualifying contestants or teams.

(3) Keep the activity moving as rapidly as possible. When the central control point sounds the whistle to stop the play, all competition ends immediately. The assistant instructor then assembles the group, forwards the team scores to the central point, and, upon a signal from the central control point, rotates the teams to their next station. It is essential that rotation and orientation of teams be quick and orderly to reduce time spent between contests.

(4) Upon completion of the final event, assistant instructors move their teams to the central control point for the announcement of winning teams, presentation of awards (if any), and final critique.

h. Scoring System. The system for determining the winner of the military field meet should be simple and efficient. At the completion of competition, the scorer totals the points that each team has scored in all contests. The scorer then subtracts the number of points scored against a team (penalty points for nonparticipation,etc.) from the total. The resulting scores are placed in a column with the highest score at the top and the lowest at the bottom. The team with the highest total is the winner. This type of scoring system encourages a team to ensure total unit participation.

Section II.Combatives

5201.DESCRIPTTON AND OBJECTIVE

Combatives are strenuous, short competitive contests in which two individuals attempt to overcome each other in a bout of skill and strength. These contests help to develop Marines' resourcefulness, confidence, strength, agility, coordination, and will to win. Any level ground area can be used. However, extremely hard ground should be avoided as some of the combatives require ground contact. A whistle is needed to control the bouts during the competition since voice commands may go unheeded. The extended rectangular formation is used for dual combatives. Combatives are conducted on an informal basis. Marines are allowed to remain at ease between activities and are allowed to brush themselves off after being on the ground. To get the most out of combatives, individuals must be urged to overcome their opponents as quickly as possible as would be required in combat.

a. Benefits for Marine Training. The possibility of close contact with an enemy in combat faces Marines at all times. They must be trained to react aggressively and violently. Combatives may be used as an introduction to such hand-to-hand contact and should be followed by hand-to-hand combat training. Marines enjoy competition, and this type of activity is a welcome change from formal conditioning activities.

b. Instructor Responsibility

(1) The instructor tells the individuals that all combatives begin and end on the whistle signal. The instructor demonstrates each activity before having it performed, explaining it in simple terms.

(2) After stopping one activity, the instructor gets everyone in place for the next.

(3) The instructor must closely supervise combatives to insure that contestants do not use unfair or unsportsmanlike tactics. To avoid unnecessary injury, instructors must see that the bouts are closely controlled and opponents equally paired. Adjustments should be made in apparent cases of mismatched abilities.

5202.COMBATIVES TABLES

There are three tables of combatives, each of which can be completed in 15 minutes. These tables become progressively more difficult from lower to higher numbers.

a.Combatives Table I

(1) Open-Hand Slap Boxing(See fig. 5-1, A. individuals assume a boxer's stance, palms open, fingers extended and joined. Each contestant tries to slap the opponent about the head and upper body with the open hand. This is a good warm- up activity.

(2) Wrist Tug-of-War(See fig. 5-1, B.)Two individuals sit on the ground with the soles of their feet in contact. Each grasps the opponent's wrists so that the hands are directly over the feet. At the whistle, each individual tries to pull the opponent from sitting to standing position.

(3) Arm-Lock Wrestling.(See fig. 5-1, C.)Two individuals sit back-to-back with legs spread and arms locked at the elbows. Each right arm is inside the opponent's left. Each tries to force the opponent's left arm or shoulder to the ground. The individual who first wins three bouts is the winner.

(4) Bulling(See fig. 5-1, D.)Two individuals assume the Westmoreland wrestling hold, each grasping the opponent's neck with the right hand and the opponent's right elbow with the left hand. Each tries to force the other to move one foot by pushing, pulling, or otherwise manipulating. The individual who first wins two boutsis the winner.

(5) Indian Wrestling(See fig. 5-1, E.)Two individuals lie on the ground, side by side, with their heads in opposite directions. They link right elbows. On the instructor's signal or by mutual agreement, each individual raises the right leg approximately straight and far enough to engage the opponent's heel. To start the contest, each person usually raises the leg three times rhythmically and on the third time engages the opponent's heel. Each tries to roll the other over backward. The right leg is used for three bouts, then the left leg for three bouts.

Figure 5-1 is available in the illustration section at the end of the book.

b. Combatives Table II

(1) Wrist Bending.(See fig. 5-2, A.) Opponents pair off, face each other, raise their arms forward, and, with palms forward, interlock their fingers. At the starting signal, each individual attempts to bend the opponent's wrist. The arms are kept up and forward and are not allowed to swing around and down to the sides. The individual who first wins two bouts is the winner.

(2) Back-to-Back Push.(See fig. 5-2, B.)Two individuals stand back-to-back with elbows locked. Each right arm is inside the opponent's left arm. At the starting signal, each pushes backward, trying to move the opponent. Opponents are not allowed to lift and carry each other; only pushing is permitted. The one who pushes the opponent the farthest wins the bout. The individual who first wins two bouts is the winner.

(3) Hop and Pull Hand.(See fig. 5-2, C.) The individuals are matched in pairs. Each grasps the opponent's right hand and, hopping on the right foot, attempts to pull the opponent off balance. Contestants automatically lose if they touch their free hand or their lifted foot to the ground. For successive bouts, they alternate hands and feet.

(4) Westmoreland Wrestling. (See fig. 5-2, D.)Each contestant grasps the back of the opponent's neck with the right hand and the opponent's right elbow with the left hand. In this position, each attempts to pull, push, or force the opponent to touch the ground with any part of the body other than the feet. The individual who first wins two bouts is the winner.

(5) Crab Fight.(See fig. 5-2, E.)Two individuals sit on the ground facing in opposite directions with their hands on the ground behind them. At the whistle, they raise their hips and push with their shoulders and bodies, each trying to make the other's hips touch the ground. The individual who first wins two bouts is the winner.

Figure 5-2 is available in the illustration section at the end of the book.

c. Combatives Table III

(1) Hand Wrestling.(See fig. 5-3, A.) Two individuals stand facing each other. Their right feet are forward and braced side by side. They grasp right hands for the first bout, left hands for the second bout. Each pulls, pushes, makes sideward movements, and otherwise maneuvers to force the opponent to move one or both feet from the original position. The contestant who first wins two bouts is the winner.

(2) Back-to-Back Tug.(See fig, 5-3, B.) Two individuals stand back-to-back with both arms linked at the elbows. Each has the right arm inside the opponent's left arm. At the starting signal, each attempts to pull the opponent. Lifting and carrying are permitted. The contestants must maintain their original direction and keep their arms linked. After a predetermined time, the player pulled or carried the farthest is the loser.

(3) Wrestling to Lift off Feet.(See fig. 5-3, C.) Contestants face each other. Each places the right arm under the left arm of the opponent and around the body. The left arm is over the opponent's right shoulder. Each tries to lift the other off the ground. The individual who first wins two bouts is the winner.

(4) Arm Pull Between Legs.(See fig. 5-3, D.) Two individuals are paired off, back-to-back. Each bends forward and, extending the right arm between the legs, grasps the opponent's right wrist. At the starting signal, each person attempts to pull the opponent After a predetermined time, the player who has pulled the opponent the farthest is the winner of the bout. The one who first wins two bouts is the winner. Repeat with the left hand and then both hands.

(5) Rooster Fight.(See fig.5-3, E.) Each contestant grasps the left foot with the right hand from behind, and the right arm with left hand. Each hops on the right foot, and, by shoulder-butting the opponent or by feinting and sudden evasions, forces the opponent to let go of the foot or arm. The contestant who first wins two bouts is winner.

Figure 5-3 is available in the illustration section at the end of the book.

Section III .Relays

5301. DESCRIPTION AND OBJECTIVE

Relays are races in which each member of a team runs one leg of the race. The team effort decides the winner. Relays provide stimulating competition and contribute to the conditioning of personnel. They also develop aggressiveness, team spirit, and the will to win. Relays should be dispersed throughout the program for short periods of time to provide a change of activity.

a. Team Organization. Relays are conducted most efficiently in platoon-size groups. Teams of equal size must be organized. Competitive spirit is encouraged and teams are organized better by basing teams on units such as squads, crews, or sections. Team captains should be designated. Extra Marines may be used as officials. The number of individuals on a relay team should be limited to squad size. If larger teams are used, the runners will spend too much time awaiting their turns and too little time actually participating. Two to six teams are ideal for relay competition. It is difficult to keep track of winners when more teams compete.

b. Administration of Relays

(1) The time spent on any one relay should be relatively short. If one team achieves a substantial lead in a long relay, the competitive spirit and enthusiastic participation of the other teams may decrease. Several short relays are generally better than one long relay.

(2) To maintain competitive spirit throughout a number of relays, determine the teams that win, place, and show in each relay and their total points for all relays. This can be done by awarding points to all teams on the basis of position at the finish of each relay. The team with the greatest number of points is the winner of the entire set of relays.

(3) Difficulties commonly encountered in conducting relays may be avoided by the following procedures:

(a) The last player in a relay race should be conspicuously identified. For example, the last runner can tie a handkerchief around the head or arm, take off the shirt, put on a hat or take it off, or use some other means.

(b) Another way to keep track of the progress of the race is to have each player sit or squat as soon as each is finished.

(c) Judges at the starting line can keep the runners from starting too soon.

(d) To prevent contestants from turning before they run the full distance, they should be required to run around a peg, pole, or assistant instructor.

(e) Batons, handkerchiefs, tent pegs, or other objects should be passed from one runner to the next when relays are run on a circular track.

(4) Before a unit's first participation in a relay, inform participants of the rules and scoring system. Violation of the rules should not result in disqualification. Instead, impose point penalties. A point penalty is imposed by subtracting a point from the team total at the conclusion of the relay.

(5) Careful administration will prevent most violations. For efficient conduct of relays, follow this procedure:

(a) Announce the name of the relay.

(b) Form the individuals in relay position.

(c) Briefly explain the relay and the rules for running it.
(d) Demonstrate.
(e) Have a definite finish line, and ensure that all know where it is.
(f) Answer questions.
(g) Conduct the relay.
(h) Determine winner and award points.

5302. RELAY TABLES

a. Events. The following relays are grouped into a table of activities. Each relay table can be completed in 15 minutes. Thus, relays can be used as a sole activity or as a part of a longer period. Each table is planned for a platoon-size group (30-60Marines). Adequate warmup for participants is provided by conducting several repetitions of exercises 1 and 2 of a conditioning drill. The recommended relay tables require an area 40 by 60 meters in size. Each table provides a variety of activity. The tables are progressive in the overload applied and should be scheduled in numerical order although not necessarily on successive days.

b. Relay Table I
(1) **60-Meter Lane Relay** (See fig. 5-4, A.) Each team is assembled in single file behind the starting line. On signal, the first individual of each team runs to the turn-around line 30 meters away, then runs back and touches the next teammate waiting at the starting line. The winning team is the first team to get its last member across the finish line. If an individual starts before being touched by the preceding runner, the team may be penalized.
(2) Wheelbarrow Race.(See fig. 5-4, B.) The players of each team pair off and line up in single file behind the starting line. The first individual walks on the hands while the partner grasps the ankles. They advance to the distance line (25 meters) behind which they exchange positions and return to the starting line. The rear individual must always hold the partner's ankles. After the first pair returns across the starting line, the next pair starts.

Figure 5-4 is available in the illustration section at the end of the book.

(3) Squad Front Relay .(See fig. 5-4, C.) The teams form in a line along the starting line with a 10-foot interval between teams. The members of each team lock elbows so that they are linked together. At the starting signal, the teams run to the distance line (20meters) where the left flank individual acts as a pivot. The team swings around on the pivot and returns to the base line. If a team breaks its links, it must reform before continuing. The first team to completely cross the base line intact is the winner.
(4) Crab-Walk Race.(See fig. 5-4, D.) The players of each team line up in single file. The first person of each team assumes the crab-walk position with feet forward on the starting line, hands on the ground behind it. At the starting signal, contestants move forward to the distance line (10 meters). They touch the line with their feet and then return to the starting line in the reverse position with the head and hands leading. The second person may not start until the first individual touches the finish line.

c.Relay Table II

(1) 100-Meter Lane Relay.(See fig. 5-5, A.)This relay is conducted exactly as the first relay of table 1 except that the start and turn-around line are 50 meters apart. This relay provides progression in sprinting.

(2) Frog-Jump Relay. (See fig. 5-5, B.) Each team lines up in single file. The first individual assumes a squatting position on the starting line. At the starting signal, the contestant progresses to the distance line (15 meters) and back by leaping forward, catching the weight on the hands, and bringing up the legs to the squat position for the next leap forward.

(3) Simple Relay. (See fig. 5-5, C.)Each team lines up in single file. Place a marker on the distance line (20 meters) in front of each team. Each team member, behind the first one, grasps the belt of the individual ahead. At the starting signal, each team runs as a unit to the marker, circles it, and returns to the starting line. The first team to completely cross the starting line intact is thewinner.

(4) Fireman's Carry Relay.(See fig. 5-5, D.)The players of each team line up in pairs, one behind the other. One individual in each pair carries the partner to the distance line (30 meters), using the fireman's carry. At the distance line, individuals exchange places and return to the starting line. As a variation, the person to be carried lies on the ground. The one carrying lifts the partner to the proper position. This relay may be performed with the other carries described in guerrilla exercises in chapter 3.

d.Relay Table III

(1) 200-Meter Circle Relay .(See fig. 5-6, A.)A course is laid out in a circular, rectangular,or oval pattern that is 200 meters around. Each team provides one runner on the starting line. On signal, the runner races around the 200-meter track and touches the next teammate, waiting at the starting line, who runs the same course. Each team member runs one lap of the course.

Figure 5-5 is available in the illustration section at the end of the book.

(2) Bear-and-Crab Race.(See fig. 5-6, B.) Each team lines up in single file. At the signal to start, the first contestant in each column assumes the bear-walk position (hands on ground in front), walks to the distance line (15 meters), and then runs back to the starting line. The returning contestant touches off the second individual and goes to the rear of the line. The second contestant assumes the crab-walk position (hands on ground behind) and crab-walks with feet leading to the distance line. This contestant also runs back to the starting line and touches off the next individual, who walks bear fashion. The rest of the members of each team alternate in this manner. The relay ends when the first person is back at the head of the line.

(3) Pilot Relay.(See fig. 5-6, C.) The players are grouped in threes, arms interlocked at the elbows, and outside players facing backwards. The middle person runs forward; the two outside individuals run backward. They run to the turning point (15meters), where they start back, this time with the middle person running backward and the two outside individuals running forward. The next set of three players starts when the first set crosses the starting line.

(4) Saddle-Back Relay.(See fig. 5-6, D.) Mark two parallel lines 15 meters apart. Each team selects a rider. The remaining members of each team count off. The even-numbered players from each team form in single files behind one line and the odd-numbered players from each team form in single files behind the other line directly across from their teammates. At the starting signal, the rider mounts the back of the first player of the team who carries the rider across the other line. There the rider changes mounts to the first player in the second line without touching the ground. This person carries the rider to the next player waiting in the first line. The relay continues until all of the mounts have carried the rider. A rider who falls off must mount again at the point of the fall. A rider who falls in changing mounts must get back on the original mount before making the change.

e. Relay Table IV

(1) 100-Meter Circle Relay (See fig. 5-7, A.)A course is laid out in either a circular, rectangular, or oval pattern that is 200 meters around. Each team is divided in half with each half positioned at starting lines on opposite sides of the track. Each runner races halfway around the track and touches a teammate who completes the lap. Each runner then waits in file at the halfway point until touched by the runner and then completes the second half of the lap. The first team to return all runners to their original starting line is the winner.

Figure 5-6 is available in the illustration section at the end of the book.

(2) In-and-Out Relay.(See fig. 5-7, B.) Each team lines up in a file with players 2 meters apart. At the starting signal, the first player runs back through the column in a zigzag fashion. He alternates, going to the right of one teammate and to the left of the next. Upon completing the run, the player lines up 2 meters behind the last one in the column. As soon as the first runner has passed the second person, the latter starts to run. This continues until all the players have realigned their original order. The team that finishes first is the winner. It may be desirable to have this relay continue until all individuals have run through their entire team two or three times in succession.

(3) Circle Race.(See fig. 5-7, **C.**) Each team forms a circle and holds hands with all individuals facing out except one who faces in and is the "driver."At the starting signal, the teams race to the distance line (20 meters) and back, keeping the circle intact. All the individuals in the circle must completely cross the distance line. The "driver" gives directions and orders. When the circle breaks, it must be reformed before it can continue. The first team completely over the starting line is the winner.

(4) Horse-and-Rider Relay.(See fig. 5-7, D.)Each team lines up in a single file. At the signal to start, the second player in each column leaps upon the back of the first who carries the rider across the distance line (30 meters). At the distance line, the rider dismounts and runs back to the starting line. There the rider picks up the third individual in the column, and carries this player to the distance line where the first player has remained. This continues until the last person is carried across the distance line.

Figure 5-7 is available in the illustration section at the end of the book.

Section IV. Team Contests and Athletics

5401. DESCRIPTION AND OBJECTIVE

Team contests are competitive activities in which Marines as a team compete with another team to win. They are guided by simple rules and organization. The function of team contests is to provide competition and an opportunity for body contact and to contribute to the development of physical readiness. In competing and working together as a team, individuals develop aggressiveness, the will to win, and teamwork. Team athletics deserve a prominent place in the physical training program because they contribute to increased combat efficiency. Because of the competitive nature of athletics and their natural appeal, individuals take part with enthusiasm. Athletic teams formed at intramural and higher levels are a strong unifying influence and provide one of the best means of developing esprit de corps.

a. Area and Equipment A level training field is sometimes the only area required. Many contests need no equipment. In contests requiring equipment, the need is for standard items such as logs, balls, nets, goals, and similar types of equipment. Specific requirements for area and equipment are listed with each contest description.

b. Progression. Team contests of a strenuous nature should be introduced after a basic period of conditioning has been completed and individuals are in the slow improvement stage of conditioning. Progression can take place from the less active to the more vigorous contests, and then from the non-contact to the contact or combative-type contests.

c. Necessity for Preconditioning Individuals must undergo conditioning prior to participation in athletics. Muscles, organs, joints, and ligaments not accustomed to stress and strain from sudden stops and starts, falls, body contact, rapid turns, prolonged running, and other rigors of athletic competition are subject to injury. Although athletics should not be introduced until players are physically prepared, there is still opportunity to engage in competition through lead-up contests. Individuals learn many of the skills required for athletics while participating in team contests.

d. Benefits. Athletics are beneficial primarily to sustain interest in the program and to maintain an achieved level of physical fitness. Athletics are a supplement, not a substitute for other types of conditioning activities which should continue. All of the desirable traits of physical fitness cannot be developed through athletics, yet their contribution is significant. For athletics to make a proper contribution to physical conditioning, the selected sports must bevigorous. The team contests recommended here are carefully selected for their simplicity, aggressiveness and applicability to squad- teaching teamwork. They are generally strenuous and involve the whole team throughout the entire duration of the contest.

5402. TEAM CONTESTS

a. Pushball

(1) **Players.** 10 to 50 players on a side. (See fig. 5-8.)

(2) Equipment A large pushball 5 to 6 feet in diameter.

(3) Area. A field 240 to 300 feet long, 120 to 150 feet wide. Mark a center line in the middle of the field parallel to the end lines. Mark a line 45 feet on either side of this center line and parallel to it, extending it across the width of the field. Mark another parallel line 15 feet from each end line, extending it across the width of the field.

(4) Game. Four 10-minute quarters are played. Give 2-minute rests between quarters and 5-minute rests between halves. The object of the game is to propel the ball over the opponent's goal line by pushing, rolling, passing, carrying, or any other way except kicking.

(a) The ball is placed on the center line with the opposing captains 3 feet from the ball. The rest of the players are 45 feet from the ball, on their half of the field. On the referee's starting whistle, the captains immediately play the ball with their respective teams coming to their assistance.

(b) At quarter time, the ball remains dead for 2 minutes at the spot where it was when the quarter ended. At half time, the teams exchange goals. The play is then started as it was in the beginning.

Figure 5-8 is available in the illustration section at the end of the book.

(c) Players may use any means of interfering with an opponent's progress except striking and clipping. Clipping is throwing the body across the back of an opponent's leg, while the opponent is running or standing. Legal use of force may be applied to all opponents whether or not they are playing the ball. For striking an opponent, the offender is removed from the game. The team penalty is half the distance to their goal. The penalty for clipping is the same.

(d) When any part of the ball goes out of bounds, it is a dead ball. The teams line up at right angles to the side lines and 3 feet apart at the point where the ball went out. The referee then tosses the ball between the teams.

(e) When, for any reason, the ball remains in one spot for more than 10 seconds, the referee declares the ball dead. The ball is then put into play as it is for an out-of-bounds situation.

(5) Scoring. A goal is scored when the ball, or any part of it, is propelled across the opponent's end line. A goal counts 5 points. The team scoring a goal has the privilege of trying for a point after the goal. To try for this extra point, the ball is placed on the opponent's 5-yard (or 15-foot) line. The teams line up on either side of this line separated by the width of the ball. Before the whistle blows, one player on each team may place hands on the ball. On the referee's signal, the ball is put into play for 1 minute. If any part of the ball is driven across the goal line in this 1-minute period, the offensive team scores 1 point. The defending team may not score during the opponent's try for the extra point.

b. Line Rush

(1) Players. Any number up to 50 on each side. (See fig. 5-9.)

(2) Equipment. None.

(3) Area. A field, 75 by 100 feet.

(4) Game. One team lines up behind one goal line and the other in midfield. On the starting signal, the team standing behind the goal line seeks to cross the goal line at the opposite end of the field within 30 seconds. The team in the center seeks to prevent it by catching and holding the runners. At the end of 30 seconds, the teams change.

(5) Scoring. Count the number of individuals who have crossed the far goal at the end of 30 seconds. After each team has had from three to five tries, the scores are added and the winner declared. A player scores 1 point when any part of the body is across the goal line.

Figure 5-9 is available in the illustration section at the end of the book.

c. Human Tug-of-War

(1) Players. 10 to 20 on a team.(See fig. 5-10.)

(2) Equipment. None.

(3) Area.40 to 60 feet.

(4) Game. Draw a line in the center of the area. Divide the players into two equal teams. Place them in single file on opposite sides of the center line facing each other. Each individual places the arms around the waist of the teammate in front. The two leaders of the opposing teams grasp each other around the waist. On signal, the teams try to pull each other over the center line within 30 seconds.

(5) Scoring. The team pulled across the center line loses. If neither team is pulled over the center line, but one team breaks its file, that team loses the match.

(6) Variation. Use a 3/4- or 1-inch rope. Space the leading players on each team 10 feet apart, each holding one end of the rope. The team pulled across the center line loses.

Figure 5-10 is available in the illustration section at the end of the book.

d. Master of the Ring

(1) Players. Any number. (See fig. 5-11.)

(2) Equipment. None.

(3) Area. A clearly marked circle large enough to contain all the players.

(4) Game. All the players stand inside the circle. At the signal, all players attempt to throw each other out of the circle. All tactics are fair except unnecessary roughness. When any part of the body touches across the line, the player is out and leaves the circle at once. Several officials are needed to spot the players who cross the line.

(5) Scoring The player who remains in the circle when all the others are out is the master of the ring.

(6) Variation l. The players are divided into two equal teams. Each team is clearly marked. On signal, each team tries to throw the opponents out of the circle. The winning team is the team that eliminates all opponents from the circle.

(7) Variation 2.The players are divided into two equal teams. Each team sends only one individual into the circle. When one person has been forced out of the circle, the losing side sends in only one player. The team which eliminates all opponents is the winner.

(8) Variation 3.A pit, approximately 4-feet deep, is used rather than a circle on level ground. The contest may then be an individual or team activity.

Figure 5-11 is available in the illustration section at the end of the book.

e. Log Pivot Circle. Space teams far enough apart so each team can pivot in a circle without colliding with any other team. Each team holds a log in the bend of the arms in front of the chest. (See fig.5-12.) At the command **CIRCLE RIGHT, MOVE**,the left-flank individual holds the pivot while the log is carried around 360 degrees in a counter- clockwise motion, back to the original position. This movement may also be performed to the left by facing in the other direction and pivoting clockwise(**CIRCLE RIGHT, MOVE**).Other commands may be used such as: **CIRCLE HALF-RIGHT, CIRCLE HALF- LEFT** and so on. The first team to complete the prescribed movement is the winner.

Figure 5-12 is available in the illustration section at the end of the book.

f. Log-Rolling Race. Each team tries to roll its log a measured distance by pushing it with the hands and driving the body forward with the legs. (See fig. 5-13.)The first team to get the entire length of the log across the finish line wins.

Figure 5-13 is available in the illustration section at the end of the book.

g. Prone Pushing Contest.. Two teams lie prone, facing each other with a log between them. (See fig. 5-14.)Both teams place their hands against the log, keeping their arms straight. Then, by driving with the legs, each team attempts to push the other a measured distance to the rear.

Figure 5-14 is available in the illustration section at the end of the book.

h. Shuttle Relay Race. Each team in this race divides into two parts, A and B. Members of part A run 50 yards with a log held under their right arms. (See fig 5-15.)At the distance line, they give the log to their teammates in part B who bring it back to the starting line. The team pair finishing first is the winner.

Figure 5-15 is available in the illustration section at the end of the book.

5403.CROSS-COUNTRY AND DISTANCE RUNNING

a. Value of Running. Long-distance running gives some benefits that cannot be obtained in the same degree from any other sport. It builds powerful leg muscles, increases lung capacity, and develops endurance. For these reasons, cross-country and distance running should be included in the physical training program. These sports require only a few miles of open space which is generally available. Short cross-country runs and middle-distance runs can be used to supplement other activities, particularly team sports or the sports that develop precision or agility rather than endurance. Short cross-country runs can be scheduled once a week, gradually increasing the distance as the physical condition of the runners improves.

b. Cross-Country Runs. A cross-country run is a distance run held on a course laid out along roads, across fields, over hills, through woods, and on any irregular ground. A flat cinder or dirt track is not a suitable surface for cross-country running. Some runs are as long as 6 miles while others are as short as 3 or 4 miles. If cross-country running is being used to supplement other activities, the 3-mile course is long enough for most. Cross-country runs should be used only after individuals reach the sustaining stage of conditioning. This type of running should then be scheduled occasionally to provide variety. Cross-country running has the advantage of allowing mass participation. Interest can be stimulated by putting the runs on a competitive basis.

c. Practice Methods. Conditioning is essential to distance and cross-country running. Championship distance running depends on stamina, which can be developed only through constant training. An individual of only average ability can become an outstanding distance runner by steady and careful training. Hiking is an excellent method for getting into condition before the season opens. Long walks build up leg muscles. During the first month of the season, training should be gradual, starting with short distances, and increasing day by day. At first the legs will become stiff, but the stiffness gradually disappears if running is practiced for a while every day. To prevent strain, it is essential to limber up thoroughly each day before running.

(1) In training a large group, leaders should be stationed at the head and the rear of the column. They should make every effort to keep the runners together. After determining the abilities of the unit in cross-country running, it is advisable to divide them into three groups. The poorest-conditioned group is started first, the best-conditioned group, last. The starting time of the groups should be staggered so that all of them come in at about the same time.

(2) In preliminary training, running is similar to ordinary road work in that it begins with rather slow jogging, alternating with walking. The speed and distance of the run is gradually increased. As conditioning improves, occasional sprints may be introduced. Atfirst, the distance run is from one-half to 1 mile. It gradually increases to 2 or 3 miles. On completing runs, require runners to walk for 3 or 4 minutes before stopping to permit a gradual cooling off.

d. Facilities and Equipment. There should be at least one timer with a stopwatch (preferably three) for timing the runners. A course 3 or 5 miles long should be measured and marked by one of the three following methods:

(1) Fasten directional arrows to the top of tall posts and place them at every point where the course turns. Such signs should also be placed at every other point where there may be doubt as to the direction of travel.

(2) Place a lime line on the ground over the entire course.

(3) Place flags, clearly visible to the runners.

- **(a)** A red flag indicates a left turn.
- **(b)** A white flag indicates a right turn.
- **(c)** A blue flag indicates the course is straight ahead.

e. Rules/Scoring. A cross-country team consists of seven individuals, unless otherwise agreed. In dual meets, a maximum of 12 on one team may be entered, but 7 or fewer are scored. First place scores 1 point, second place 2, third place 3, and so on. All who finish the course are ranked and tallied in this manner. The team score is then determined by totaling the points scored by the first five of each team to finish. The team scoring the least number of points is the winner. If fewer than five (or the number determined prior to the race) finish, the places of all members of that team are disregarded. If two or more teams score the same number of points, the event is called a tie.

5404.TEAM ATHLETICS

Team athletics add variety to the combat readiness program. At no time should they be allowed to replace intramural activities or to provide practice time for intramural or base teams. If team athletics are integrated into the program, the goal should be total participation and competition. Therefore, the games selected can be played by the skilled or the unskilled. They require limited equipment, yet will teach competitiveness, aggressiveness, and teamwork while they improve conditioning.

a.Soccer. Soccer is one of the best athletic activities for developing endurance, agility, leg strength, and great skill fulness in using the legs. The most popular sport in the world, soccer is the national game of many European, Asian, and Central and South American countries. In recent years, it has become popular in the United States. A soccer ball is the only equipment needed for the game, and Marines can learn to play it easily. Players do not need much skill to participate, but the amount of skill they can develop is almost unlimited.
(1) Place in the Program. Soccer should be introduced during the latter part of the slow improvement stage of physical training. It can be used as a competitive activity in the sustaining stage. It is primarily a spring or fall sport. Any level field is suitable for competition. The boundaries for the soccer field are similar to the dimensions for a football field. (See fig. 5-16.) Goal posts are essential, but they are easily constructed and usually temporary so that they may be removed when not in use.

Figure 5-16 is available in the illustration section at the end of the book.

(2) Basic Skills

(a) Passing. Passing with the feet is the primary means of moving the ball. Short passes are easier to control. Emphasis should be continually placed on the skill of passing.

(b) Dribbling. The ball is dribbled by a series of kicks with the inside or outside of the foot. Do not kick with the toe. Keep the head over the ball when kicking and propel it only a short distance at a time. Keep it close to the feet. If a player lets a ball move far from the feet, an opposing player can easily take it away.

(c) Instep Kicking. The instep kick is the basic soccer kick. The toe does not come in contact with the ball. The toe is pointed downward, and the instep (the area under the shoe laces) is applied to the ball with a vigorous snap from the knee.

(d) Inside-of-the-Foot Kicking. The ball is kicked with the inside of the foot, and the leg is swung from the hip. The toe is turned outward, and the sole of the foot is parallel with the ground as the foot strikes the ball. This kick is used for short passes and for dribbling.

(e) Foot Trapping. The foot trap is the method of stopping the ball by trapping it between the ground and the foot. Place the sole of the foot on top of the ball at the instant it touches the ground, keeping the foot relaxed. This is an effective way to stop a fast-moving ball.

(f) Shin Trapping. .The shin trap is a method of stopping the ball with the shins. Stand just forward of the spot where the ball should strike the ground and allow it to strike the shins in flight or on the bounce. Use either one or both legs from the knee down so as to absorb the bounce.

(g) Body Trapping. .Intercept the ball with any part of the upper body except the arms and hands. Keep the body relaxed and inclined toward the ball. To keep the ball from bouncing, move backwards from it as it strikes the body. This drops the ball at the feet in position for dribbling or passing.

(h) Heading. Heading is a technique for changing the direction of the flight of a ball by butting it with the head. Tense the neck muscles and jump up to meet the ball. Butt the ball with the forehead at about the hairline to reverse its direction. Use the side of the head to deflect it to the side.

(3) Offensive and Defensive Positions. Forwards usually play on the offensive half of the field. Fullbacks usually play on the defensive half of the field. Halfbacks are the backbone of the team. They move forward on the offense and back on defense. The goalkeeper almost always remains within a few feet of the goal.

(4) Abridged Rules

(a) A soccer team is composed of 11 players. (See fig. 5-16.)

(b) The player propels the ball by kicking it with the feet or any part of the legs, by butting it with the head, and by hitting it with any portion of the body except the arms or hands.

(c) The goalkeeper is the only individual allowed to place hands on the ball, but only in the goalkeeper's area. The term "hands" includes the whole arm from the point of the shoulder down.

(d) A goal is made by causing the ball to cross completely the section of the goal line lying between the uprights and under the cross bar.

(e) Each goal scores 1 point for the team scoring the goal.

(f) The penalty for a foul committed anywhere on the playing field (except by the defensive team in its penalty area) is a free kick awarded to the opposing team.

(g) All opponents must be at least 10 yards from the ball when a free kick is taken.

(h) The penalty for a foul committed by the defensive team in its penalty area is a penalty kick.

(i) A penalty kick is a free kick at the goal from the spot 12 yards directly in front of the goal. The only players allowed within the penalty area at the time of the kick are the kicker and the defending goalkeeper.

(j) An official game consists of two 30-minute halves,

(k) Teams change goals at the end of every quarter.

(l) In the event of a tie, an extra quarter may played. This may be followed by a sudden death period. If the tie still exists, the game can be decided by penalty kick.

(m) After a team has propelled the ball across a side line and out of play, the ball is put back into play by a throw-in from the side line by a member of the opposing team. The ball is thrown in from the point where it crossed the side line as it went out of bounds. The ball must be thrown in with both hands using an over-the-head motion, keeping both feet on the ground.

(n) When the offensive team propels the ball across the defensive team's goal line but not in the goal, the defensive team is awarded a goal kick-a free kick taken from within its own goal area. The ball must come out of the penalty area to be in play.

(o) When the defensive team causes the ball to go behind its own goal line, excluding the portion between the goal posts, the offensive team is awarded a corner kick—a free kick taken by a member of the offensive team at the quarter circle, at the cornerflag post nearest to where the ball went behind the goal line. The flag post must not be removed.

(p) The game is started and, after a goal has been scored, is resumed by placing the ball in the center of the midfield line. Players must be on their side of the line until the ball is kicked. The ball must be kicked forward and must move at least 2 feet to be legal. The first kicker may not touch the ball twice in succession at the kick off. The opposing team must be 10 yards from the ball until it moves.

b.Speedball. Speedball offers vigorous and varied action with plenty of scoring opportunities. It is easy to learn and provides spontaneous fun. Little equipment is needed-a ball is all that is absolutely necessary. Speedball combines the kicking, trapping, and intercepting elements of soccer; the passing game of basketball; and the punting, drop kicking, and scoring pass of football.

(1) Place in the Program.Speedball, like soccer, should be introduced into the physical training program during the latter part of the toughening stage and used as a competitive activity in the sustaining stage. Speedball may be played any time weather permits but is primarily a spring or fall activity.

(2) Method of Play.The game is played as follows:

(a) Two teams of 11 each play the game under official rules, but any number of players may successfully constitute a team.An inflated leather ball, usually a soccer ball, is used. The playing field is a football field with a football goal post at each end. (See fig. 5-17.)

(b) The game starts with a soccer-type kickoff. The kicking team tries to retain possession of the ball and advance it toward the opposite goal by passing or kicking it. Running with the ball is not allowed, with the result that there is no tackling or interference. After the ball touches the ground, it cannot be picked up with the hands or caught on the bounce but must be played as in soccer until it is raised into the air directly from a kick. Then the hands may be used.

(c) When the ball goes out of bounds over the sidelines, it is given to a player of the opposing team and is put into play with a basketball throw- in. When it goes over the end line without a score, it is given to a player of the opposing team who may either pass or kick it onto the field.

(d) When two opposing players are contesting possession of a held ball, the official tosses the ball up between them as in basketball.

(e) Points are scored by kicking the ball under the crossbar of the goal post, drop kicking the ball over the crossbar, or completing a forward pass into the end zone for a touchdown.

Figure 5-17 is available in the illustration section at the end of the book.

(3) Basic Skills

(a) Shared Skills. Skills include kicking, passing, heading, and trapping (from soccer); punting, dropkicking and forward passing (from football); and passing, receiving, and pivoting (from basketball).

(b) Kickups. The kickup is a play in which a player lifts the ball into the air with the feet so as to legally play the ball with the hands. The kickup generally makes the transition from ground play to aerial play. The technique of making the play depends upon whether the ball is rolling or stationary. If the ball is rolling or bouncing toward the player, the foot is held on the ground with the toe down until the ball rolls onto the foot. Then the foot is raised, projecting the ball upward. If the ball is stationary, the player rolls it backward with one foot. Then the player places a foot where the ball will roll onto it. The player then raises the foot, projecting the ball as before. If a ball is rolling away, the player should stop it with a foot and play it as a stationary ball. The player can also raise the ball by standing over it with a foot on either side. The player presses the feet against the ball and jumps into the air, propelling the ball into the hands.

(4) Offensive and Defensive Play. The positions of players in speedball are much the same as in soccer. However, some of the positions have different names. Each team has 11 players. The forward line is composed of five players, the right end, right forward, center, left forward, and left end. The second line consists of right halfback, fullback, and left halfback. In the next line is the right guard and left guard. The player who defends the goal is the goal guard. The strategy employed in speedball during offensive play is very similar to that of soccer. There are two types of defensive formations in speedball: man-for-man and position defense. Man-for-man defense is recommended for beginning players.

(5) Abridged Rules

(a) Field. A speedball field is 100 yards long with additional 10-yard end zones on each end and 55 yards wide, the same as a regulation football field.

(b) Players. Eleven players are on a team. The goal guard has no special privileges.

(c) Time. Ten-minute quarters. Two minutes between quarters. Ten minutes between halves. Five minutes for extra overtime periods.(Begin the first overtime by a jump ball at center, same goals; change goals in the event of a second overtime period.)

(d) Winner of Toss. The winner of the toss has the choice of kicking, receiving, or defending a specific goal.

(e) Kickoff. A kick off starts the first and second half of a game and starts play after a score. The kick off is made from any place on the 50-yard line. The kicking team must be behind the ball when it is kicked. The receiving team must stay back of its restraining line (10 yards distance) until the ball is kicked. The ball must go toward the opposing team's goal before it can be played. Kick off out of bounds goes to the opponent at that spot. A kick off touched by the receiving team before going out of bounds, with no impetus added, still belongs to the receiving team. A kick off, in possession and control of the receiving team and then fumbled out of bounds, belongs to the kicking team at that spot.

(f) Starting Second and Fourth Quarters. The ball is given to the team that had possession at the end of the previous quarter, from out of bounds, as in basketball.

(g) Second Half. The first-half receiving team kicks off at the beginning of the second half.

(h) Fly and Ground Ball. The most characteristic playing rule of speedball is the difference between a fly ball (or aerial ball) and a ground ball. A player is not permitted to touch a ground ball with the hands and must play it as in soccer. A fly ball is one that has risen into the air directly from the foot of a player (for example, a punt, drop kick, place kick, or kickup).Such a ball may be caught with the hands provided the catch is made before the ball strikes the ground again.

(i) Kickup and Overhead Dribble. A kickup is a ball that is so kicked by a player that he can catch it himself. A bounce from the ground may not be touched with the hand because it has touched the ground since being kicked. This rule prohibits the ordinary basketball dribble, but one overhead dribble (throwing the ball into the air and advancing to catch it before it hits the ground) is permitted.

(j) Out of Bounds. If a team causes the ball to go out of bounds over the sidelines, a free throw-in (any style) is given to the opposing team. When the ball goes over the end line without scoring, it is given to the opponents who may pass or kick from out of bounds at that point.

(k) Tie Ball. In case two players are contesting the possession of a held ball, even in the end zone, a tie ball is declared. The ball is tossed up between them.

(6) Scoring

(a) Field Goal(3 points).A soccer-type kick, in which a ground ball is kicked under the crossbar and between the goal post from the field of play or end zone, is a field goal. (A punt going straight through is not a field goal for it is not a ground ball. The ball must hit the ground first.) A drop kick from the field of play that goes under a crossbar does not count as a field goal. A drop kick from the end zone that goes under the crossbar counts as a field goal;if it goes over the crossbar, it is ruled as a touch back, and the other team takes possession of the ball.

(b) Dropkick(2 points). A scoring drop kick must be made from the field of play and go over the crossbar and between the uprights. The ball must hit the ground before it is kicked (usually with the instep).

(c) End goal(1 point). This is a ground ball which receives its impetus (kicked or legally propelled by the body) from any player, offensive or defensive, in the end zone and which passes over the end line but not between the goal posts.

(d) Penalty Kick(1 point).A ball kicked from the penalty mark that goes between the goal posts and under the crossbar is a penalty kick. The penalty mark is placed directly in front of the goal at the center of the goal line.

(e) Touchdown(1 point). A touchdown is a forward pass from the field of play completed in the end zone. The player must be entirely in the end zone. If the player is on the goal line or has one foot in the field of play and the other in the end zone, the ball is declared out of bounds. If a forward pass is missed, the ball continues in play, but it must be returned to the field of play before another forward pass or drop kick may be made.

(7) Substitutions. Substitutions may be made any time when the ball is not in play. A player who is withdrawn may not return during that same period.

(8) Timeout. Three legal timeouts of 2 minutes each are permitted each team during the game.

(9) Fouls

(a) Personal Foul. Kicking, tripping, charging, pushing, holding, blocking, and unnecessary roughness of any kind, such as running into an opponent from behind, are personal fouls. Kicking at a fly ball and thereby kicking an opponent is a personal foul. Four personal fouls disqualify a player from the game.

(b) Technical Foul. Illegal substitution, more than three timeouts in a game, unsportsmanlike conduct, and unnecessarily delaying the game are technical fouls.

(c) Violation. Traveling with the ball, touching a ground ball with the hands or arms, double overhead dribble, violating tie ball, and kicking or kneeing a fly ball before catching it are violations.

(d) Penalties. The offended player shall attempt the kick.

(e) Officiation of Fouls

1 Fouls in the field of play allow no follow-up while fouls in the end zone always allow follow-up.

2 On penalty kicks, with no follow-up, only the kicker and goalie are involved.

3 On penalty kicks, with a follow-up, the kicking side is behind the ball and the defending side behind the end line or in the field of play. No one is allowed in the end zone or between the goal posts except the goal guard. The kicker must make an actual attempt at goal and cannot play the ball again until after another player plays it.

c. Volleyball. Volleyball is a popular sport. The game entails much physical activity, yet it is not strenuous. It is, therefore, a game for young and old alike, for beginners and for skilled players. It may be played indoors or outdoors on any type of terrain. As an informal activity, volleyball can be played by any number; as an organized activity, it provides, as few other sports do, a game for 12 to play in a limited area. While volleyball requires no great skill to play, it permits a high degree of proficiency. Players naturally get more enjoyment when they know the game and play it well. For this reason, instruction in the basic skills should be provided.

(1) Place in the Program Volleyball may be used occasionally as a competitive activity during the sustaining stage. Though it is a year- round sport, it should be included in the physical training program only when a more strenuous activity is impractical. Volleyball is an excellent off-duty activity.

(2) Instruction. Usually during the first three or four classes, a 10- to 15- minute period of instruction, followed by scrimmage, is enough to teach the basic skills, rules, and techniques of volleyball. More time, if available, can be given to teaching basic skills, but emphasize competitive play rather than formal instruction. Divide the unit into 6-person teams, organized on the basis of ability .Teams should be as nearly equal as possible.

(3) Basic Skills

(a) Passing the High Ball. The chest pass is the most effective method of playing the ball. To receive the ball, the knees are fixed, and the body is tilted forward. The elbows are raised sideward to a point in line with the shoulders. The wrists are extended in line with the forearm. The arms, wrists, and hands are rotated inward. To pass the ball, the hands are chest-high, thumbs pointing inward. The fingers are flexed, forming a cup, allowing them to contact the ball. On contact with the ball, the wrists are snapped while the fingers and elbows are pushed upward, sending the ball upward. A high ball is much easier to handle than a low one.

(b) Passing the Low Ball. A ball that is lower than the waist is one of the easiest to hit, but it is also a frequent cause of fouls (holding or carrying the ball). The best position for handling a low ball is to have the knees flexed and arms flexed at the elbows and rotated so the thumbs are pointing outward, the palms up. When the fingers contact the ball, the entire body reacts in a lifting motion. The arms and hands swing upward in a scooping action. It is important that the fingers, not the palms, contact the ball, and that the ball is batted, not thrown.

(c) Underhand Serve. Take a position behind the back line facing the net, left foot forward, holding the ball in the palm of the left hand. The left knee is flexed; the right knee is straight. Swing the right arm back and at the same time move the left hand (holding the ball) across the body in line with the right hip. Then swing the right arm forward hitting the ball off of the left hand with the palm of the right hand, raising the hips and arching the back in the same motion. Be certain to swing the right arm in a straight line, or the ball will be difficult to control. When the opposition is in formation, the server should try to place the ball in the right or left back area, and not near the net.

(d) Setup. A setup is a ball hit into the air near the net by one player, so that a teammate may hit or "spike" it sharply downward into the opponent's court. The chest pass is the best pass to use. The ball is sent approximately 10 feet into the air toward the spiker so it will descend from 4 to 20 inches from the net.

(e) Spiking. The spike is a leap into the air and a sharp downward hitting of the ball into the opponent's court. A spiker must be able to spring easily from the floor, judge the movement of the ball, and strike it with a downward movement of the arm. To jump from the floor, step off with one foot and jump with the other. Stand with the right or left side to the net, facing the setup player. Much depends upon the setup player to place the ball in the proper position. The spiker jumps into the air and strikes the ball above its center so as to drive it downward. A snapping movement of the arm and wrist will drive the ball forward and downward with power and control. Aim for a weak spot in the opponent's defense.

(f) Blocking. The block is a defense technique used to prevent a spiker from driving the ball across the net. It is an attempt by one or more defensive players at the net to block a hard-hit shot by using the force of the ball to send it immediately back into the opponent's court. In an effective block, forwards on the defensive team spring into the air at the time of the spike, placing both hands and arms in the expected path of the ball. An effective block tends to upset the offense and presents another element for the spiker to worry about. To be effective, the blocker must anticipate the path of the ball and time the block with the spike.

(4) Abridged Rules

(a) The volleyball court is 30 feet wide by 60 feet long. (See fig. 5-18.)

(b) The top of the net is 8 feet high.

(c) A volleyball team consists of 6 players.

(d) A match consists of the best two out of three games.

(e) The first team scoring 15 points wins the game, provided that they have 2 points more than their opponents.

(f) A deuce game is a game in which both teams score 14 points. The game is continued until one team obtains a 2-point advantage over the other.

(g) Only the serving team can score. If the serving team commits a fault, it loses the serve to the opposing team.

(h) The team members who receive the served ball rotate one position in a clockwise direction.

(i) The ball is put into play by serving from behind the back line.

(j) A served ball touching the net results in the loss of the serve. At any other time during play, a ball touching the net is still in play.

Figure 5-18 is available in the illustration section at the end of the book.

(k) The ball is out of play when it touches the ground or goes outside one of the boundary lines.

(l) All line balls are good.

(m) The players must hit or bat the ball; they may not throw, lift, or grasp it.

(n) A player may not touch the ball with any part of the body below the knees.

(o) A player may not play the ball twice in succession. In receiving a hard-driven spike, a defensive player may make several contacts with the ball even if they are not simultaneous. All such contacts, however, must constitute one continuous play, and all must be above the knees.

(p) The ball may be touched no more than three times on one side of the net before being returned across the net to the opposing team.

(q) A player must not touch or reach across the net.

(r) A player may touch the line under the net, but may not cross it.

Chapter 6
EVALUATION OF PERFORMANCE DURING TRAINING

6001.OBJECTIVES OF COMBAT READINESS

The objectives of the Marine Corps physical fitness program are-
a. To contribute to the health and well-being of every Marine through regular exercise and health education.
b. To develop Marines who are physically capable of performing their duties in garrison and in combat.
c. To develop Marines with a reserve level of physical fitness that will help them win in combat.
d. To provide a medium for developing the self-confidence of the individual Marine and thereby enhancing overall discipline, morale, esprit de corps, unit efficiency, and the desire to excel within the Marine Corps.

6002. THE PHYSICAL FITNESS TEST

The physical fitness test (PFT) has distracted attention from the proper goal of the physical fitness program. Simply stated, the goal of the physical fitness program is the success of Marines in combat. Unit commanders must not allow the PFT to become an end in itself. The PFT should be used within the commander's program of combat readiness training as a standard, an index of the physical fitness of individual Marines at a given time and place. The PFT is not difficult, and it must be kept in perspective. All athletes should be first class, and first class is attainable for all Marines. The conduct of the PFT is described in Marine Corps Order 6100.3_.

6003. OTHER METHODS OF EVALUATION

The commander should use other ways of evaluation to provide variety and imagination. Intangible objectives, such as confidence and aggressiveness, are hard to measure. The use of either inspection or observation is necessary to evaluate objectives of this nature. The commander has several other methods available when evaluating the physical condition of the command. The additional methods of inspection, observation, medical examination, and testing are summarized as follows:
a. Formal Inspections. Formal inspections, using inspection officers and standardized rating criteria, may help evaluate unit physical fitness.
b. Daily Observation. Routine observation of physical performance and appearance can serve as an indicator of an individual's or unit's physical readiness. However, mere observation is not a totally reliable or accurate means of evaluation.
c. **Physical Examination.** Medical examination may detect any individual disability or detrimental physical condition. It may guide in application of remedial, therapeutic, or limited exercise programs.

d. Other Physical Fitness Tests. Some other tests are described in detail in this manual for information purposes. Commanders are cautioned, however, not to place too great an emphasis on testing.

6004. RESPONSIBILITY

The commander is responsible for the physical fitness of the command and for the measurement and evaluation of its physical readiness. The goal of combat readiness training remains to prepare Marines and units for the physical demands of combat. The bulk of the commander's combat readiness training program should be such activities as obstacle and confidence courses, endurance courses, and progressive load-bearing marches which will prepare Marines for combat. Group activities which are competitive and combative in nature are advantageous and add imagination and enthusiasm to the program. Activities such as orienteering, which is not only physically demanding but also competitive and which teaches essential subject skills, are excellent and motivate all concerned. A commander's goal should be to provide an integrated training program that is so aggressive and demanding of Marines and leaders alike that a PFT administered at any time will reflect that each and every Marine is both physically and mentally prepared for the demands of combat.

6005. COMMANDER'S PHYSICAL FITNESS PROGRAM

The Commandant of the Marine Corps has directed that the physical fitness test shall be the universal measure of adequate individual physical fitness. The focus of a commander's physical fitness training program, however, should not be to prepare Marines to pass the physical fitness test, but to prepare them for the physical rigors of combat.

a. Administration. The physical fitness program requires each Marine to spend a minimum of 3 hours per week in physical fitness training, to be tested semiannually, and to obtain a minimum standard of third class. The attainment of a higher score is a laudable individual goal which should be encouraged but should not become a unit objective. Overemphasis could be detrimental to training required to develop the complete Marine. All commanders should exercise sound judgment as well as positive, aggressive leadership in striving toward the training objective.

b. Variety of Activities. The program must include the following elements:

(1) Physical Readiness Training. physical readiness training is a complete physical training program which develops and maintains the strength, endurance, and physical skills needed to sustain the individual during combat.

(2) Remedial Physical Conditioning. Remedial physical conditioning is a process by which physically substandard Marines are conditioned to meet prescribed standards.

(3) Competitive Conditioning Activities. Competitive conditioning activities consist of teams or individuals competing against an opponent to win. This involves a combination of sports and military skills designed to foster competitive attitudes and develop unit pride and esprit de corps.

(4) Occupational Conditioning. Marching long distances, dry net training, etc., should not be ignored as a means to fulfill the time requirement.

c.Time. In order to develop the desired level of physical fitness, unit commanders are advised that a minimum program of 3 hours per week is mandatory. The program must encompass at least three exercise periods weekly. Each period should include calisthenics, running, or other forms of vigorous activity .The minimum desirable length for each exercise period is 60 minutes. Shorter periods may be authorized by the unit commander if dictated by the local training situation. Unit commanders are encouraged to use the normal working day to satisfy this training requirement. However, commanders are authorized to conduct required physical fitness training during off-duty hours when the mission, workload, personnel status or other significant factors preclude fulfilling it during duty hours.

d.Action. Marine Corps Order 6100.3_ requires that commanders-

o Establish and maintain an effective physical fitness program which ensures that all Marines maintain an acceptable level of physical fitness.

o Ensure that all individuals are medically qualified prior to participating in the physical fitness program.

o Establish a minimum physical fitness program of at least 3 hours per week. Physical training may be authorized on an individual basis at the discretion of the unit commander.

o Ensure that all Marines participate in physical fitness conditioning activities commensurate with their medical qualifications and limitations.

o Conduct physical fitness testing for all Marines.

o Place all Marines who fail to pass the PFT on a daily command-supervised remedial physical conditioning program until they pass the test.

o Ensure that results of physical fitness testing are entered on section A of the fitness report. Comments are placed in section C in connection with attainment of superior physical performance or a medical excuse from the PFT.

6006. OTHER FORMAL TESTS

Commanders must realize that the best index of combat readiness is personal observation of performance on conditioning marches of 15 miles while under load. This manual presents three additional indicators of unit fitness. These indices are the physical readiness test, which was formerly the Marine Corps standard and is now routinely administered at Officer Candidate School; the airborne trainee physical fitness test which many Marines who attend U.S. Army formal schools are required to take; and the battle fitness test, the ultimate test of the combat readiness of Marines. Commanders may wish to include these tests or portions thereof within their combat readiness programs.

6007. PHYSICAL READINESS TEST

a. Purpose. The purpose of this test is to measure the Marine's ability to meet the minimum standard of physical readiness for duty in the field. This test is currently administered to all male officer candidates at Officer Candidate School, MCCDC, Quantico, VA. Female candidates do not take this test.

b. Test Area. Administration of this test requires an area suitable for crawling, jumping, and running 160 yards to include a trench that is 8 feet wide.

c. Equipment. The equipment required for this test includes—
o A platform, rail on a ledge, or a box 18 inches high,
o A 20-foot climbing rope,
o A stopwatch.

d. Conduct of Test. Adequate timers and other supervisory personnel must be provided by the company staff. All events should be conducted in a single session of one morning or afternoon. The events may be run in any sequence.

e. Uniform. The uniform for the physical readiness test is as follows: full utilities, cartridge belt, belt suspender straps, two canteens (full of water), rifle, and helmet. Depending on the heat condition, the PT shirt may be worn in place of the utility jacket. All equipment will remain secure throughout the test. Except for emergencies, canteens are not to be emptied.

f. Events
(1) Event 1: Climbing Uphill (Step-ups)This event simulates marching uphill at a rapid and steady pace. The Marine stands in front of the step. On the command GO, the Marine places either foot on the step and steps up. Hand pressure on the knee may be used if desired. The Marine then stands erect and steps backward and down, on one foot at a time. The same leg may be used for each repetition or the legs may be alternated. For 100 points, Marines will complete 60 repetitions in the maximum allotted time of 1 minute and 20 seconds (80 seconds).
(2) Event 2: Rope Climb. This event resembles entering and leaving a hovering helicopter, using ropes in house-to-house fighting, and ascending and descending landing nets. The Marine assumes a sitting position on the ground at the bottom of the rope with hands at the highest point that can be reached. On the command GO, the Marine jumps up and climbs to the top. The Marine must touch the 20-foot mark with one hand, then descend without sliding or dropping. Marines must reach the 20-foot mark in the maximum allotted time of 30 seconds or less.
(3) Event 3: Evacuation. This event simulates reaching and evacuating a wounded Marine under fire. The Marine assumes the prone position at a distance of 50 yards from a casualty who is of approximately equal weight. On the command GO, the Marine springs up and covers the distance in a sprint. The Marine lifts the casualty to a fireman's carry and returns to the starting point. Assistance by the casualty is permitted. The tested individual must carry all equipment; the casualty has none. Marines must negotiate the course in a maximum time of 47 seconds or less.
(4) Event 4.'Advance by Fire and Maneuven. This event simulates advancing and assaulting during an attack. The Marine assumes the prone position at the starting point. On the command GO, the Marine then alternately crawls, runs in a zigzag manner, rolls, jumps, and assumes different firing positions for the next 130yards. Before reaching the objective, the Marine must "hit the deck," roll over, and assume the firing position three times. While assuming the firing position, the Marine must pull back the charging handle, take aim, and squeeze the trigger before rising from the ground. After rising from the ground for the third time,the Marine continues running and jumps to clear the 8-foot trench in one leap. The Marine then finishes the course in

the maximum allotted time of 2 minutes. The Marine is allowed multiple attempts only within the specified time limit.

(5) Event 5: Forced March. This event is a combat run of 3 miles without halts and against time. The Marine must double-time. Marines must complete the 3-mile course in a maximum allotted time of 30 minutes or less.

g. Event Standards

(1) To successfully pass the physical readiness test, the Marine must pass each event with a minimum of 80 points for each. Points are then assessed for each event in accordance with the scoring tables. (See fig. 6-1.)

(2) If a Marine fails one event, the highest possible score that the Marine may receive for the entire test is 79 percent. (See example 1 in fig. 6-2.) If two events are failed, the highest possible score for the entire test is 69 percent, regardless of the total points accumulated. (See example 2 in fig. 6-2.) If three events are failed, the highest possible score is 59 percent. Failure in four or more events will result in a score of zero for the entire test.

(3) To convert the total points for all five satisfactorily passed events to a final percentage score, total the accumulated points for all five events and divide by five. (See example 3 in fig. 6-2.)

Figure 6-1 is available in the illustration section at the end of the book.

Figure 6-2 is available in the illustration section at the end of the book.

6008. AIRBORNE TRAINEE PHYSICAL FITNESS TEST

a. Use and Composition of the Test.

(1) Use. The airborne trainee physical fitness test is a means of determining the physical ability of the applicant for acceptance to and retention in the airborne course of instruction.

(2) Test Events The test battery consists of five events as follows: chinups, knee bender, pushups, situps, and an endurance run.

b. Method of Scoring and Standards

(1) Scoring. The Marine will be scored by a trained scorer who is thoroughly familiar with the minimum standards for the test events. The Marine will be scored on a pass or fail basis. The performance on each event may be recorded on the reverse side of DA Form 705 (Physical Fitness Testing Record).The examinee records the personal information on the face of the card as specified on the first, second, and third lines. The face of the card is identified by writing diagonally across the lower half of the card "Airborne Trainee PPT—See reverse side."Using line 7 on the reverse side of the scorecard, the scorer enters the test title in the block entitled "(Other (Specify))" and completes the required information on the remainder of that line. The "Remarks" section can be used to record the test event titles and the applicant's performance on each test event.

(2) Standards. To successfully pass the test, the examinee must reach the standard in each test event. The standards follow:

(a) Chinups

(b) Knee bender 80(2-minute period)

(c) Pushups 45(2-minute period)

(d) Situps45(2-minute period)

(e) Endurance runs:2 miles in 15:54 minutes or less in athletic gear and 4 miles in 32 minutes in utilities and running shoes.

c. Uniform for Testing

(1) Examinees. The prescribed uniform for test participation is boots and the work uniform of the season. No headdress is worn. When climatic conditions permit, jackets or outer shirts may be removed.

(2) Officials. Scorers and other test officials should be uniformly and distinctively dressed for contrast with Marines being tested.

d. Test Administration Procedure

(1) Preparation for Test. The administration of the test to a large group makes it mandatory that the test effort be organized and efficiently operated. All testing is not completed with large test groups; on certain occasions, individuals and small groups are tested. Care must be exercised to administer the test uniformly and to standardize the conduct of all elements of the test. Regardless of the size of the test group, the following elements of sound test administration should be included:

(a) An orientation to include the purpose, method of administration, scoring of the test, preparation of the score card, and required standards.

(b) A correct demonstration of each event to ensure that there is no misunderstanding of the proper form and required standards.

(c) Completion of all five test events in one test period with all Marines taking the events in the same sequence.

(d) Adequate rest periods between the test events to allow for recovery before the next event.

(2) Method of Administration with a Large Group. With a minimum of 12 lanes per test event, 14 officials can administer the test battery to 150 or 200 Marines in 2 hours. The officials are designated as follows: one officer in charge, one demonstrator,and 12 scorers. If more or fewer Marines are to be tested, a greater or lesser number of officials will be required. The following procedure is recommended:

(a) Conduct an orientation and ensure the examinees have properly completed their score cards.

(b) Assign Marines to lanes and caution them to remain in the same lane order throughout the test.

(c) Explain and demonstrate the chinup event, administer it, and score it. Then proceed to the knee bender and pushup events, and administer them in the same manner.

(d) Grant a 5- to 10-minute rest period after the pushup event. Advise against excessive consumption of water during the break period.

(e) Explain and demonstrate the situp event, administer it, and score it as prescribed. Then move to the run area, explain the running event, and complete it.

(f) Retain the scorecards at the completion of the running event.

(3) Method of Administration with a Small Group. A similar procedure is followed for the testing of individuals and small groups. The informality usually associated with small groups must not conflict with sound test administration. With fewer examinees, fewer officials are required.

e. Description and Explanation of Test Events

(1) Test Event l:Chinups

(a) Purpose. This event is devised to test arm and shoulder flexor strength.

(b)Equipment There is one horizontal bar per lane, made of plumber's pipe or a gymnasium horizontal bar 1 1/2 inches in outside diameter. The bar must be rigidly supported at a height of 8 feet above the ground. The upright supports must be 5 feet apart. There must be a movable stand at each bar for short Marines to stand on to reach the bar.

(c) Officials There is one scorer per lane.

(d) Organization. The examinees, holding their scorecards, stand in order behind the restraining line in their respective lanes. The scorers take each scorecard when the Marine is called forward for the test.

(e) Starting Position The bar is grasped with the palms turned toward the face, the thumbs underneath the bar. The body is fully extended in a "dead" hanging position with the arms straight and the feet above the ground.

(f) Movement. Pull the body directly upward until the chin is placed over the bar. Lower the body until the elbows are completely straight and the body is again in the "dead" hanging position. Repeat as many times as required.

(g) Instructions. Explain and demonstrate the fully extended "dead" hanging position with the proper grasp. Show that the chin is placed over the bar at the top of the movement and that the arms are fully extended, the elbows completely straight, at the bottom of the movement (the hanging position). Explain that the body must be kept from swinging and that it is permissible to raise the legs and flex the hips when pulling up, but that any kicking, bicycling, or jerking motion with the trunk or legs is not acceptable. Inform the examinees that no penalty is exacted for hanging on the bar to rest in the bottom position but that this is not to their advantage. Tell them that half-completed chinups are not counted, and that the scorer will repeat the number of the last correct chinup when incorrect execution is detected.

(h) Administration and Scoring. Caution the examinees to assume the "dead" hanging position and wait for the scorer's command to begin. The scorer is at the examinee's left with a clear view of the bar. If the examinee begins to swing widely, the scorer should stop the swinging by extending the left arm across the front of the examinee's body, being sure not to hinder the execution of the chinup. The scorer counts aloud the number of chinups correctly executed. When a chinup is not correctly executed, the scorer repeats the number of the last correct one. The scorer records the number of correct chinups on the scorecard and returns the card to the examinee.

(2) Test Event 2:Knee Bender

(a) Purpose. This event measures the strength and endurance of the leg muscles.

(b) Equipment. None.

(c) Officials There is one scorer per lane.

(d) Organization. Marines stand in numerical order behind the restraining line in their respective lanes. The scorer takes each scorecard when the Marine is called forward for the test.

(e) Starting Position. The feet are spread less than shoulder width apart, hands on hips, thumbs in the small of the back, elbows back.

(f) Movement. Do a knee bend and at the same time bend slightly forward at the waist and thrust the arms between the legs until the extended fingers touch the ground. The hands are about 6 inches apart. The bend is approximately a three-quarters bend. From this knee bend position, recover to the starting position by moving the body upward, straightening the knees, and returning the hands to the hips. Repeat as many times as required.

(g) Instructions. Explain and demonstrate the correct starting position. Be certain examinees understand the correct knee bend and that only the tips of the fingers touch the ground. Tell them the scorer will repeat the number of the last correct knee bender when incorrect execution is detected. Some of the common errors are failure to correctly bend the knees, failure to touch the ground, and failure to assume the completely erect position after the bend has been executed.

(h) Administration and Scoring. The scorer stands to one side so as to see that the knees are properly bent and the fingers touch the ground as prescribed. From this position, the scorer can view the examinee to see that a properly erect position is assumed after each knee bend. The scorer counts aloud the number of correctly executed knee bends. When a knee bend is done incorrectly, the scorer repeats the number of the last correct one. The scorer records the number of correct knee bends on the scorecard and returns it to the examinee.

(3) Test Event 3:Pushups

(a) Purpose. Pushups measure arm and shoulder extensor strength.

(b) Equipment. None.

(c) Official. There is one scorer per lane.

(d) Organization. Marines stand behind the restraining line in their respective lanes until the scorer calls on them to perform. The scorer takes each scorecard when the Marine comes forward.

(e) Starting Position. The front-leaning rest position is the starting position. The body is straight from head to heels, palms are flat on the ground directly underneath the shoulders, and elbows are straight and locked. The body weight is supported on the hands and toes throughout the event.

(f) Movement. Bending only the elbows, lower the body in one straight plane until the chest touches the scorer's hand. Straightening and locking the elbows, raise the body in one straight plane, returning to the original front-leaning rest position. Repeat as many times as required, keeping the body in a straight line from head to heels.

(g) Instructions. Explain and demonstrate that the arms are straight at the beginning and completion of the movement and that the chest must touch the scorer's hand, but the stomach and thighs must not touch the ground. Also explain that the whole body must be maintained in a straight line as it is lowered and raised; that is, there is to be no breaking at the hips or shoulders so that any body part is lowered or raised in advance of the other or as a separate segment .Dipping or rolling through the shoulders is illegal, as is lowering or raising the body with one arm or shoulder at any time. Resting is not permitted during repetitions. Instruct

Marines that the scorer will repeat the number of the last correct pushup when incorrect execution is detected.

(h) Administration and Scoring. It is recommended that Marines assume a prone position while placing their feet and hands in the proper positions. This permits them to rest while the scorer gets into position and, at the same time, provides a feel of the body in a straight plane from head to heels. The scorer lies on the right hip and side to the right of the examinee. This gives the scorer a clear view of the examinee's body to see any errors. The palm of the scorer's right hand rests flat on the ground underneath the lowest part of the examinee's chest. By keeping the right forearm flat on the ground at an angle in front of the examinee's right arm, the scorer's position will not prevent the examinee from lowering the body completely. The scorer's left hand is free to test the straightening of the elbow at the completion of the movements and to point out body segments being lowered or raised separately. When in position and ready, the scorer has the examinee assume the starting position and begin doing pushups. The scorer counts aloud the repetitions done correctly and repeats the number of the last correct pushup if one is done incorrectly. There is no penalty if the contour of the examinee's body causes the hips to protrude slightly out of line, provided that the whole body is raised and lowered simultaneously. The scorer enters the number of repetitions on the scorecard and returns it to the examinee.

(4) Test Event 4: Situps

(a) Purpose. This event primarily measures abdominal strength.

(b) Equipment. None.

(c) Official. There is one scorer per lane.

(d) Organization Marines stand behind the restraining line in their respective lanes until the scorer calls on them to perform. The scorer takes each scorecard when the Marine comes forward.

(e) Starting Position. The examinee lies flat on the back with knees flexed, both feet flat on the ground. The correct angle of the thighs to the ground is 45 degrees. If the heels are too near the buttocks, the applicant will not be able to sit up. Legs are spread shoulder width apart. The examinee interlaces fingers and places them behind the head in contact with the ground. The feet are not held by another person.

(f) Movement. Bend forward at the waist and raise the upper body until the head is directly over the knees. Heels are not to leave the ground. Elbows remain in the same plane to the head and body throughout the event. The upper body is slowly lowered to the starting position until the head touches the ground. Repetitions are done at a slow cadence with no rest periods.

(g) Instructions. Explain and demonstrate the correct starting position and the proper execution of the situps to be sure that Marines understand the movement. Warn them that their knees must remain flexed during each situp, the heels cannot leave the ground at any time, and they may not roll up on one side and push up with one elbow. Tell them they must do the repetitions at a slow cadence with no rest periods. Instruct Marines that the scorer will repeat the number of the last correct situp when incorrect execution is detected.

(h) Administration and Scoring. When the performer is in position and ready, the scorer has the performer assume the starting position and begin doing situps. The scorer counts aloud the correct executions. When a situp is improperly done, the scorer repeats the number of the last correct one. No situp is credited if the hands are unclasped from behind the head, if the back is used to bounce up from the ground (which means the shoulders would not touch the ground), or if one shoulder or elbow is used to push up. The examinee is not penalized if the heels slide forward slightly so long as the knees remain flexed and the heels maintain contact with the ground. The scorer enters the number of repetitions on the scorecard and returns it to the examinee.

(5) Test Event 5:Endurance Run

(a) Purpose. This event measures cardiovascular endurance.

(b) Equipment One stopwatch or watch with a sweep second hand.

(c) Area. A large training field on which a quarter-mile track has been staked out or a level road over flat terrain may be used as a running surface. A 1-mile route is designated with wooden stakes marking the start point, finish point, and one-quarter mile intervals.

(d) Officials For large groups, a scorer times the event and controls the conduct of the run, and a guide runs with the group and sets the pace.

(e) Organization The run is conducted with groups of Marines in a column formation. Company-sized units may run at the same time with platoons serving as running groups. The scorer issues the command to assume double-time.

(f) Starting Position. Marines are assembled in proper column formation (column of twos, threes, or fours, as appropriate to the size of the group) with short men to the front. When all is ready, the column is moved forward a short distance before the running period is started.

(g) Movement. At the command DOUBLE-TIME, MARCH, the examinees retain their places in the column formation and execute the command. Length of steps is about 40 inches. The scorer has the group execute the run. The formation is maintained during therun.

(h) Instructions. Marines are instructed to maintain formation while running and are informed that the guide will set the proper pace. They will be instructed in the command used to control the column in the execution of the test. The scorer should announce the 4-minute, 2-minute, 1-minute, and 30-second remaining time intervals.

(i) Administration and Scoring. The event may be administered as previously prescribed to a large group, to several Marines, or to an individual. An individual examinee usually does not require a guide or pacer. If the event is administered on a training field, the scorer may stand in the center as the runner(s) circles about the field. This method of administration relieves the scorer of running with each group to be tested. Scoring is based on successful completion of the run as prescribed. The scorer should announce the remaining times as prescribed in (h).

6009. BATTLE FITNESS TEST

The battle fitness test is a physical fitness evaluation which is currently under development as part of the basic warrior training concept plan.

Chapter 7
THE HUMAN BODY

7001.GENERAL

To implement an effective physical conditioning program, a leader must understand how the human body functions and how exercise affects the body .This chapter provides only an introductory discussion of anatomy and functioning.

a. Body Functioning During the Stages of Conditioning. For more on the stages of conditioning, see chapter 1.

(1) During the toughening stage, the waste products of muscle activity (lactic acids) collect more rapidly than the blood can remove them. This acid waste builds up in the muscle tissue and irritates the nerves in the muscle fiber, causing pain and stiffness. As the exercise program continues, more blood is carried through the muscle, removing the waste materials more rapidly and eventually causing the soreness to disappear.

(2) During the slow improvement stage, the blood circulation in the muscles increases, and the body as a whole becomes a more efficient machine. The improvement is rapid in the first few weeks, but as a higher level of skill and conditioning is reached, the improvement becomes less noticeable. The body reaches its maximum level of performance after 6 to 10 weeks and should then be maintained at this peak.

(3) During the sustaining stage, it is possible to maintain this state of conditioning through 15 to 20 minutes of exercise a day, but the exercise must be quite strenuous.

b. Diet. Regular exercise increases the appetite. If the desire for greater amounts of food is satisfied by a balanced diet, the body benefits.

(1) There are two main types of foods: body-building and energy- producing.

(a) Body-building foods consist of proteins, which build up tissue and repair wear and tear.

(b) Energy-producing foods include carbohydrates and fats. Carbohydrates provide a quick source of energy, while fats act as a reserve store of energy.

(2) In addition, food contains vitamins, mineral salts, and water. During hot weather and strenuous training periods, the body requires greater fluid intake.

(3) Diet should be supplemented with proper rest to allow the digestive system to digest the lood.

(4) Occasionally, especially during early stages of conditioning, strenuous exercise may cause vomiting. Although vomiting is not a frequent occurrence, it is not usually a cause for concern.

7002. SYSTEMS OF THE BODY

The systems of the body include the skeletal, muscular, circulatory, respiratory, endocrine, digestive, genitourinary, and nervous systems. Each has a different function, but all must work in cooperation with one another to insure a sound body. Of these systems, the first four are the most affected by exercise and are discussed in succeeding paragraphs.

7003. THE SKELETON

a. Bones. The skeleton (see fig. 7-1) is composed of about 206 bones, which-

o Provide structure for the body and support for the attachment of muscles,

o Protect vital organs, such as the brain, lungs, and heart,

o Manufacture red blood cells, which carry oxygen through the body.

Figure 7-1 is available in the illustration section at the end of the book.

b. Joints. A joint is a place of union between two or more bones. Joints can be-

o Immovable, such as in the face and head.

o Slightly movable, such as in the vertebrae or sternum.

o Freely movable, such as in the shoulder, hip, knee, ankle, and elbow. The bones in a freely movable joint are held in place by ligaments. Freely movable joints are of the greatest importance in physical training because exercise improves their mobility and stability.

c. Cartilage and Ligaments. The joints in the body, where bones connect, are supported by cartilage and ligaments.

o Cartilage is a tough, elastic, translucent tissue that acts as a shock absorber or buffer between bones. Examples are the discs between the vertebrae, the tissue attaching the ribs to the sternum, and buffers in the knee joints. Damaged cartilage does not heal.

o Ligaments are connective tissue that bind bones together. They contract and extend but are not elastic. Consequently, ligaments can be torn or strained. Damaged ligaments will heal to some extent.

7004. THE MUSCLES

a. General. Muscles are tissue, or an organ composed of tissue, which contract or extend to produce movement. Muscles are of three classifications: involuntary, voluntary, and cardiac. The voluntary and cardiac muscles are the most affected by physical training; physical training has little effect on involuntary muscles.(Seefigs. 7-2, 7-3, 7-4.)

o People have no control over the involuntary muscles, an example of which is the diaphragm,

o Voluntary muscles are the larger skeletal muscles which cause movement and which the individual can control.

o Cardiac muscle is found only in the heart and for all practical purposes is an involuntary muscle. Physical training has a significant effect on the fitness of cardiac muscles.

b. Muscle Structure. Muscle is composed of cells which are grouped into bundles called fibers. Groups of fibers form muscles. These bundles of fibers are held in place by sarcolemma, a thin, sheath-like material that surrounds the muscle bundles and secretes a fluid that lubricates the muscle tissue. The fused ends of the sarcolemma form the tendons which attach the muscles to the bones.

c. Attachment of Muscles. The arrangement of muscles on the skeleton provides the proper angle of pull to make movement possible. Voluntary muscles are usually attached to the skeleton in two places.

d.Action of Muscles. To produce motion and to do work a muscle usually shortens its fibers. The movement may be flexion such as bending the arm at the elbow, or extension, such as the straightening the arm. Some muscles raise the arms or legs, others depress the raised limb. Some muscles have the primary function of rotating the trunk from side to side. In addition, muscles act as stabilizers as well as prime movers.

Figure 7-2 is available in the illustration section at the end of the book.

Figure 7-3 is available in the illustration section at the end of the book.

Figure 7-4 is available in the illustration section at the end of the book.

7005. FUNCTIONING OF THE SKELETON AND MUSCLES

a. General. The skeleton and muscles function in close coordination. Muscles move the body, and bones support the muscles.

b. Effect of Exercise on Bones. Continuous exercise, particularly among younger people, usually strengthens the bones, allowing them to withstand greater stress and strain. Bones which are not used lose much of their strength, a factor which should be considered when individuals return to physical training after a prolonged period of inactivity. People in this category should refrain from activities which might result in bone breakage before the bone is exercised back to normal condition. The condition known as "march fracture" is usually a result of this condition.

c. Muscular Strength. When exercised vigorously to improve strength, muscle grows in size. In general, the larger the muscle, the stronger the muscle. Furthermore, trained muscle functions more efficiently than untrained muscle. Trained muscles contract more vigorously and with less effort. Regular and strenuous exercise also toughens muscle, making it firmer and able to stand more strain.

d. **Muscular Endurance.** Muscles naturally become fatigued during continuous, repeated work. Through specific exercises, such as chinups or situps, local muscle groups become exhausted long before an individual fatigues. Training to develop muscular endurance enables people to continue a relatively heavy load of exercise over a long period of time. Lengthened exercise periods improve muscular endurance.

e. Muscular Coordination. Improved speed and strength result in part from improved muscular coordination. An unskilled individual may use irrelevant muscles to perform a particular activity, increasing the amount of work without increasing the mechanical output. This increase in skill is highly desirable, but it should be offset by greater effort in duration or intensity to compensate for the loss in overload due to increased skill.

f. Muscular Fatigue. When the rate of work is increased, the energy required is proportionately much greater than the increase in rate. For example, if an individual doubles running speed, the amount of power demanded to do this is increased eight times.

g. Circulation in Muscles Regular, strenuous exercise causes the creation of new capillaries and the opening of inactive, latent capillaries, increasing blood circulation within the muscle by as much as 400 percent. This increases the supply of food materials and oxygen to the muscle, improving its endurance. This process takes about 8 to 12 weeks of regular conditioning in young adults and longer as age increases.

7006. THE CIRCULATORY AND RESPIRATORY SYSTEMS

a. The Circulatory System. The circulatory system transports blood to all parts of the body, removes waste products for disposal, and delivers protecting and repairing substances where needed. The circulatory system includes the heart and blood vessels.

(1) The Heart. The heart is a pump which forces blood through the blood vessels to the parts of the body. The heart is a little larger than a fist and is located in the left center of the chest between the lungs.

(2) Blood Vessels. The vessels carrying blood away from the heart are the arteries. They eventually divide into capillaries, the very small vessels through which diffusion and osmosis take place. The capillaries gradually increase in size, forming the veins which carry blood back to the heart.

b. The Respiratory System. The respiratory system performs the function of breathing, whereby oxygen is inhaled and carbon dioxide is exhaled. The respiratory system consists of the mouth, nose, trachea, lungs, and diaphragm.

o Trachea. The trachea, or windpipe, is a hollow, tube-like structure that carries air from the mouth to the lungs.

o Lungs. The lungs are elastic bags in the chest where the exchange of oxygen and carbon dioxide takes place.

o Diaphragm. The diaphragm is a thin, sheet-like muscle just below the lungs. During inhalation, the diaphragm flattens out and lowers, allowing the lungs to expand and fill with air. During exhalation, the diaphragm raises into a dome shape, helping to expel air from the lungs.

7007. CARDIOVASCULAR FUNCTIONING

a. General. Cardiovascular functioning is the combined functioning of the circulatory and respiratory systems. The chief organs of these systems, the heart and lungs, function together to provide oxygen to the body.(See fig. 7-5.)

Figure 7-5 is available in the illustration section at the end of the book.

b. Heart Action. The heart, the organ which propels blood through the blood vessels, is the chief organ of cardiovascular endurance. A person tires quickly if the heart lacks the capacity to circulate the blood through the body. The heart is a muscular organ and is developed by exercise just like any other muscle. The heart cannot be exercised alone; any exercise that puts a sufficient load on the heart, such as running or cycling, will also exercise other body parts. A conditioned heart tends to beat more slowly but pump more blood when at rest. This is known as an increase in "stroke volume,"a desirable condition because it enables the heart to pump more blood with a slower contraction rate.

c. Functions of the Lungs. Strenuous and regular exercise can improve the efficiency with which the lungs transmit oxygen to the blood by as much as 25 percent In a poorly conditioned person, some of the alveoli (air sacs) within the lungs are closed or collapsed. The forced breathing created by exercise over a period of weeks causes the air sacs to slowly expand, increasing the ability to absorb oxygen.

d. Relationship of Heart and Lungs. The heart pumps carbon dioxide-laden blood through the pulmonary artery to the lungs. In the lungs the carbon dioxide is exchanged for oxygen and the purified blood is returned to the heart by way of the pulmonary vein. The heart then pumps the blood through the aorta for circulation throughout the body. Blood moving into the muscles exchanges oxygen for carbon dioxide, after which time it is ready for the return trip to the heart.

e. Cardiovascular Functioning in High Altitudes. Marines to be employed in high altitudes should be acclimated in a similar area for 10 to 14 days prior to employment. Persons not accustomed to the rarefied air of higher altitudes tire more quickly and may collapse after rapid physical exertion. Air is much less dense at high altitudes than at sea level. Persons accustomed to sea level inhale only about 80 percent of the oxygen they are used to no matter how hard they breathe. Further, people accustomed to sea level or moderate altitude simply do not have enough red corpuscles in the blood to fulfill their needs at high altitudes. The red blood cells will increase over time.

f. Symptoms of Cardiovascular Malfunctioning. The brain is the first organ to react to a lack of oxygen. Unconsciousness results when the brain is denied sufficient oxygen. "Blacking out" is actually a defense mechanism to keep the body alive.

APPENDIX A
INSTRUCTOR HINTS AND INSTRUCTOR TRAINING

1. General. This appendix informs instructors on administration of exercise periods, commands, the extended rectangular and circle formations, methods of instruction and conduct of exercise activities, explanation of basic positions associated with the various activities, and instructor training.

2. Aim of the Physical Fitness Program. The aim of the physical fitness program is to prepare Marines physically for war. This can be conveniently split into two phases.

a. Preparation Phase

(1) Development of all-around physical fitness.

(2) Development of purposeful physical skills.

(3) Development of mental alertness.

(4) Development of character and leadership.

b. Maintenance Phase. Once the preparation is completed, the program must progress to a maintenance program. This phase revolves around the need to maintain the trained Marine's physical fitness for a particular role. This will include—

(1) Maintenance of a high standard of all-around physical fitness.

(2) Application of purposeful physical skills to combat skills.

(3) Development of leadership and self-confidence.

(4) Development of mental toughness and endurance.

3. Physical and Mental Development

a. Marines need to be developed physically and mentally. The physical training (PT) instructor must ensure that the program accomplishes this.

b. The system of the body functions best in response to progressive exercise. This means that Marines should master the easier exercises before they attempt the more difficult ones.

c. The body and mind are linked inseparably and must be considered as one in physical fitness training. Physical exercises assist in developing the mind.

d. The confidence in oneself and in one's comrades attained by physical achievement helps to promote the indomitable will to win. The Marine must be taught to realize this. It will help during arduous training, which at all times must be imaginative, realistic, and filled with enthusiasm. All physical fitness training must be associated with developing the qualities required of a Marine.

4. Planning and Preparation of Programs The following factors must be considered when compiling a program (see also ch. 2)~

a. Aim or desired skill level.

b. Duration of the complete training program and number of physical training lessons.

c. Purpose of included exercises,drills, or practical applications.

d. Beginning/inventory fitness levels.

e. Medical facilities, medical personnel, and Marine Corps or local command regulations governing their proximity to the training.

f. Number of Marines to be trained.

g. Availability of instructors, assistants, and facilities.

h. Apparatus and equipment available.

i. Alternative training areas for varying weather conditions,

j. Stages of progression and tests of achievement.

5. Five Basic Principles of Program Design

a. Regularity. Regularity of exercise is far more important than the amount of exercise performed on a given day. A Marine should exercise a minimum of three times per week. Human muscle begins to atrophy after 3 days without exercise.

b. Balance. The program must be balanced in terms of total physical fitness. Imbalance is created when, for example, an individual merely runs. While the cardiovascular system is improved, general upper body strength may be lacking. Conversely, a person who just does weight training may develop strength, but not cardiovascular endurance.

c. Overload. If total body fitness is to be developed, then the muscular and cardiovascular systems must be made to perform more work than previously. This is done by increasing the frequency, intensity, and duration of the exercise. The body adapts to the amount of stress placed upon it and becomes stronger and more efficient.

d. Progression. To be successful, any type of training program must be progressive. This is an area where untrained persons conducting PT can cause injury to Marines resulting in failure of the PT program. Close supervision must be made of untrained instructors. The result of progressing too fast is that the class begins to reject PT because it is no longer enjoyable or motivating. The program should start at a slow pace, and the pace should increase as proficiency improves. Average persons will encounter three stages of progressiom the toughening stage, the slow improvement stage, and the sustaining stage. (See ch. 1.)

e. Variety. Some programs fail because the training becomes boring. Variety is essential and perhaps presents the greatest challenge to the commander and staff.

6. Grouping. Providing for different levels of physical fitness is particularly recommended in the early stages of conditioning. Older Marines and those in poorer physical condition should be expected to attain a group level of fitness, but they should be given more time to do it.

a. Homogeneous Grouping One simple method of providing for the difference in levels is to group them according to their condition. A two-group classification would divide Marines into highly conditioned and average groups. A finer classification could be obtained by dividing them into three groups—a highly conditioned, a moderately conditioned, and an unconditioned group.

b. Performance or Age Grouping. The segregation of Marines into different exercise groups should be based on physical fitness test scores or on the level of fitness they demonstrate. They may also be grouped at first according to age. A common classification by ages is under 30, 30 to 34, and 35 and above.

7. Preparatory Commands and Commands of Execution. The prepreparatory command describes and specifies what is required. The command of execution calls into action what has been prescribed. All preparatory commands are given with a rising inflection. The interval between commands is long enough to permit the average Marine to understand the first one before the second one is given.

8. Extended Rectangular Formation. The traditional formation for carrying on many physical training activities is the extended rectangular formation (fig. A-1). This formation is more compact than any other. It is the best type to use for large numbers of Marines because it is simple and easy to assume.

a. For the formation of oneplatoon, the base man paces off five paces from the stand, faces left and moves five paces, halts, and again faces left. With the base man positioned facing the stand, the platoon leader then commands. **FALL OUT AND FALL IN ON THE BASE MAN**. At this command, all Marines run to the designated area and reform. This procedure is preferred to marching the unit into position. If more control is desired, the unit may march at double time to the vicinity of the base man and then be directed to fall out and fall in on him. Much valuable time is wasted in the field by needless maneuvering of Marines at quick time in an effort to position the platoon or unit on the exact spot for the exercises.

b. A company-size unit assumes the extended rectangular formation from a column of threes or fours at normal intervals between squads. This extension can also be executed from a company mass without interval between platoons. In extending either a platoon or company-size unit, take your place at the head of the column and command:

(1) TAKE INTERVAL TO THE LEFT, MARCH. At this command, the Marines in the right flank file stand fast with arms extended sideward. All others turn to the left and run forward at double time. After taking a sufficient number of steps, all Marines face the front with both arms extended sideward. The distance between finger-tips is about 12 inches, and dress is right.

Figure A-1 is available in the illustration section at the end of the book.

(2) ARMS DOWNWARD, MOVE. At this command, the arms are lowered smartly to the sides.

(3) LEFT, FACE

(4) EXTEND TO THE LEFT, MARCH. At this command, those in the right flank file stand fast with arms extended sideward. All others turn to the left and run forward at double time. Spacing is the same as in (1) above, and dress is right.

(5) ARMS DOWNWARD, MOVE. Same as in (2) above.

(6) RIGHT, FACE.

(7) FROM FRONT TO REAR, COUNT OFF. At this command, the leading Marine in each column turns the head to the right rear, calls of **ONE** and faces the front. Successive Marines in each column call off in turn **TWO, THREE, FOUR, FIVE** in the same manner.

(8) EVEN NUMBERS TO THE LEFT, UNCOVER. At this command, each even-numbered Marine stride jumps sideward to the left, squarely in the center of the interval. In doing this, each one swings the left leg sideward, jumps from the right foot to the left foot, and smartly brings the right into position against the left.

c. To assemble the unit, you commandASSEMBLE TO THE RIGHT, MARCHAt this command, all return to their original position in the column at double time and reform on the base man.

d.It is recommended that the area for grounding equipment and arms be at the edge of, or well away from, the area to be used for exercising. To conserve time and ensure proper position of the unit, the base man or, if the unit is composed of several platoon-sizegroups, the various base men may precede the unit and establish their positions in relation to the instructor's stand.

9. Circle Formation.The circle formation is effective for conducting various exercise activities (fig. A-2). This formation has an advantage in that supervision of all Marines is facilitated and a moving formation is available which provides control. Guerrilla exercises, grass drills, and some forms of running are examples of activities which are more easily conducted in the circle formation than in the extended rectangular formation.

a. It is not advisable to have more than 60 Marines in a circle. When more must be accommodated, separate circles should be used.Concentric circles may be employed to reduce the size of the circle or to accommodate more persons. If concentric circles are formed, the different circles are made by designating squads for each circle.Each additional circle requires more Marines than the one inside it.For example, one squad of a platoon may form the inner circle and theremaimng three squads the outer circle.

b. When a platoon is to form a circle, the commands ai€IRCLEFORMATION, MARCH, FOLLOW MEUpon this command, the right flank squad of the column moves forward at double time with the leader of the platoon group gradually forming a circle in a counterclockwise direction. Each succeeding file falls in behind that on the right. After the rough outline of the circle is formed, the leader commandsPICK UP A **5**-YARD INTERVAL.

Figure A-2 is available in the illustration section at the end of the book.

c.The group may be halted and faced toward the center, or, if instruction is not necessary, the exercise activity may be executed without stopping the platoon.

lO.Leadership Techniques

a. Unless you experience all the exercises, you cannot appreciate how arduous they are, what movements are most strenuous and difficult, where the errors in performance are likely to occur, and what the proper cadence should be.

b. You must give everyone careful supervision and participate in the exercises to show that you can do them. When you participate, your assistant instructors should supervise because it is difficult for you to supervise and exercise at the same time.

c. Marines should never be kept too long in one position, especially a constrained one. They should never have to perform an exercise more times than they can do it without losing the proper form. You should insist upon proper form in the execution of all exercise activities.Even slight deviations from the proper form reduce the value of the exercise.

d. Avoid long explanations.As a rule, it should be necessary to give a full explanation of new exercises only. Explain the most essential features of an exercise first; add details later. Too many details at one time are more likely to confuse than to assist them.Minor corrections should be made to the entire class while the exercise is in progress (for exampHEADS **UP,** KNEES STRAIGHT). If necessary, follow this direction by the name of the person who is particularly at fault. If a Marine requires special attention, givethat person separate instruction by an assistant instructor to avoidwasting the time of the group.

e. The heavy demand on your voice can be lightened by training assistant instructors to assume some of the instruction. Using mass cadence is also an effective method of lessening the demand on your voice.

“Illustrations”

DAY	TIME	ACTIVITY
MONDAY	60 MIN	RUNNING ACTIVITY
TUESDAY	60 MIN	COMBATIVES
WEDNESDAY	60 MIN	CONDITIONING & LOG DRILLS
THURSDAY	60 MIN	CONDITIONING & RELAYS
FRIDAY	60 MIN	RUNNING & CONDITIONING
SATURDAY	60 MIN	RUNNING ACTIVITY

Figure 2-1. Sample A- -Single Activity.

DAY	TIME	ACTIVITY	REMARKS
MONDAY	20 MIN 20 MIN 20 MIN	RIFLE DRILL COMBATIVES RUNNING	PROGRESSIVE ACTIVITY SYSTEM
TUESDAY	30 MIN	GUERRILLA EXERCISES TABLE I	SINGLE ACTIVITY
WEDNESDAY	30 MIN	RIFLE DRILL	SINGLE ACTIVITY
THURSDAY	30 MIN	CIRCUIT INTERVAL TRAINING	SINGLE ACTIVITY
FRIDAY	20 MIN 20 MIN 20 MIN	STRENGTH CIRCUIT CONDITIONING DRILL CROSS-COUNTRY RUN	ROTATION ACTIVITY
SATURDAY	60 MIN	WARM UP RUN AND GRASS DRILL	SINGLE ACTIVITY

Figure 2-2. Sample B- -Various Time Periods, Activities, and Systems.

DAY	TIME	ACTIVITY
MONDAY	45 MIN	RUNNING – CROSS-COUNTRY
TUESDAY	75 MIN	TEAM ATHLETICS – TOUCH FOOTBALL
WEDNESDAY	45 MIN	RUNNING – CROSS-COUNTRY
THURSDAY	20 MIN 40 MIN	LOG DRILL DUAL COMBATIVES
FRIDAY	90 MIN	SPEED MARCH
SATURDAY	90 MIN	TEAM CONTESTS

Figure 2-3. Sample C--Sustaining Activities.

Route	Kilometers from SP	Recommended rate of march (kmph)	Remarks
SP: RJ 8th Div Rd—Superhighway	......	4	Bridge; hard surface; two guides.
RJ Jamestown—Hersey Rd	2,25	4	Traffic heavy; two guides; Jamestown Rd bears to right.
RJ Jamestown—Yankee Rd	7.10	4	Two guides
RJ Jamestown—Lightning Rd	10.50	4	Two guides
RJ Jamestown—Sunshine Rd	12.40	4	Light traffic; one guide Sunshine Rd; gravel, poor traction when wet.
RP; RJ Sunshine—Sedan Rd	14.80	4	Two guides

Figure 3-1.Example of Route Reconnaissance Report.

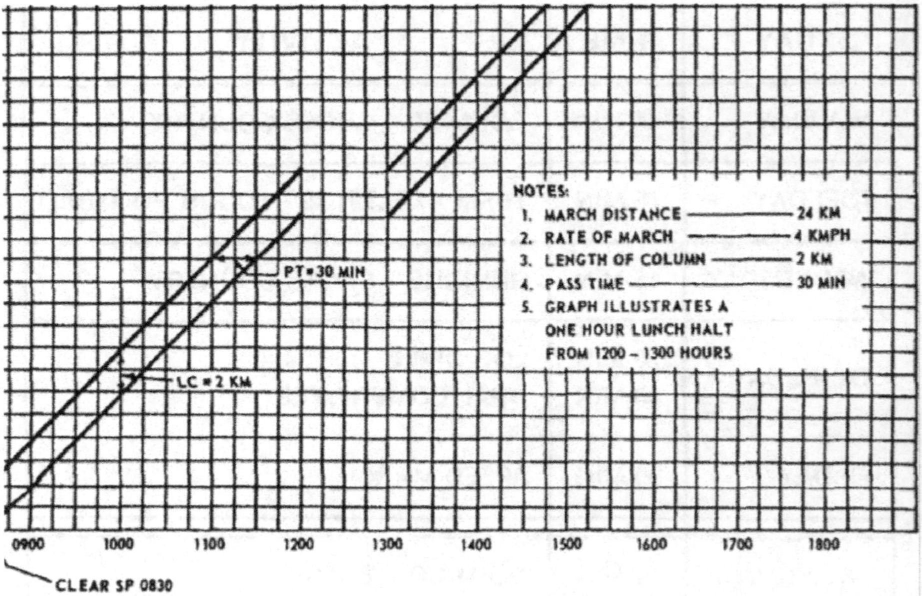

Figure 3-2.Road Movement Graph.

Figure 3-3.The Pace-Setter.

	SINGLE FILE	COLUMN OF TWO
2 m/MAN APART	24	12
5 m/MAN APART	54	27

GAP DISTANCES			
SERIAL:	(3-1) × 100 m	=	200 m
MARCH UNIT:	(12-1-2-) × 50 m	=	450 m
	TOTAL COLUMN GAP		650 m

Figure 3-4.Conditioning Drill One.

Figure 3-5.Conditioning Drill Two.

Figure 3-6.Conditioning Drill Three.

Figure 3-7. Rifle Drill.

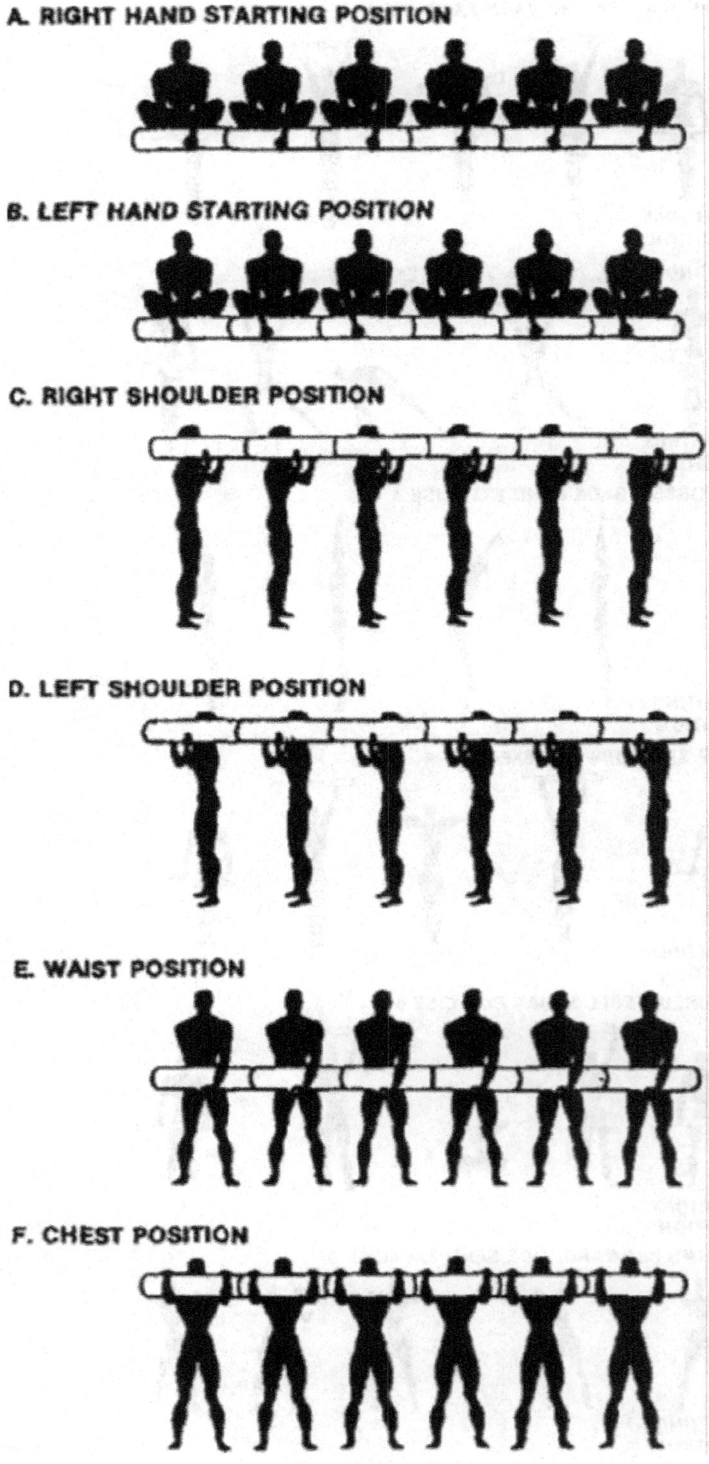

Figure 3-8. Starting Positions.

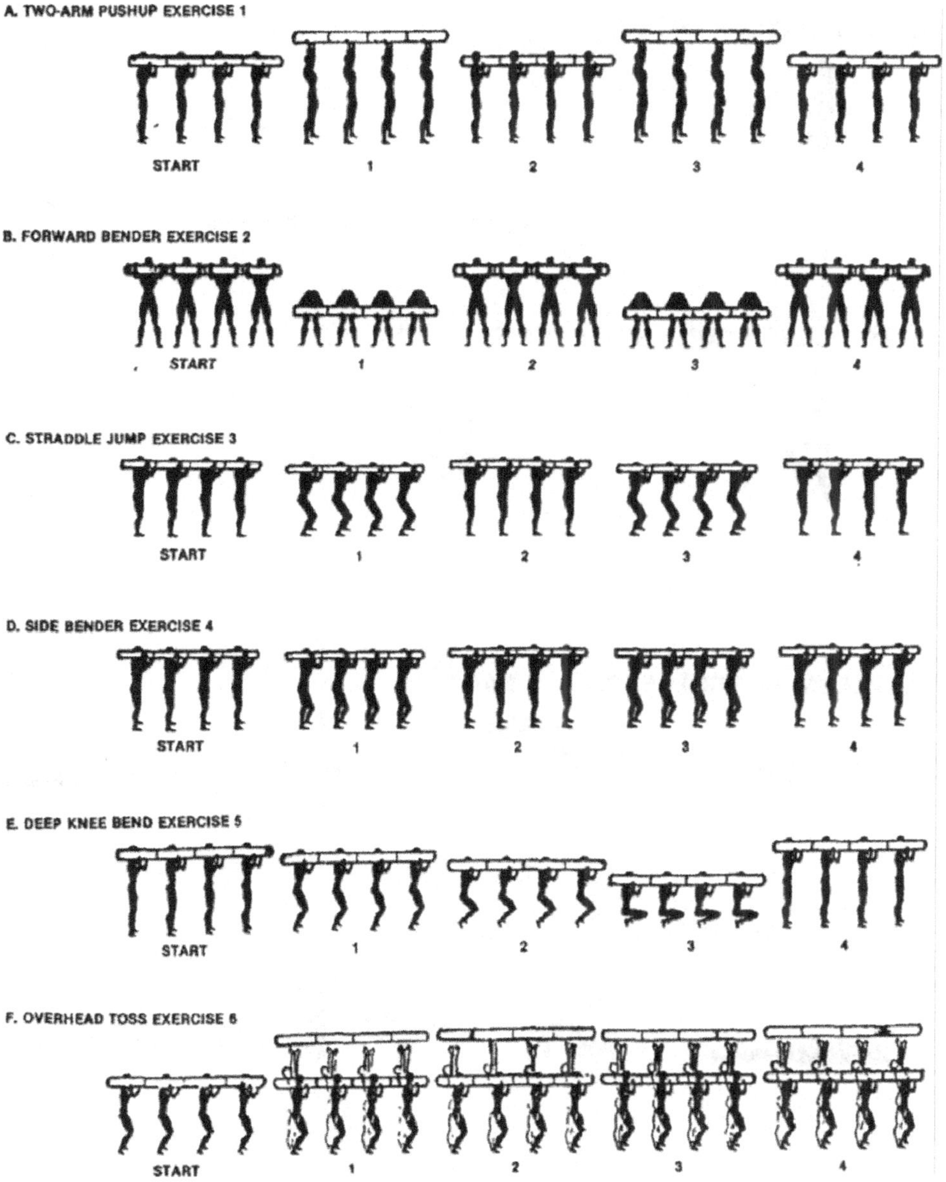

Figure 3-9. Log Drill

Figure 3-10.Basic Positions for Grass Drill.

Figure 3-11.Grass Drill One.

Figure 3-12.Grass Drill Two.

Figure 3-13.Guerrilla Table I.

Figure 3-14.Guerrilla Table II.

Figure 3-15.Proper Running Form.

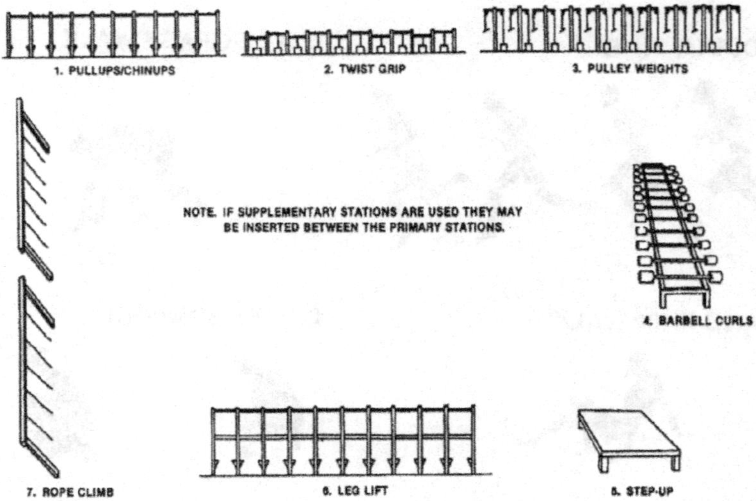

Figure 3-16. Fixed Strength Circuit.

STATION	ITEM	NO.	SPECIFICATION
1	BARBELL		1¼-INCH PIPE 5 FEET LONG WITH CONCRETE FILLED NO. 10 CANS
2	JUMP ROPE		¼- OR ¾-INCH ROPE, 10 FEET LONG
3	TWIST GRIP		HANDLE 12 INCHES LONG, ROPE 4 FEET LONG, NO. 10 CAN CONCRETE FILLED.
4	INCLINE PLANE		¾-INCH PLYWOOD PLATFORM 2 FEET WIDE AND 6 FEET, 6 INCHES LONG, ELEVATED 10 INCHES AT ONE END. STRAP TO HOLD FEET DOWN.
5	WAR CLUB		HEAD IS 6 BY 12 INCHES, HANDLE IS 14 INCHES LONG BY 1¼ INCHES IN DIAMETER, ABOUT 20 POUNDS.
6	BICYCLE RIDE		PLYWOOD BOARD OR PLATFORM 2 BY 3 FEET WITH 2 BY 2 RUNNERS.
7	STEP—UP		A BOX OR STURDY PLATFORM 18 INCHES HIGH, 18 INCHES WIDE, 24 INCHES LONG.
8	ISOMETRIC PULL		TWO HANDLES 12 INCHES LONG WITH 4 FEET (BETWEEN HANDLES) OR LIGHT WIRE CABLE OR ¼-INCH ROPE

Figure 3-17.Movable Strength Circuit.

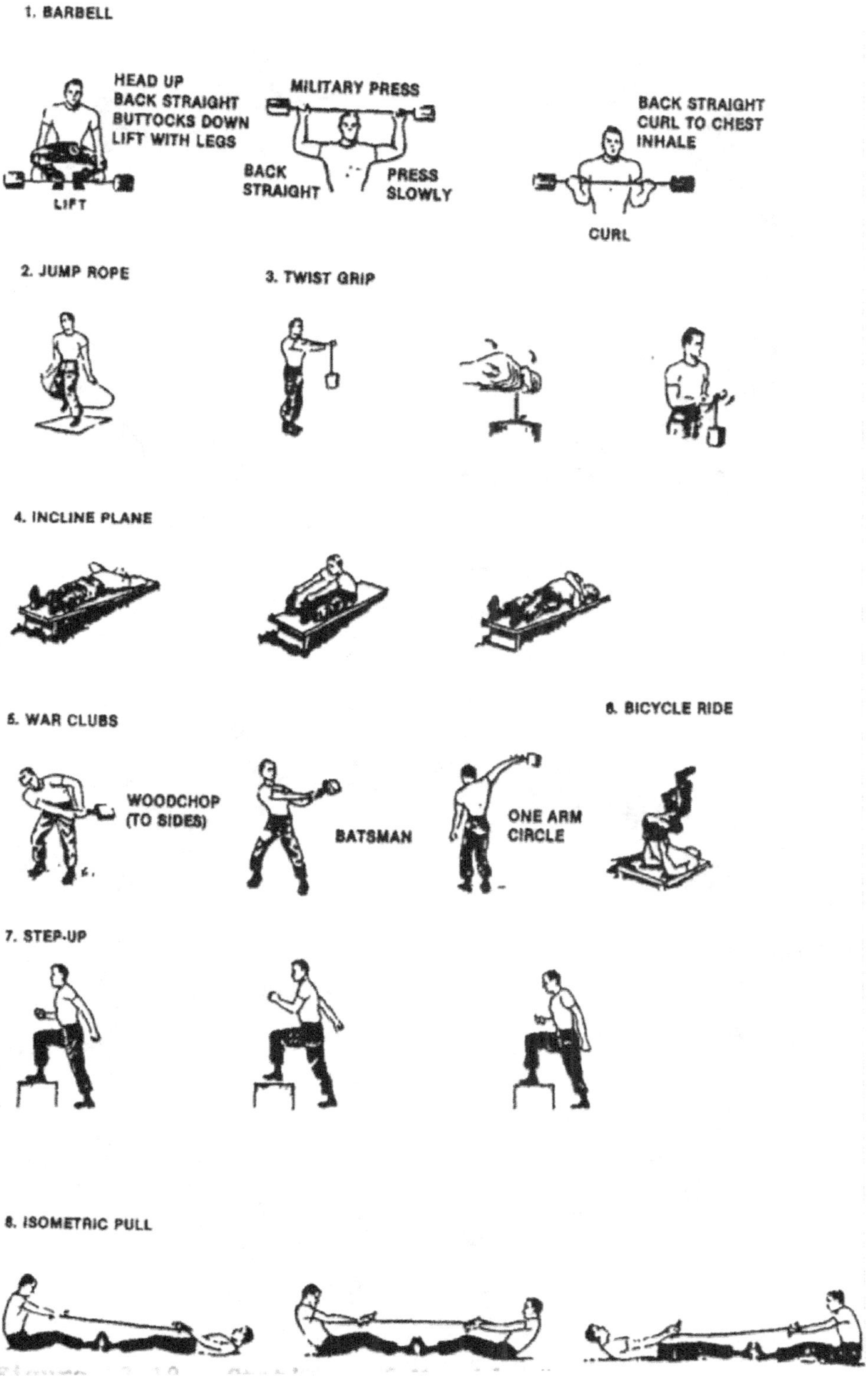

Figure. 3-18.Stations of Movable Strength Circuit.

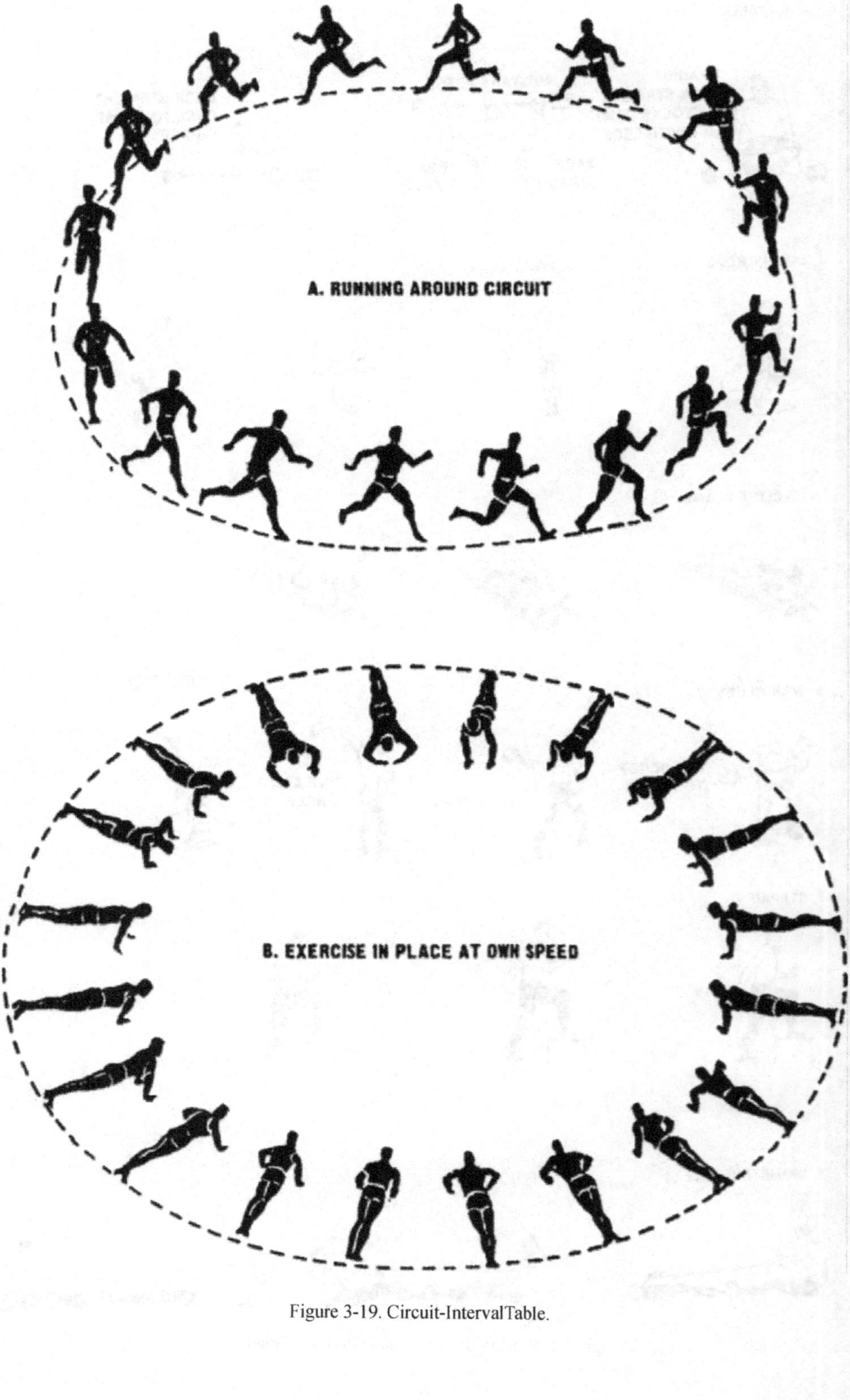

Figure 3-19. Circuit-IntervalTable.

Figure 3-20. Jumping.

Figure 3-21.Dodging.

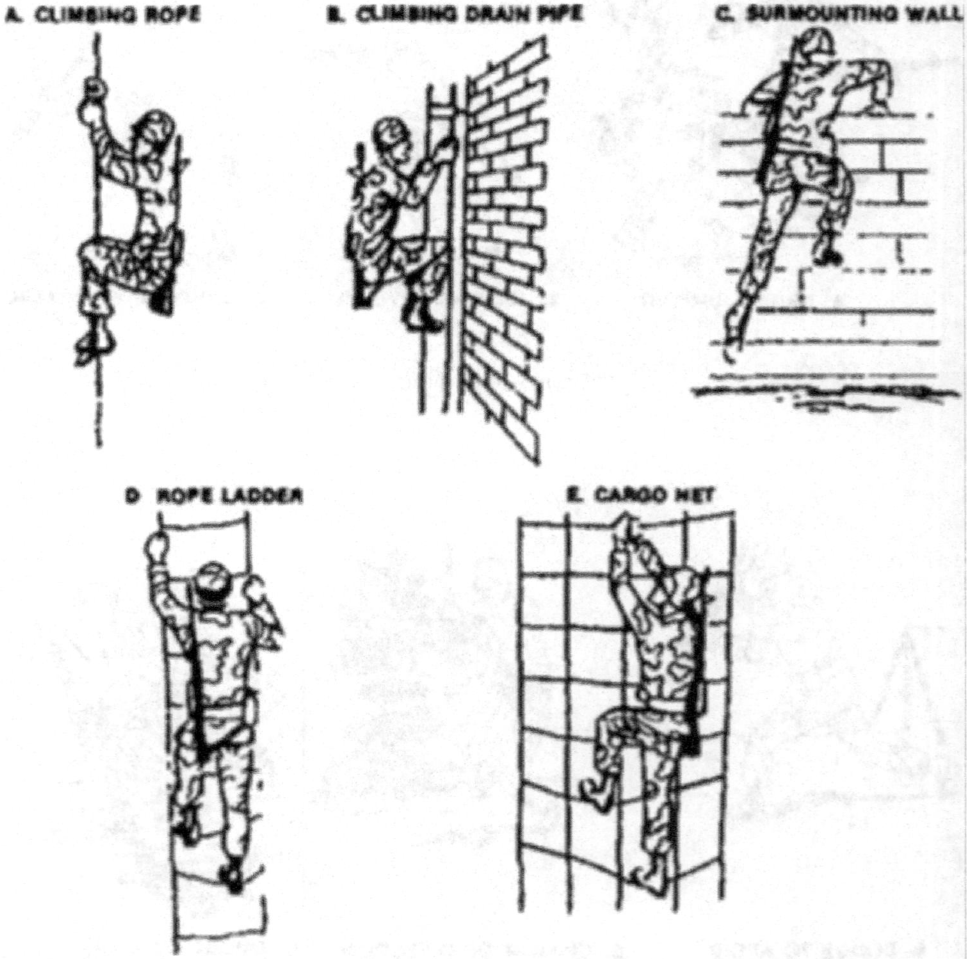

Figure 3-22.Climbing and Surmounting Vertical Objects.

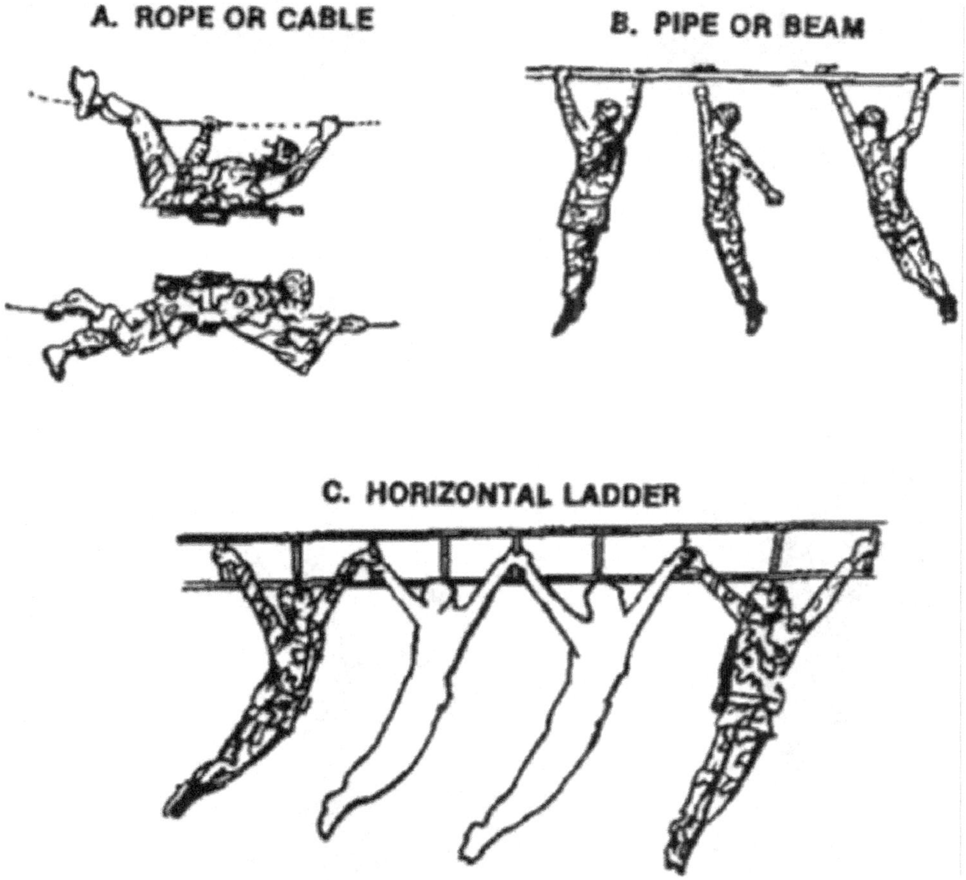

Figure 3-23. Traversing Horizontal Objects.

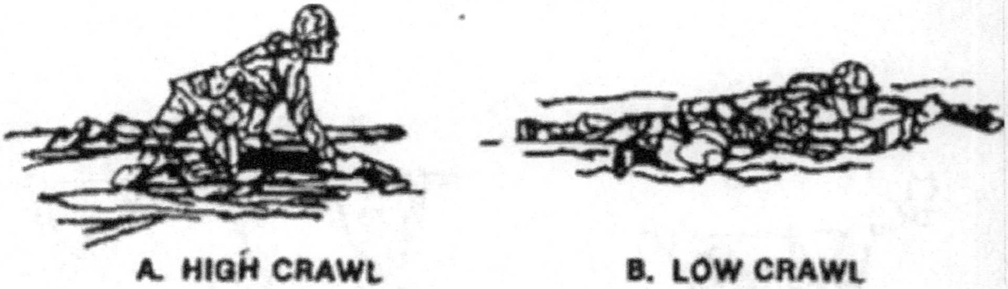

Figure 3-24.Crawling.

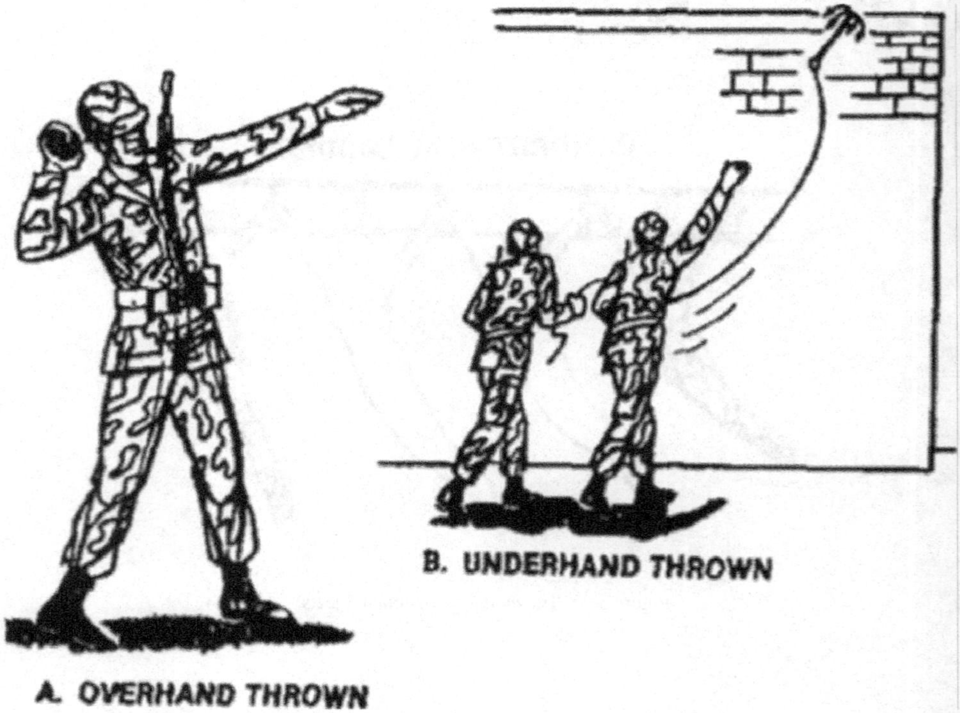

Figure 3-25.Throwing.

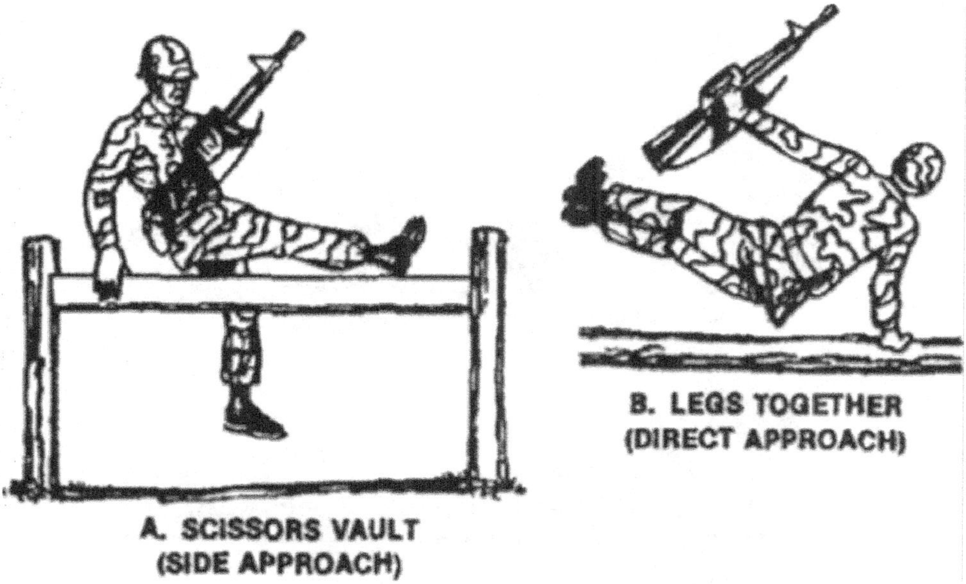

Figure 3-26. Vaulting

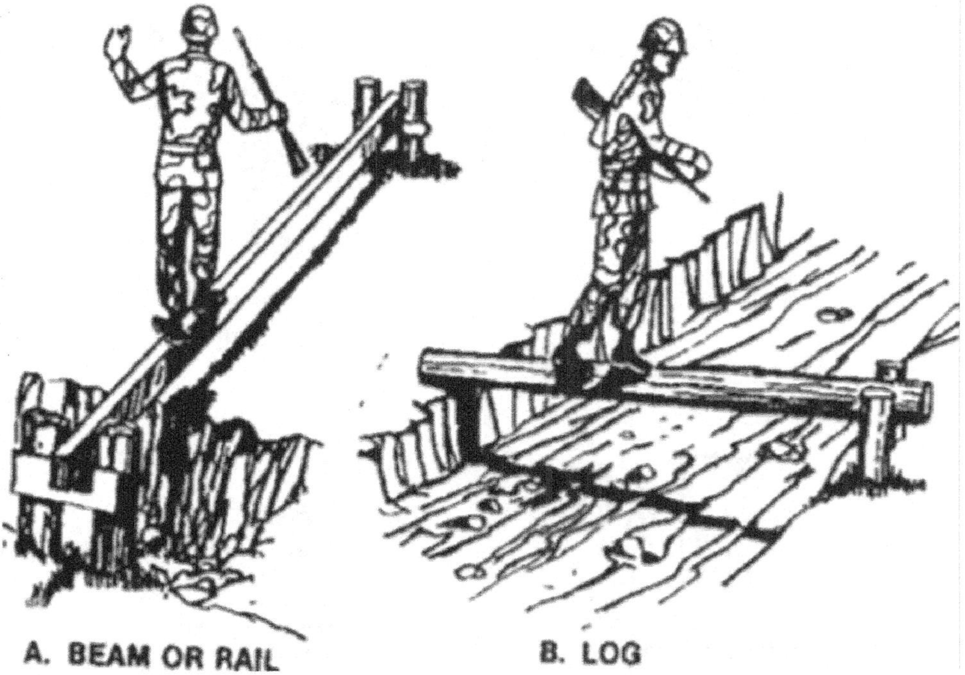

Figure 3-27.Balancing.

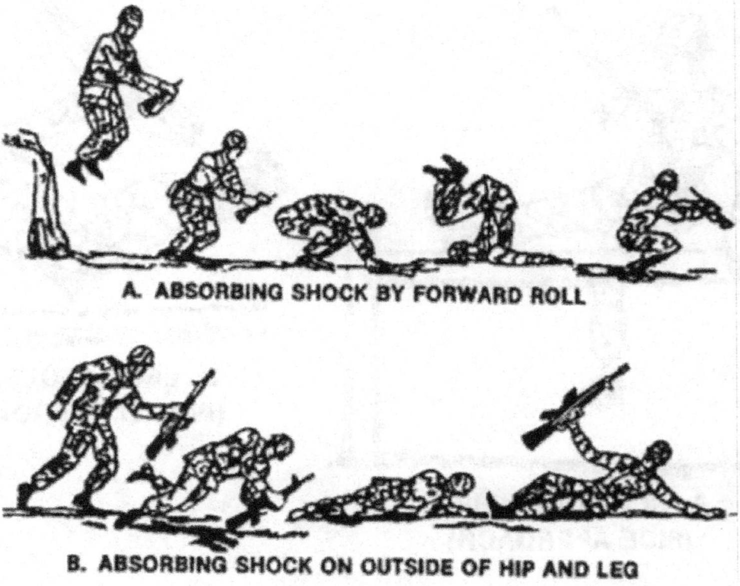

Figure 3-28.Falling.

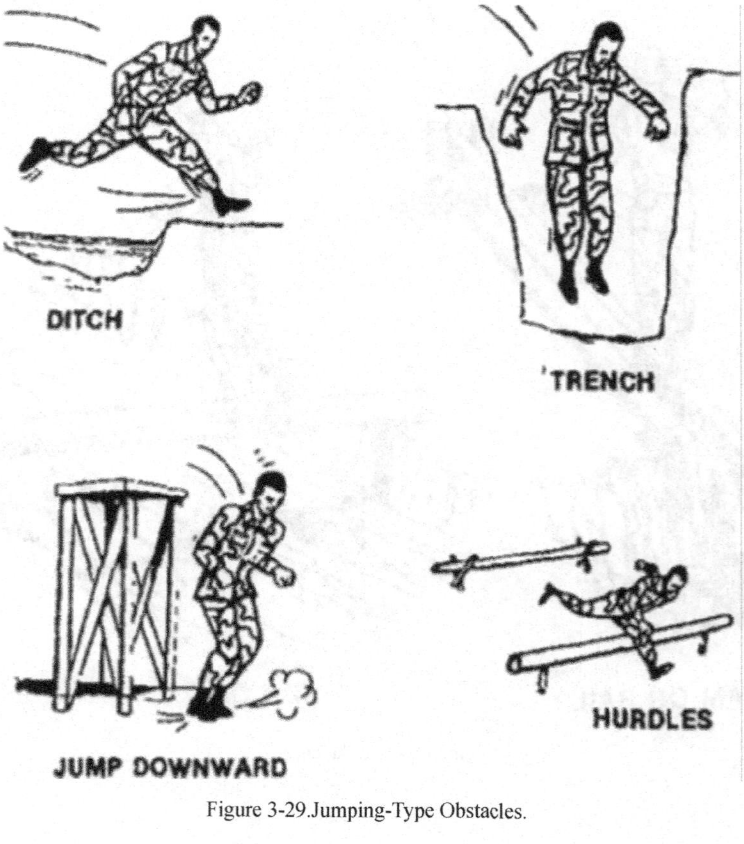

Figure 3-29.Jumping-Type Obstacles.

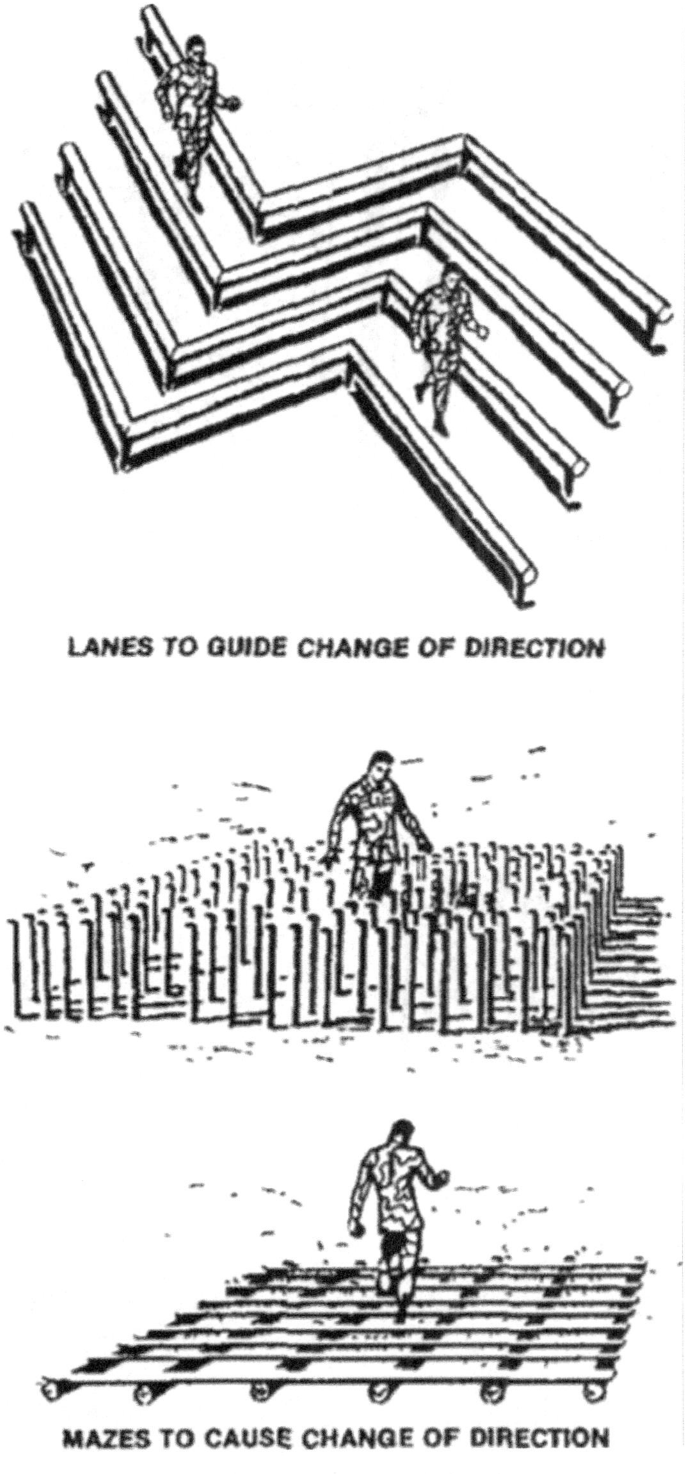

Figure 3-30.Dodging-Type Obstacles.

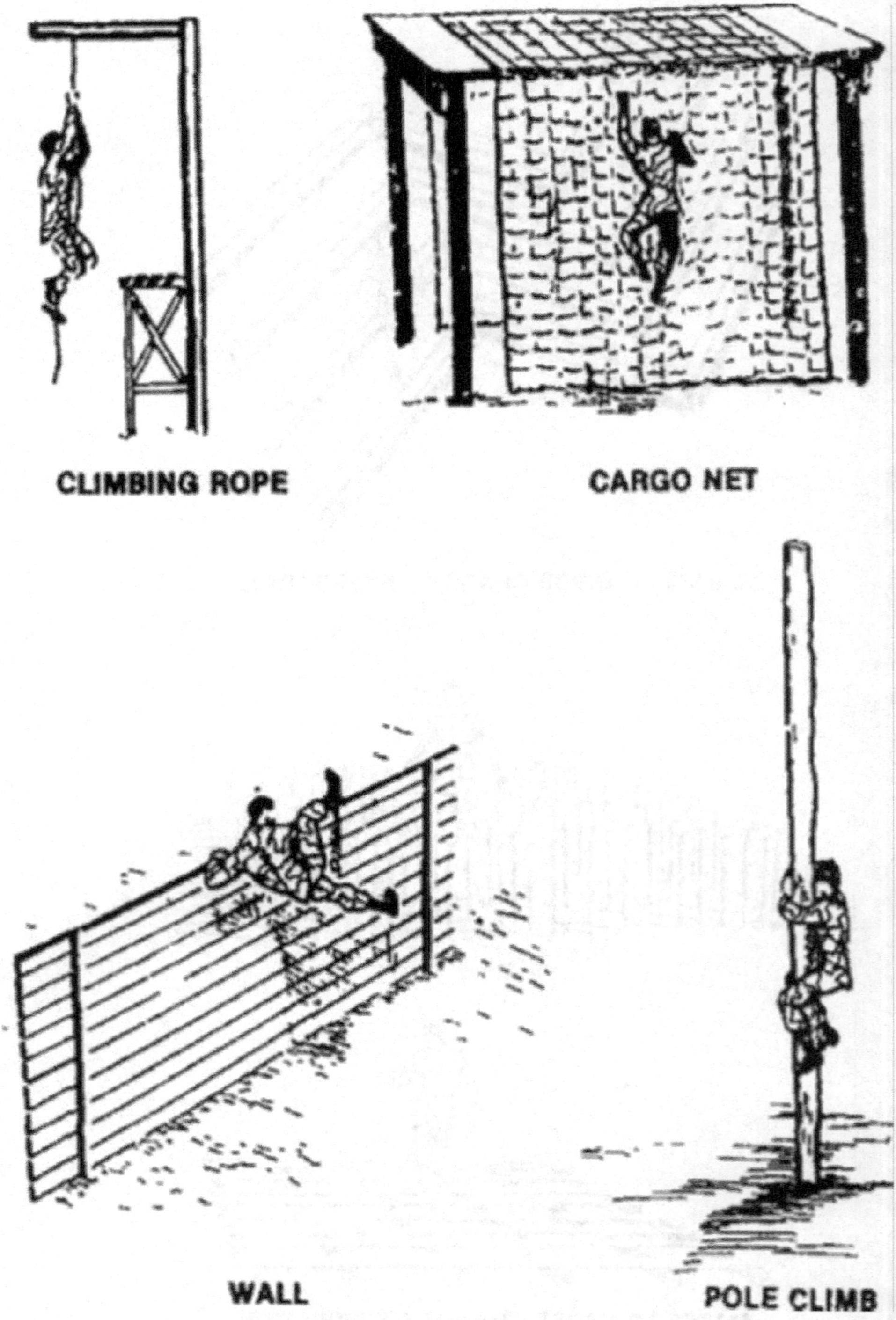

Figure 3-31. Vertical Climbing Obstacles.

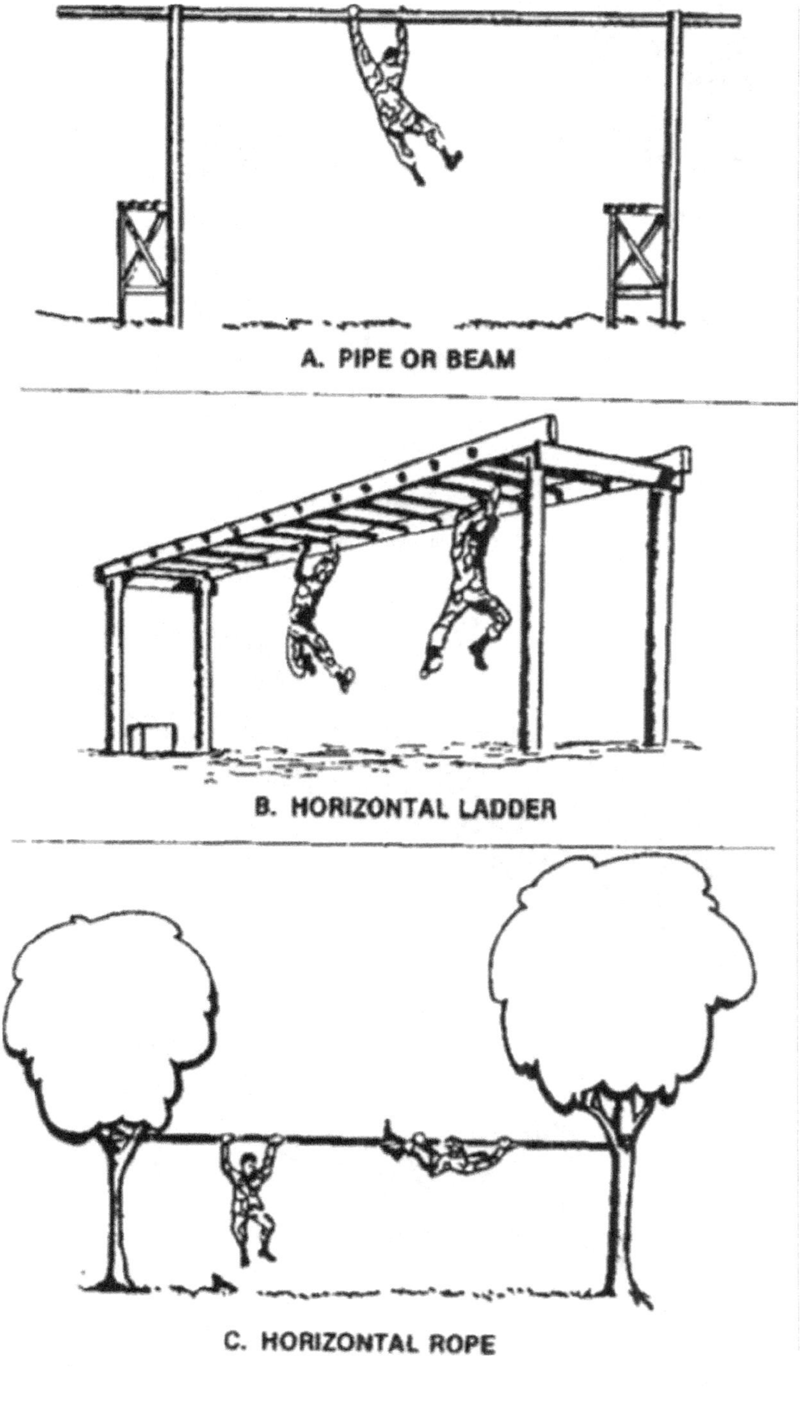

Figure 3-32.Horizontal Type Obstacles.

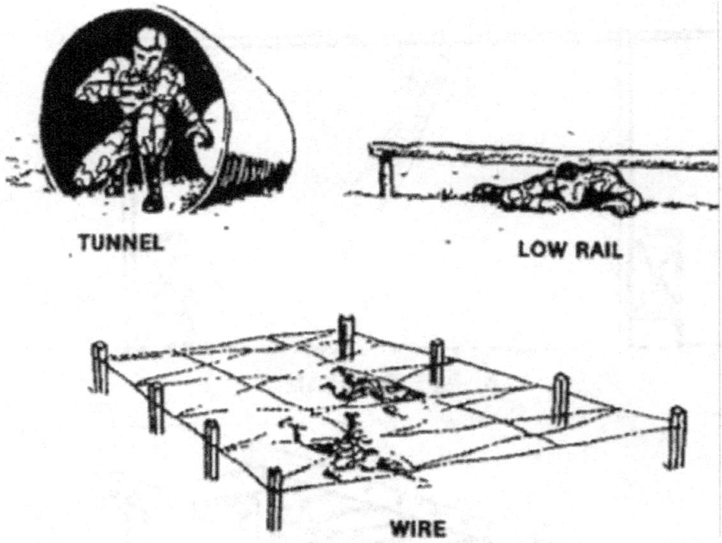

Figure 3-33.Crawling-Type Obstacles.

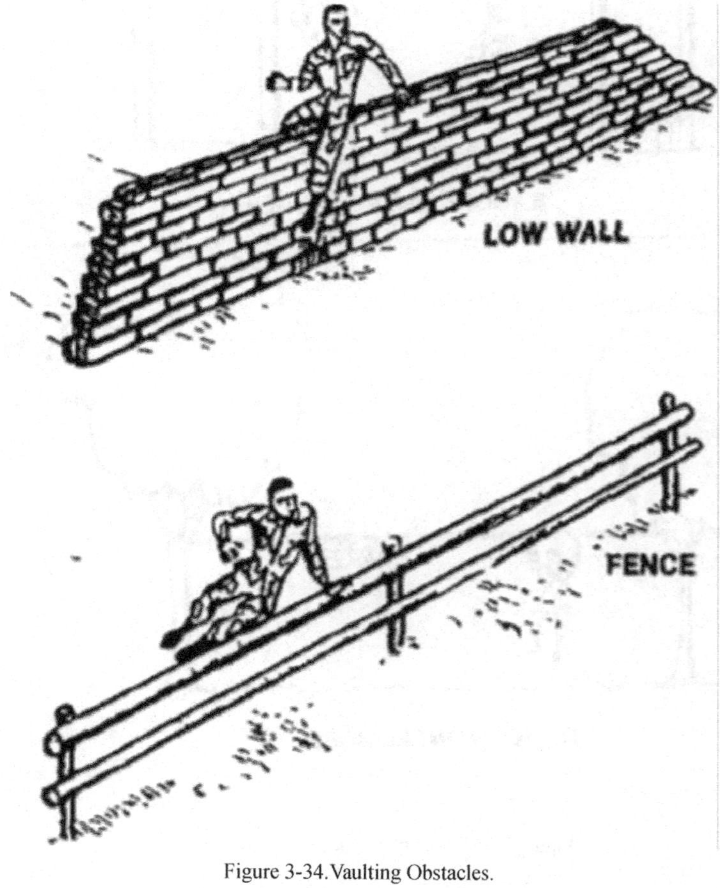

Figure 3-34.Vaulting Obstacles.

Figure 3-35.Balance-Type Obstacles.

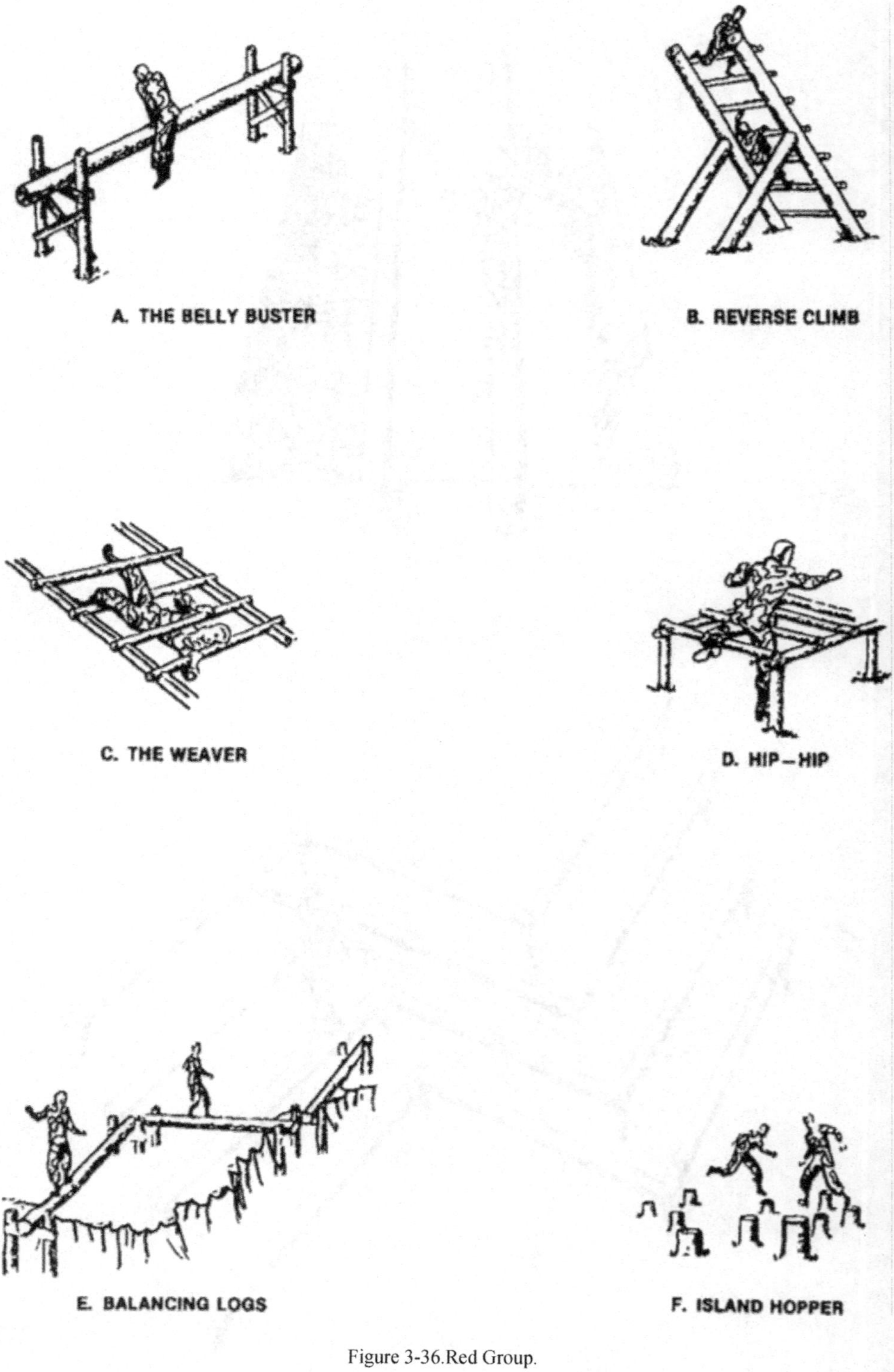

Figure 3-36.Red Group.

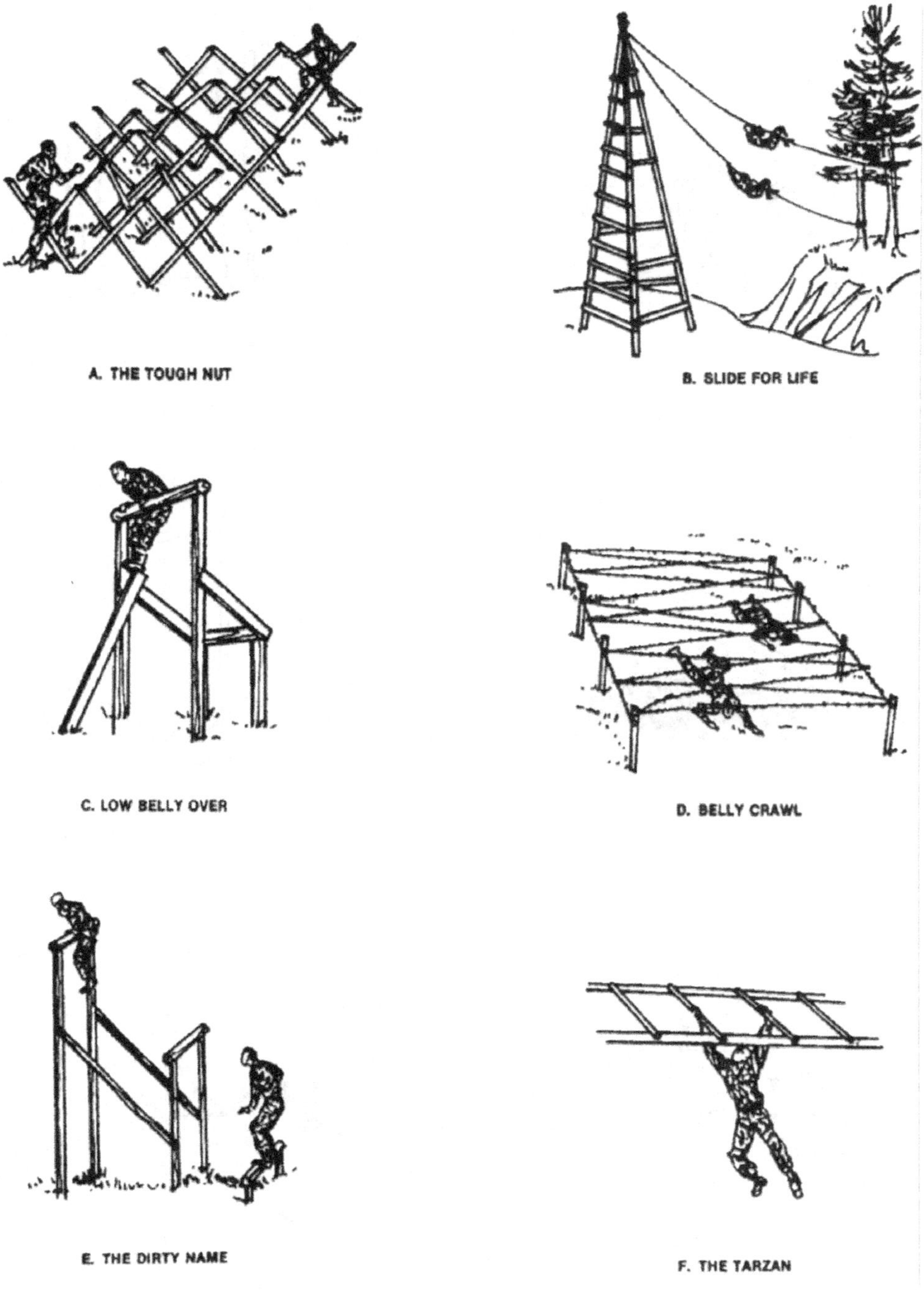

Figure 3-37.White Group.

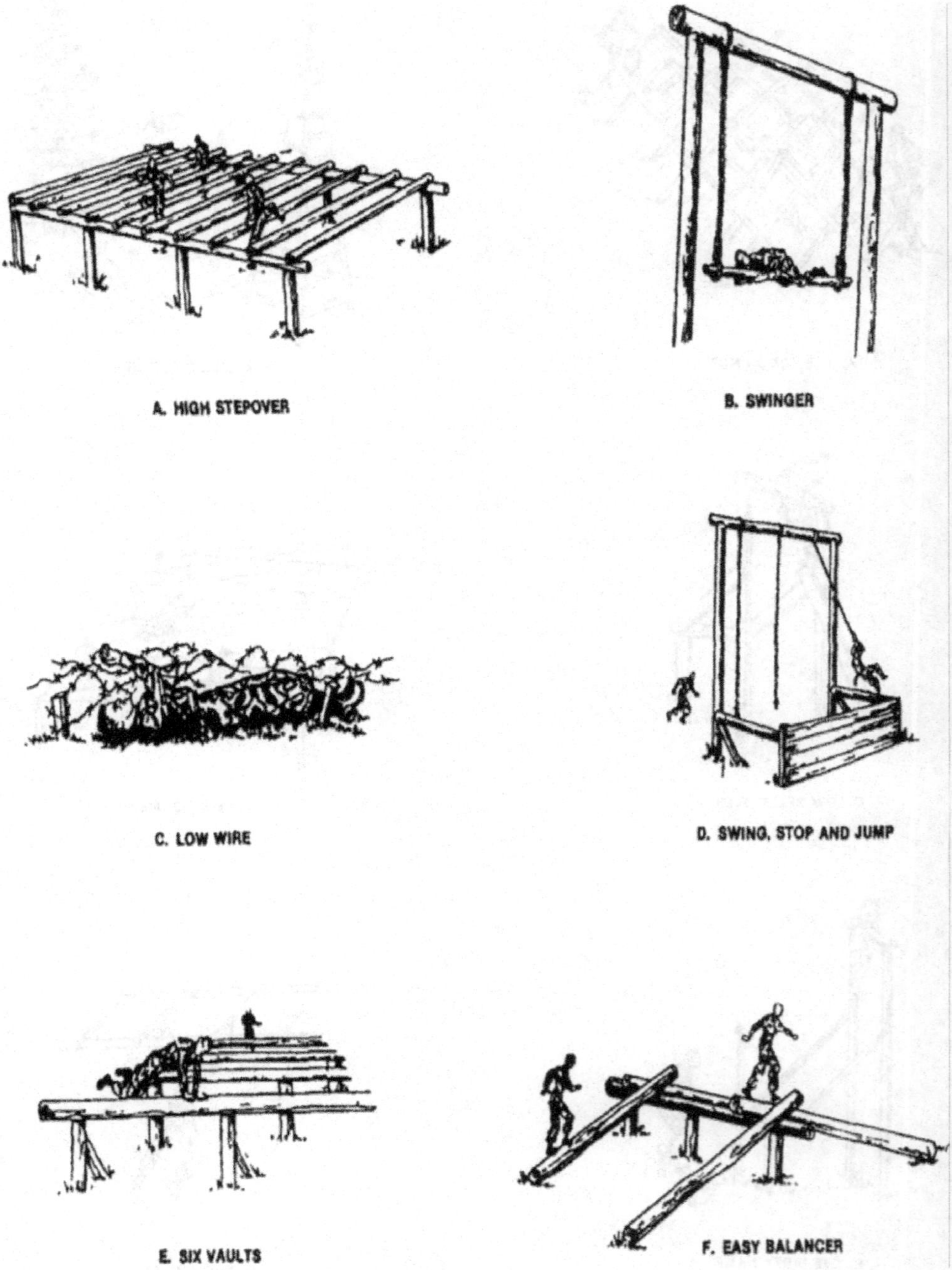

Figure 3-38.Blue Group.

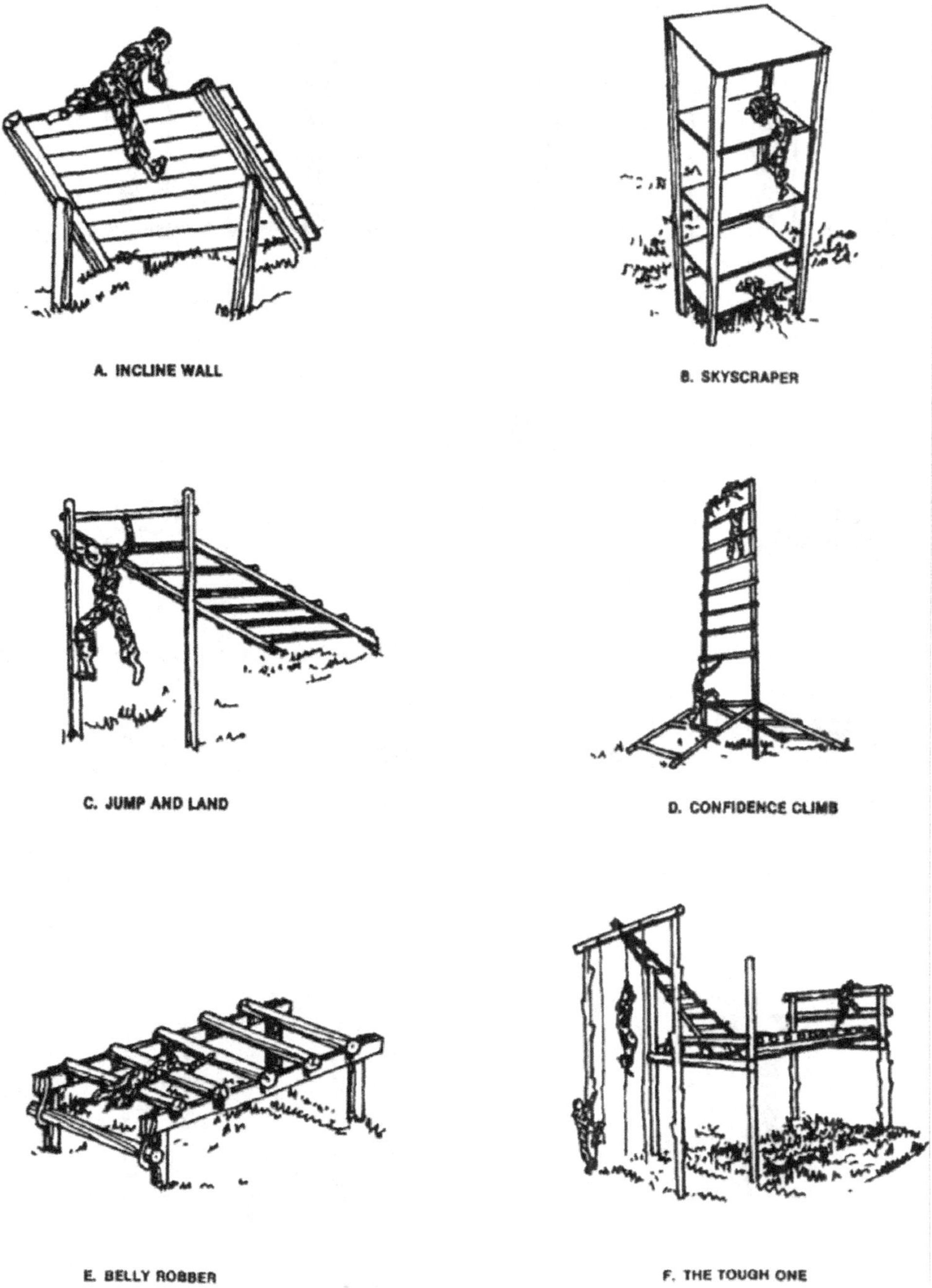

Figure 3-39.Black Group.

Figure 3-40. Bench Conditioning, Table I.

Figure 3-41. Bench Conditioning, Table II.

PROGRESSION GUIDE

AGE GROUP	LEVEL	EXERCISES 1	2	3	4	5	6
17	A	15	18	14	15	15	250
to	B	13	16	13	13	13	235
29	C	11	14	12	11	11	[illegible]
30	A	13	14	12	13	13	200
to	B	11	13	11	11	11	185
39	C	9	12	10	9	9	165
40	A	11	11	10	11	11	150
to	B	9	10	9	9	9	135
44	C	7	9	8	7	7	120
45	A	9	8	8	9	9	100
to	B	7	7	7	7	7	90
49	C	5	6	6	5	5	80
50	A	7	6	6	7	7	75
to	B	5	5	5	5	5	70
59	C	3	4	4	3	3	60
60	A	4	5	4	4	4	50
and	B	3	4	3	3	3	40
over	C	2	3	2	2	2	30
Minutes for each exercise		2	1	1	1	2	5

Table I. Progression Guide.

Table I. Exercises.

PROGRESSION GUIDE

AGE GROUP	LEVEL	EXERCISES 1	2	3	4	5	6
17	A	17	17	17	9	19	300
to	B	15	15	15	8	17	270
29	C	13	1[illegible]	13	7	15	245
30	A	15	15	15	8	17	235
to	B	13	13	13	7	15	210
39	C	11	11	11	6	13	190
40	A	13	13	13	7	15	175
to	B	11	11	11	6	13	155
44	C	9	10	9	5	11	135
45	A	11	11	11	6	13	125
to	B	9	9	9	5	11	110
49	C	7	7	7	4	9	100
50	A	9	9	9	5	11	95
to	B	7	7	7	4	9	85
59	C	5	5	5	3	7	75
60	A	6	7	7	4	9	70
and	B	5	5	5	3	7	60
over	C	4	4	4	2	5	50
Minutes for each exercise		1	1	1	1 1/2	1 1/2	6

Table II. Progression Guide.

Table II. Exercises.

PROGRESSION GUIDE

AGE GROUP	LEVEL	EXERCISES 1	2	3	4	5	6
17	A	10	19	19	16	10	350
to	B	9	17	17	15	9	315
29	C	8	15	15	14	8	280
30	A	9	17	17	14	9	270
to	B	8	15	15	13	8	240
39	C	7	13	13	12	7	210
40	A	8	15	15	12	8	200
to	B	7	13	13	11	7	180
44	C	6	11	11	10	6	160
45	A	7	13	13	10	7	150
to	B	6	11	11	9	6	135
49	C	5	9	9	8	5	120
50	A	6	11	11	8	6	115
to	B	5	9	9	7	5	105
59	C	4	7	7	6	4	95
60	A	5	9	9	7	5	90
and	B	4	7	7	6	4	80
over	C	3	5	5	4	3	70
Minutes for each exercise		1 1/2	1	1	1 1/2	1	6

Table III. Progression Guide.

Table III. Exercises.

PROGRESSION GUIDE

AGE GROUP	LEVEL	EXERCISES 1	2	3	4	5	6
17	A	12	9	12	24	25	400
to	B	11	8	11	22	23	380
29	C	10	7	10	21	21	360
30	A	11	8	11	23	23	305
to	B	10	7	10	21	21	290
39	C	9	6	9	20	20	275
40	A	10	7	10	20	21	225
to	B	9	6	9	18	18	215
44	C	8	5	8	16	14	205
45	A	8	6	8	16	16	175
to	B	7	5	7	14	14	165
49	C	6	4	6	12	12	155
50	A	6	5	6	13	13	135
to	B	5	4	5	11	11	130
59	C	4	3	4	10	10	120
60	A	5	4	5	10	10	100
and	B	4	3	4	9	9	95
over	C	3	2	3	8	8	90
Minutes for each exercise		1	2	1	1	1	6

Table IV. Progression Guide.

Table IV. Exercises.

PROGRESSION GUIDE

AGE GROUP	LEVEL	EXERCISES 1	2	3	4	5	6
17	A	14	13	28	14	30	450
to	B	13	12	27	13	28	430
29	C	12	11	26	12	26	410
30	A	12	12	25	12	26	350
to	B	11	11	24	11	24	330
39	C	10	10	23	10	22	310
40	A	11	11	23	11	23	250
to	B	10	10	21	10	21	240
44	C	9	9	19	9	19	230
45	A	9	9	20	9	20	200
to	B	8	8	18	8	18	190
49	C	7	7	16	7	16	180
50	A	7	7	16	7	16	170
to	B	6	6	14	6	14	155
59	C	5	5	12	5	12	140
60	A	6	6	12	6	12	115
and	B	5	5	11	5	10	110
over	C	4	4	9	4	9	105
Minutes for each exercise		2	1	1	2	1	5

Table V. Progression Guide.

EXERCISE 1

EXERCISE 2

EXERCISE 3

EXERCISE 4

EXERCISE 5

EXERCISE 6

Table V. Exercises.

PROGRESSION GUIDE

AGE GROUP	LEVEL	EXERCISES 1	2	3	4	5	6
17	A	17	15	32	32	35	500
to	B	16	14	30	30	33	480
29	C	15	13	28	28	31	460
30	A	15	13	30	30	31	400
to	B	14	12	28	28	29	380
39	C	13	11	26	26	27	360
40	A	13	10	27	27	27	310
to	B	12	9	25	25	25	285
44	C	11	8	23	23	23	265
45	A	11	9	23	23	23	250
to	B	10	8	21	21	21	230
49	C	9	7	19	19	19	210
50	A	9	8	19	19	19	200
to	B	8	7	17	17	17	190
59	C	7	6	15	15	15	175
60	A	8	7	15	15	17	140
and	B	7	6	13	13	15	130
over	C	5	5	10	10	12	120
Minutes for each exercise		2	1	1	1	1	6

Table VI. Progression Guide.

Table VI. Exercises.

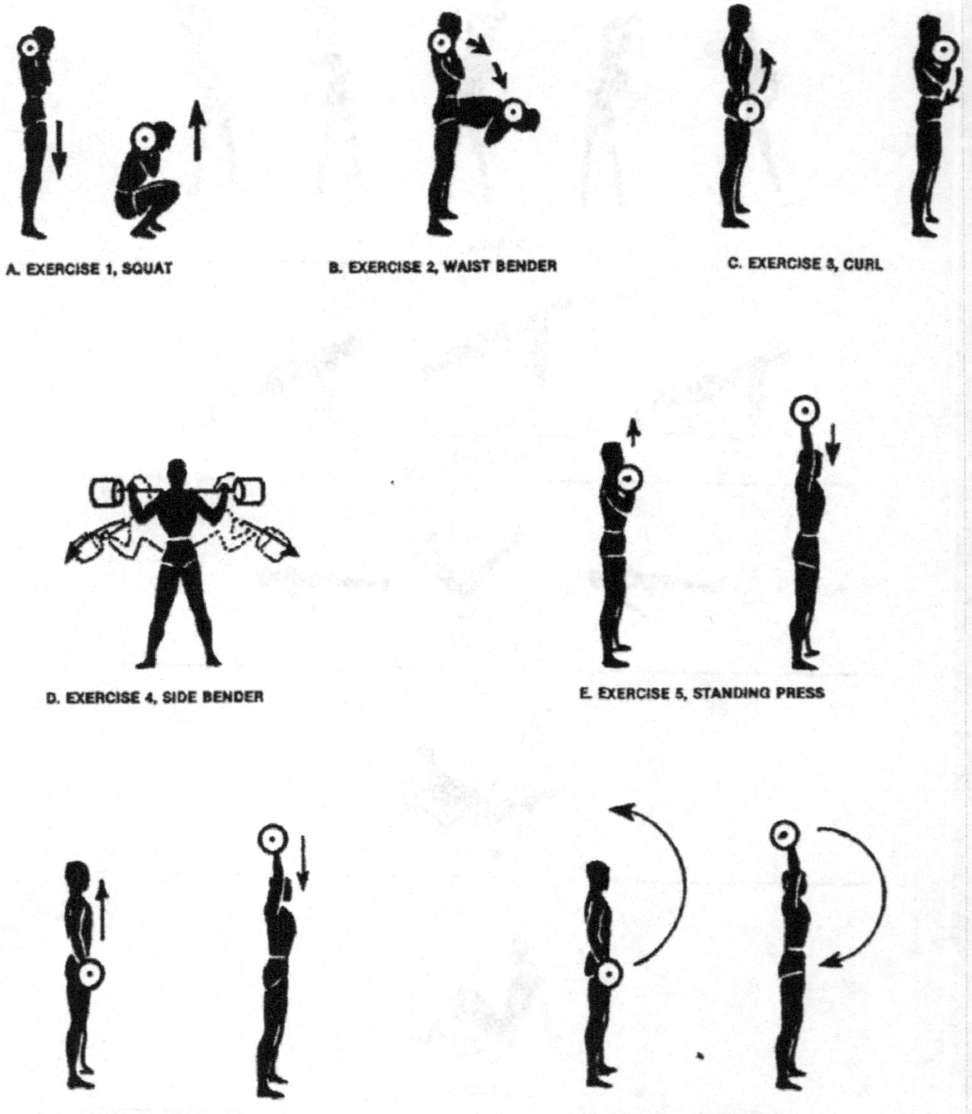

Figure 3-42. Weight Training.

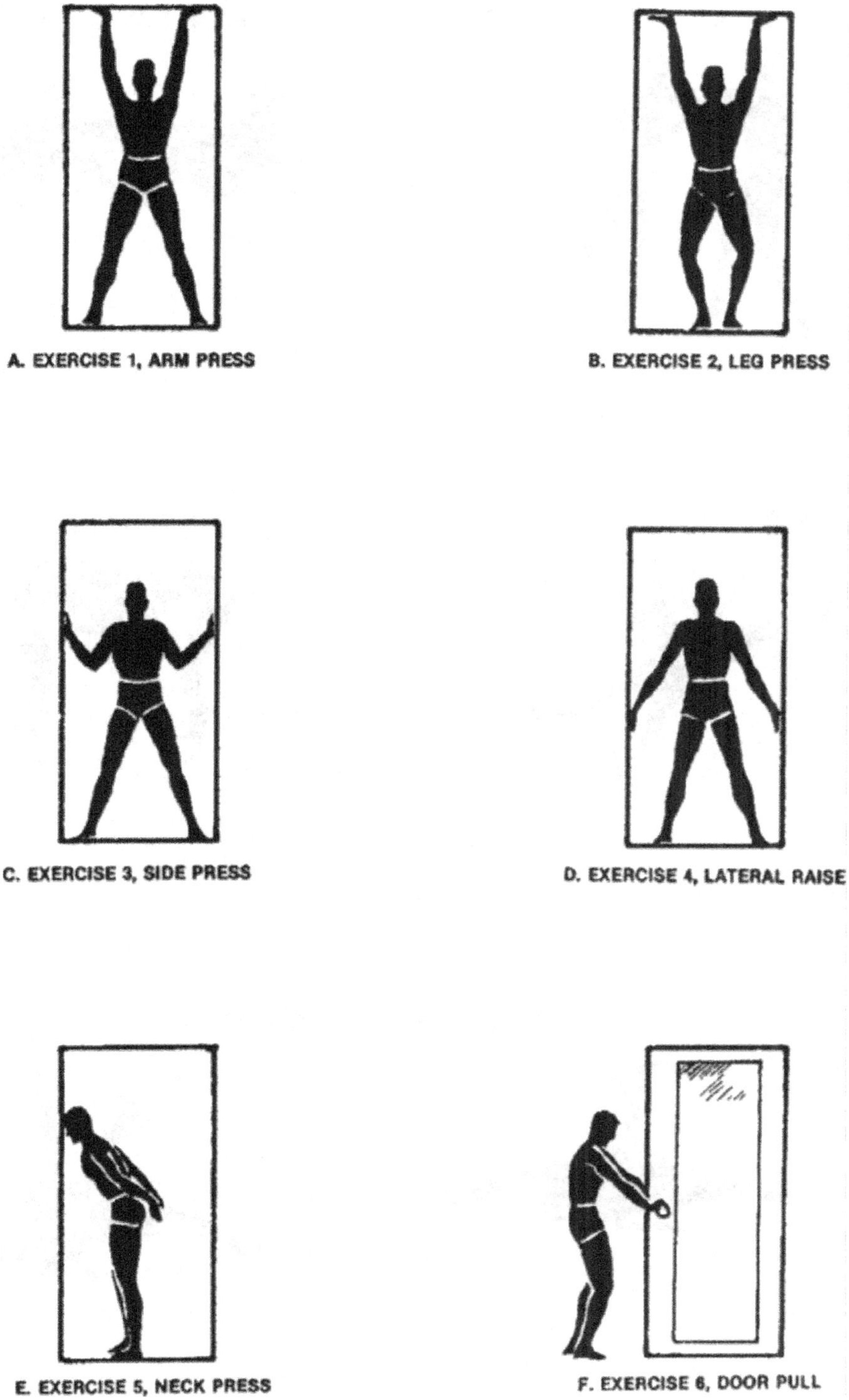

Figure 3-43. Isometric Exercises.

Figure 5-1.Combatives Table I.

Figure 5-2.Combatives Table II.

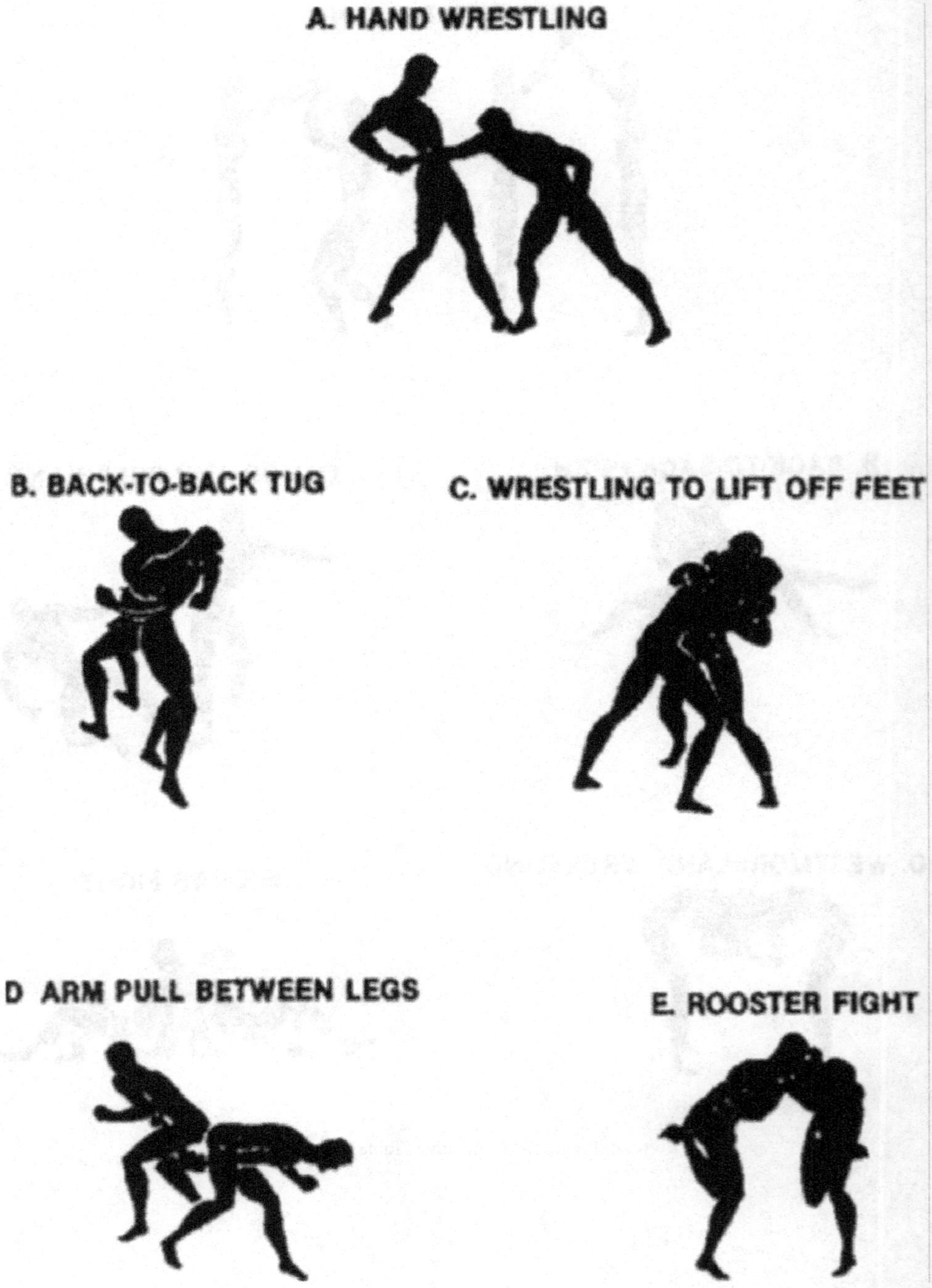

Figure. 5-3.Combatives Table III.

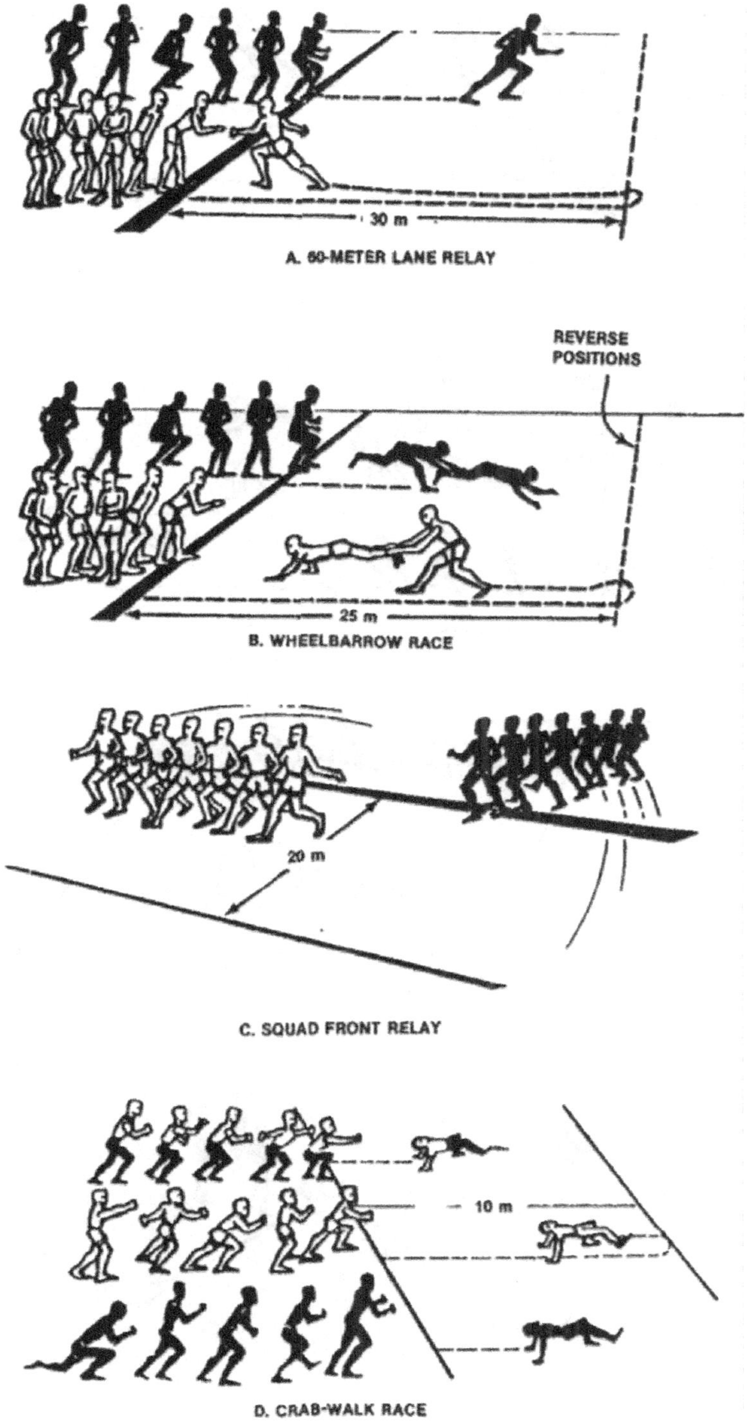

Figure 5-4.Relay Table I.

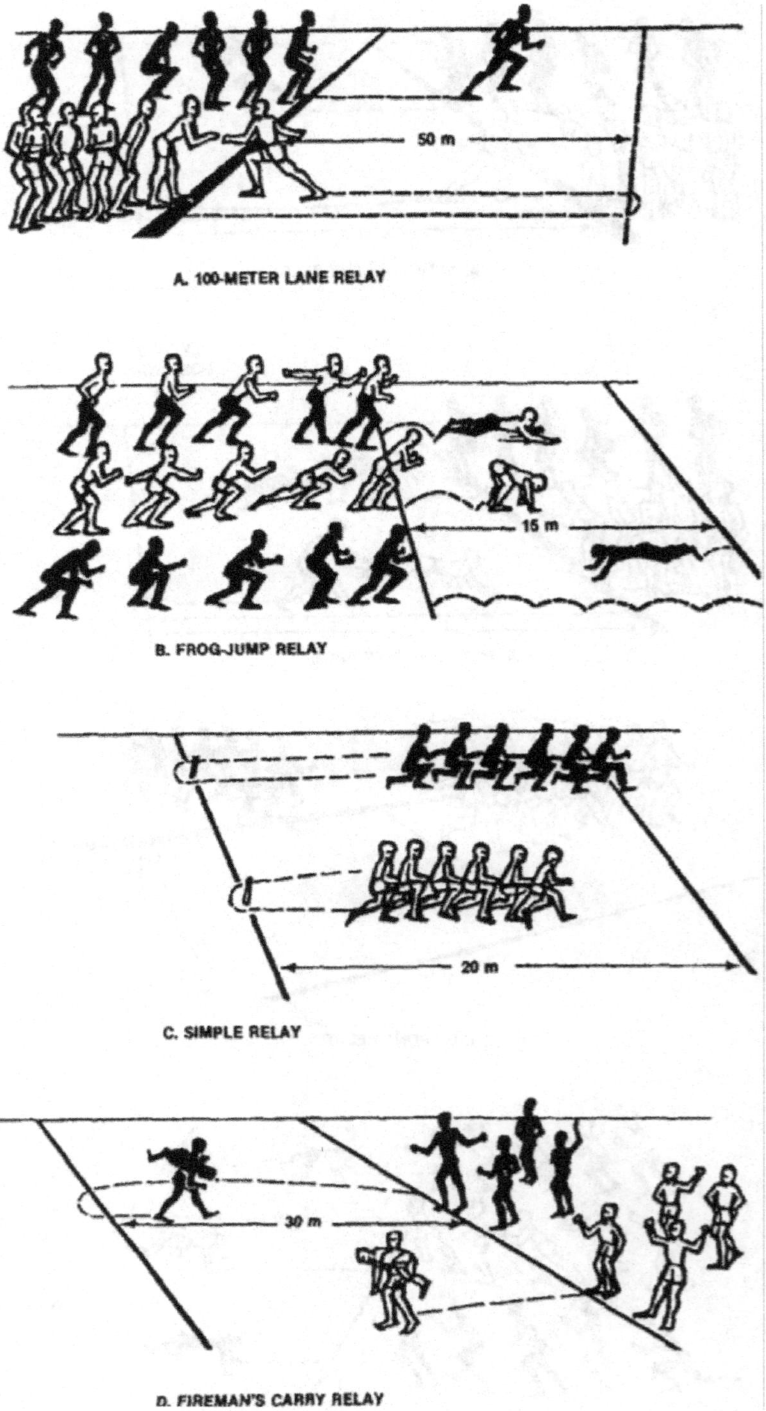

Figure 5-5.Relay Table II.

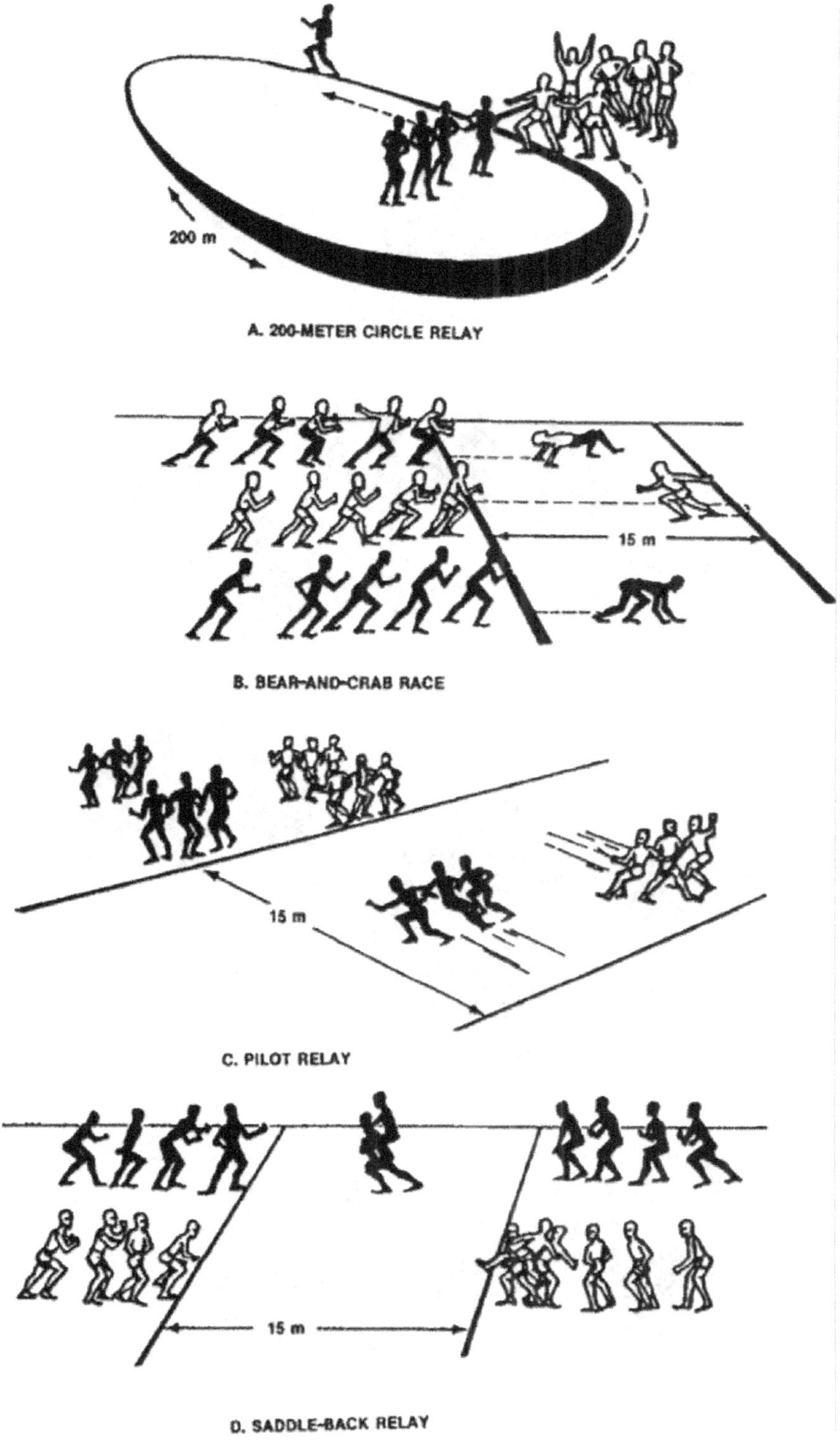

Figure 5-6.Relay Table III.

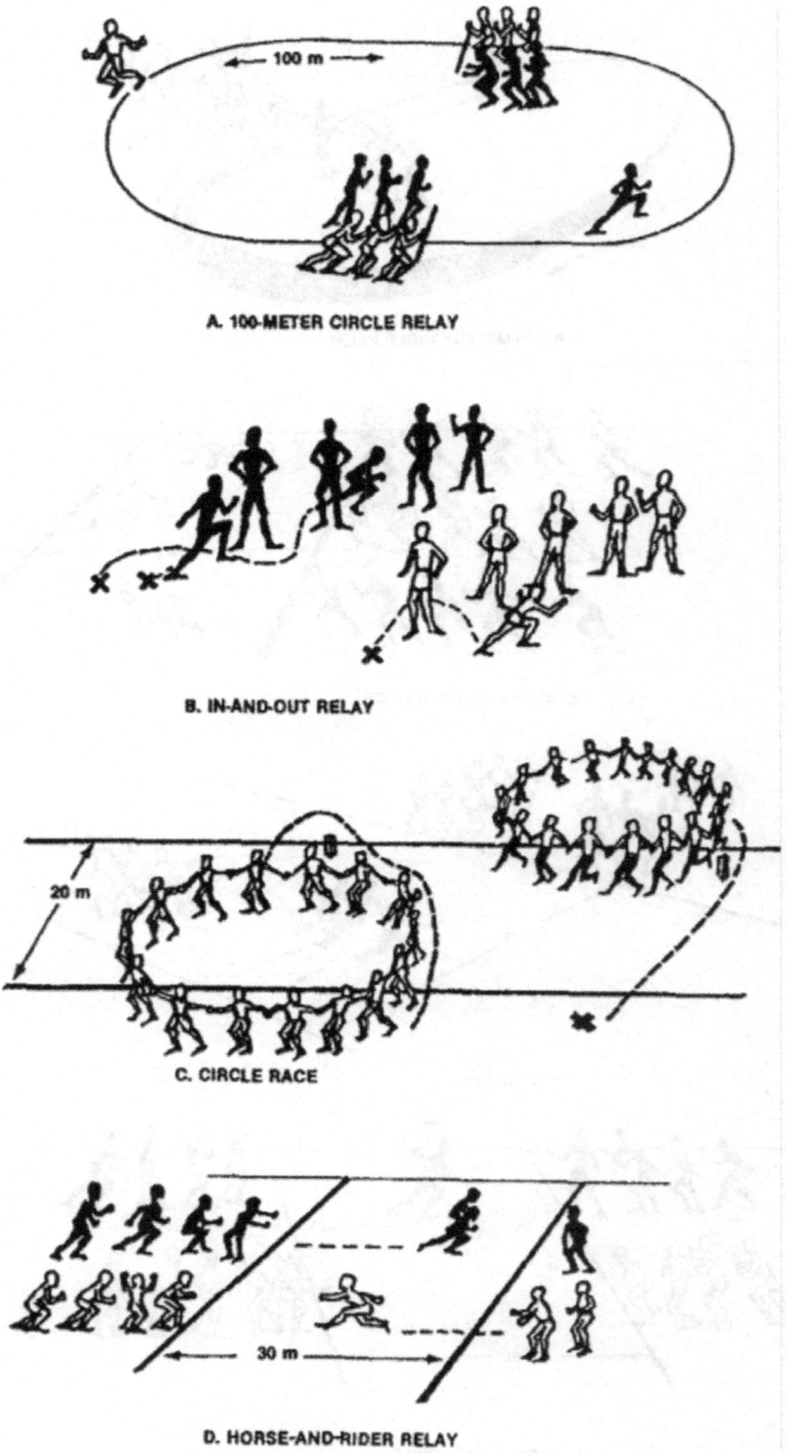

Figure 5-7.Relay Table IV.

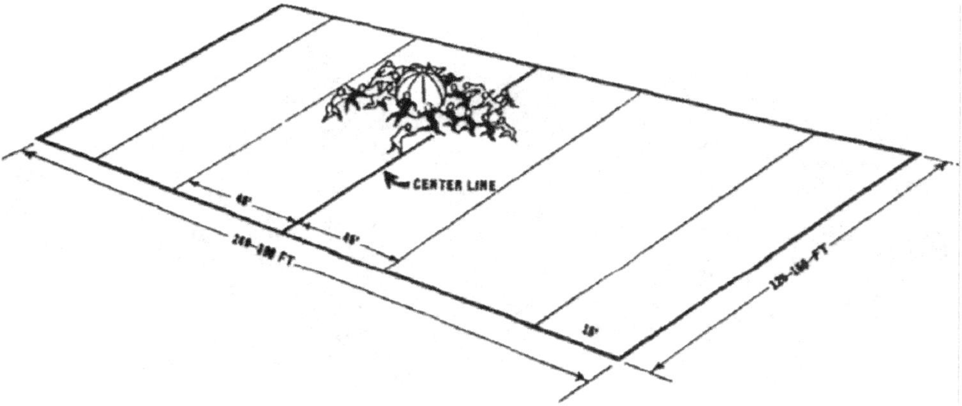

Figure 5-8.Pushball.

Figure 5-9.Line Rush.

Figure. 5-10. Human Tug-of-War.

Figure 5-11. Master of the Ring.

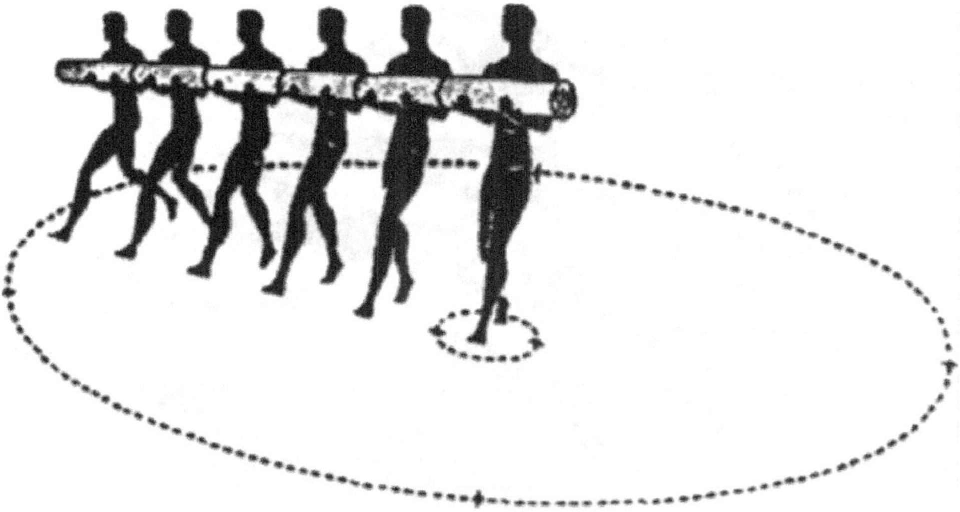

Figure 5-12.Log Pivot Circle.

Figure 5-13.Log-Rolling Race.

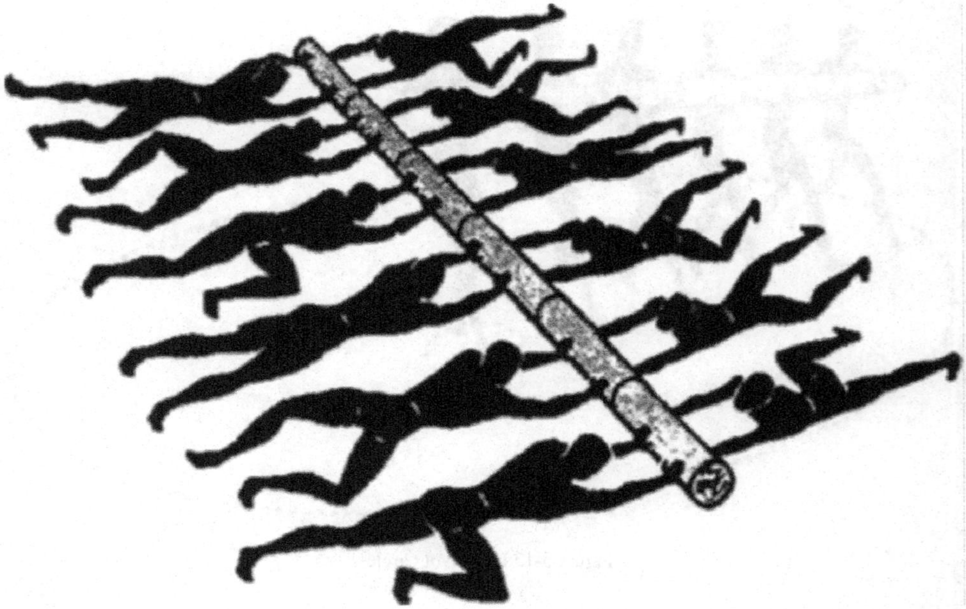

Figure 5-14. Prone Pushing Contest.

Figure 5-15. Shuttle Relay Race.

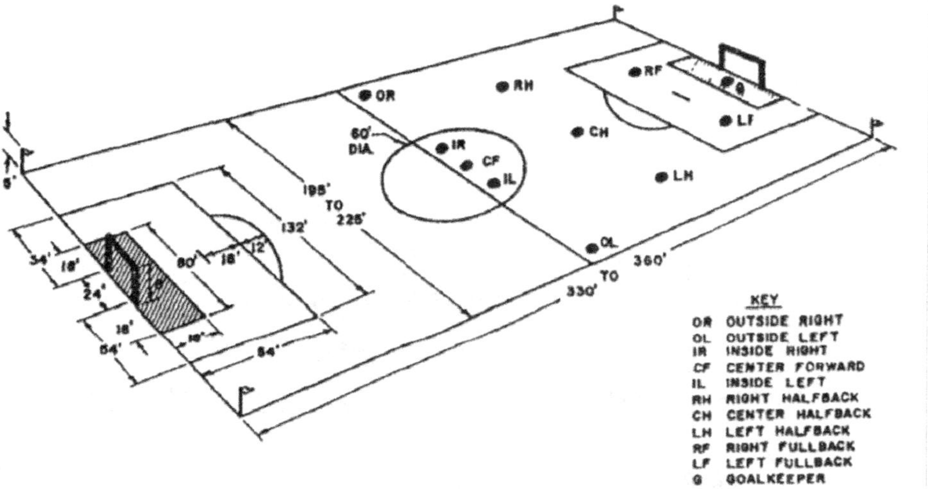

Figure 5-16. Soccer.

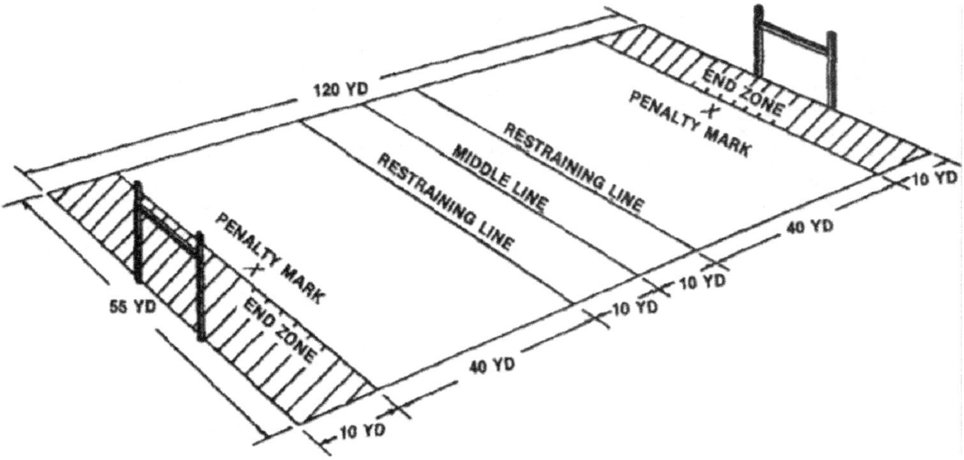

Figure 5-17.Speedball Field.

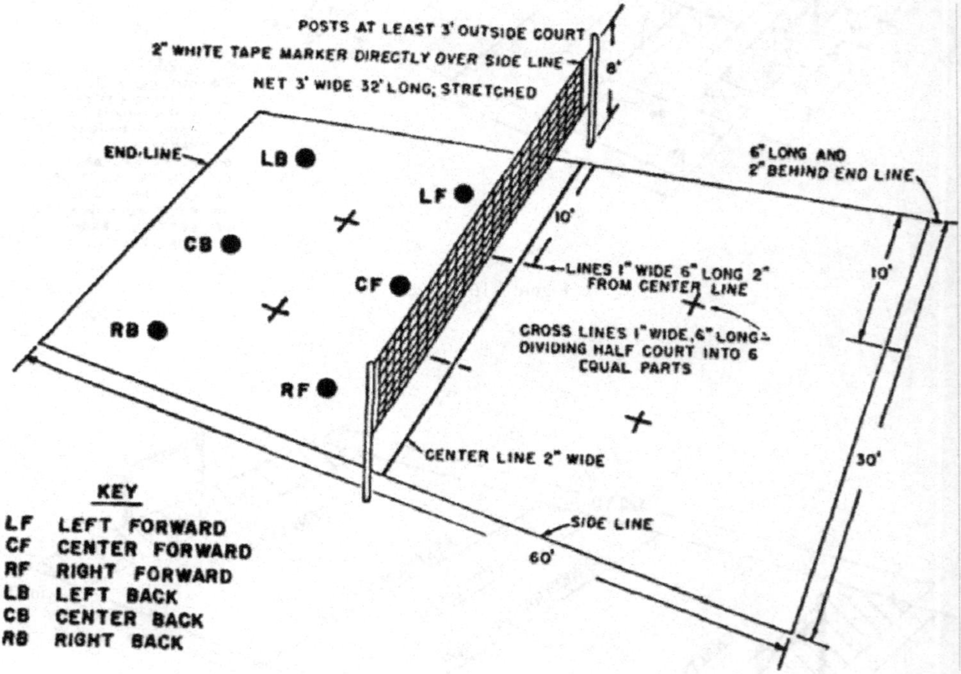

Figure 5-18. Volleyball.

PTS	UPHILL CLIMB	ROPE CLIMB	EVACUATION	FIRE AND MANEUVER	FORCED MARCH
100	1:20-BELOW	0:10-BELOW	0:25-BELOW	1:20-BELOW	26:00-BELOW
99	1:21-1:22	0:11		1 21-1:22	26:01-26:12
98	1:23-1:24	0:12	0:26	1:23-1:24	26:13-26:24
97	1:25-1:26	0:13		1:25-1:26	26:25-26:36
96	1:27-1:28	0:14	0:27	1:27-1:28	26:37-26:48
95	1:29-1:30	0:15		1:29-1:30	26:49-27:00
94	1:31-1:32	0:16	0:28	1:31-1:32	27:01-27:12
93	1:33-1:34	0:17		1 33-1.34	27:13-27:24
92	1:35-1:36	0:18	0:29	1:35-1.36	27:25-27:36
91	1:37-1:38	0:19		1:37-1:38	27:37-27:48
90	1:39-1:40	0:20	0:30	1:39-1:40	27:49-28:00
89	1:41-1:42	0:21	0:31	1:41-1:42	28:01-28:12
83	1:43-1:44	0:22	0:32	1:43-1:44	28:13-28:24
87	1:45-1:46	0:23	0:33	1:45-1:46	28:25-28:36
86	1:47-1:48	0:24	0:34	1.47-1:48	28:37-28:48
85	1:49-1:50	0:25	0:35	1.49-1:50	28:49-29:00
84	1:51-1:52	0:26	0:36	1:51-1:52	29:01-29:12
83	1:53-1:54	0:27	0.37	1:53-1:54	29:13-29:24
82	1:55-1:56	0:28	0:38	1:55-1:56	29:25-29:36
81	1:57-1:58	0:29	0:39	1:58-1:58	29:37-29:48
80	1:59-2.00	0:30	0:40	1:59-2:00	29:49-30:00
79	2.01-2:02	0:31	0:41	2:01-2:02	30:01-30:12
78	2:03-2:04	0:32	0:42	2:03-2.04	30:13-30:24
77	2:05-2:06	0.33	0:43	2:05-2:06	30:25-30:36
76	2:07-2:08	0.34	0.44	2:07-2:08	30:37-30:48
75	2:09-2:10	0:35	0:45	2:09-2:10	30:49-31:00
74	2:11-2:12	0:36	0:46	2:11-2 12	31.01-31:12
73	2:13-2:14	0:37	0.47	2.13-2:14	31:13-31:24
72	2:15-2:16	0:38	0:48	2.15-2:16	31:25-31:36
71	2:17-2:18	0:39	0:49	2:17-2:18	31:37-31:48
70	2:19-2:20	0:40	0:50	2:19-2.20	31:49-32.00
69	2.21-2:22	0:41	0:51	2:21-2:22	32:01-32:12
68	2:23-2:24	0:42	0:52	2:23-2:24	32:13-32:24
67	2:25-2.26	0:43	0:53	2:25-2:26	32:25-32:36
66	2:27-2:28	0:44	0:54	2.27-2:28	32.37-32:48
65	2:29-2:30	0.45	0.55	2.20-2 30	32.49-33:00

Figure 6-1.Physical Readiness Test Scoring Table for Male.

	EXAMPLE 1	EXAMPLE 2	EXAMPLE 3
Event #1	100 pts	100 pts	99 pts
Event #2	100 pts	99 pts	86 pts
Event #3	85 pts	75 pts (F)	80 pts
Event #4	90 pts	79 pts (F)	100 pts
Event #5	79 pts	85 pts	90 pts
Pts	454 pts	438 pts	455 pts
Final Score	79.0% (F)	69.0% (F)	91.0%
(F) = Failure			

Figure 6-2.Scoring Examples.

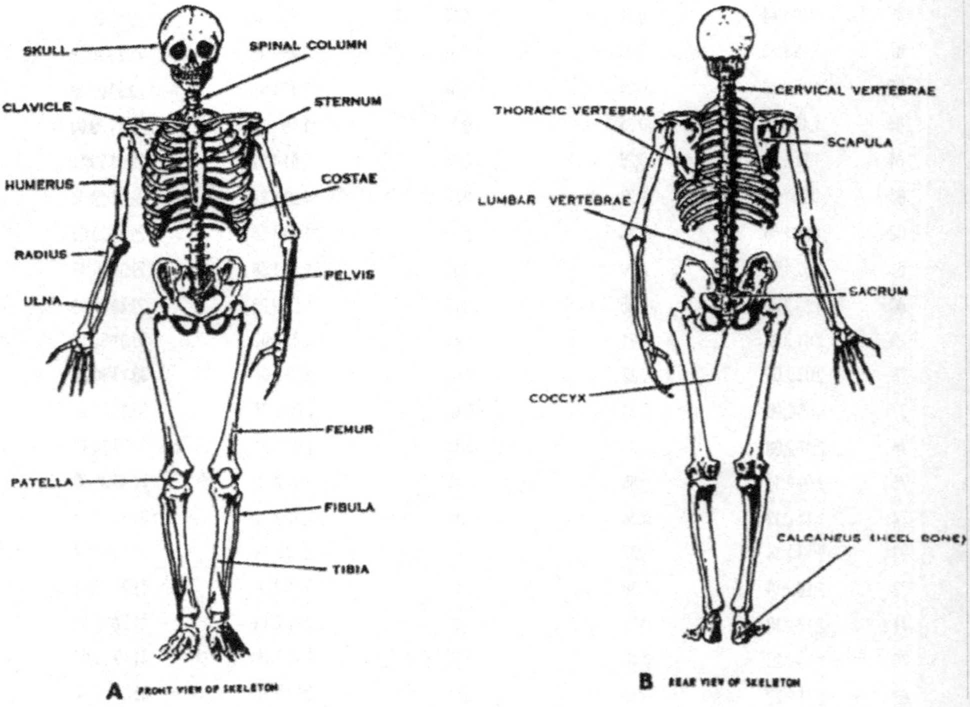

Figure 7-1.Front and Rear Views of Skeleton.

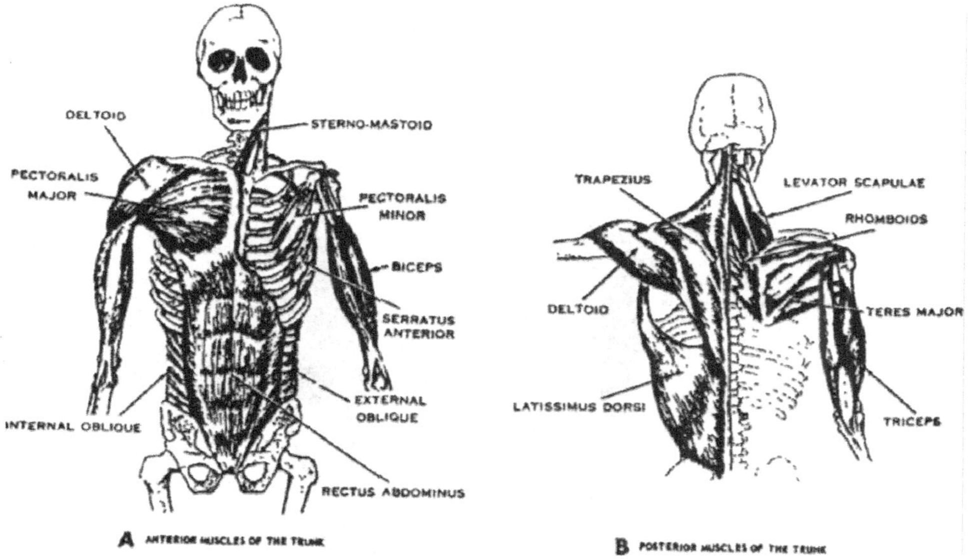

Figure 7-2.Muscles of the Trunk.

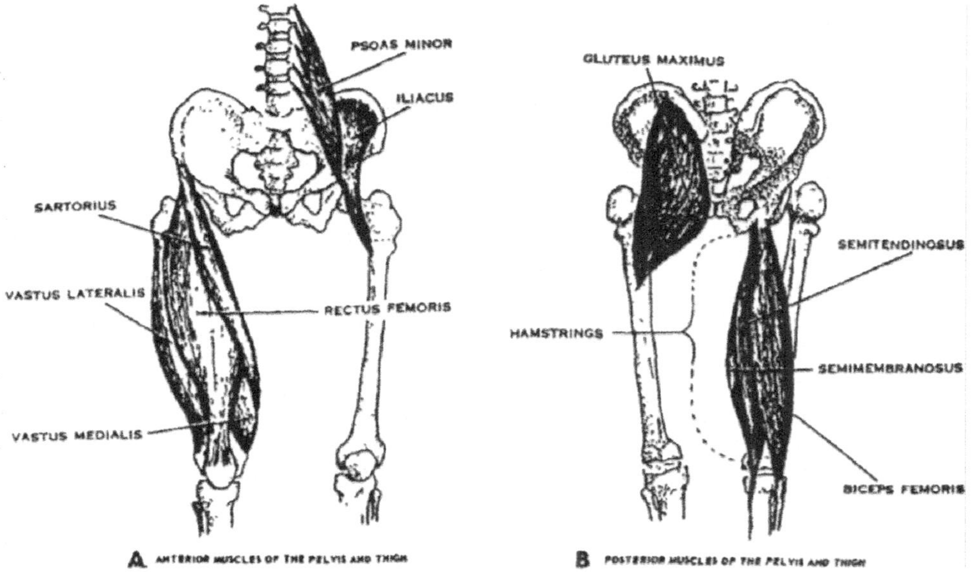

Figure 7-3.Anterior and Posterior Muscles of the Pelvisand Thigh.

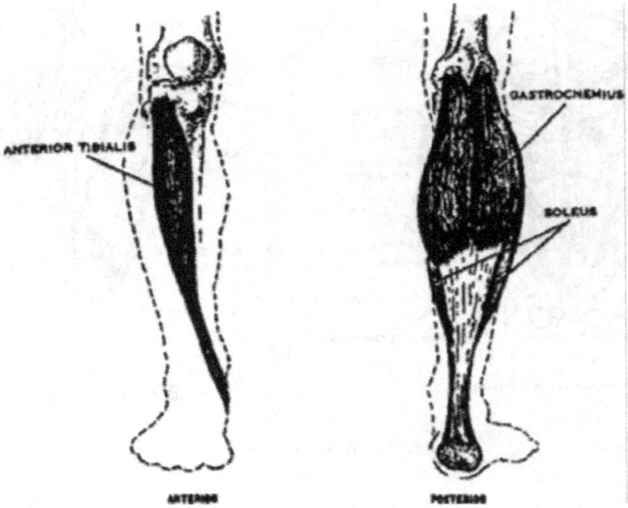

Figure 7-4. Muscles of Lower Leg.

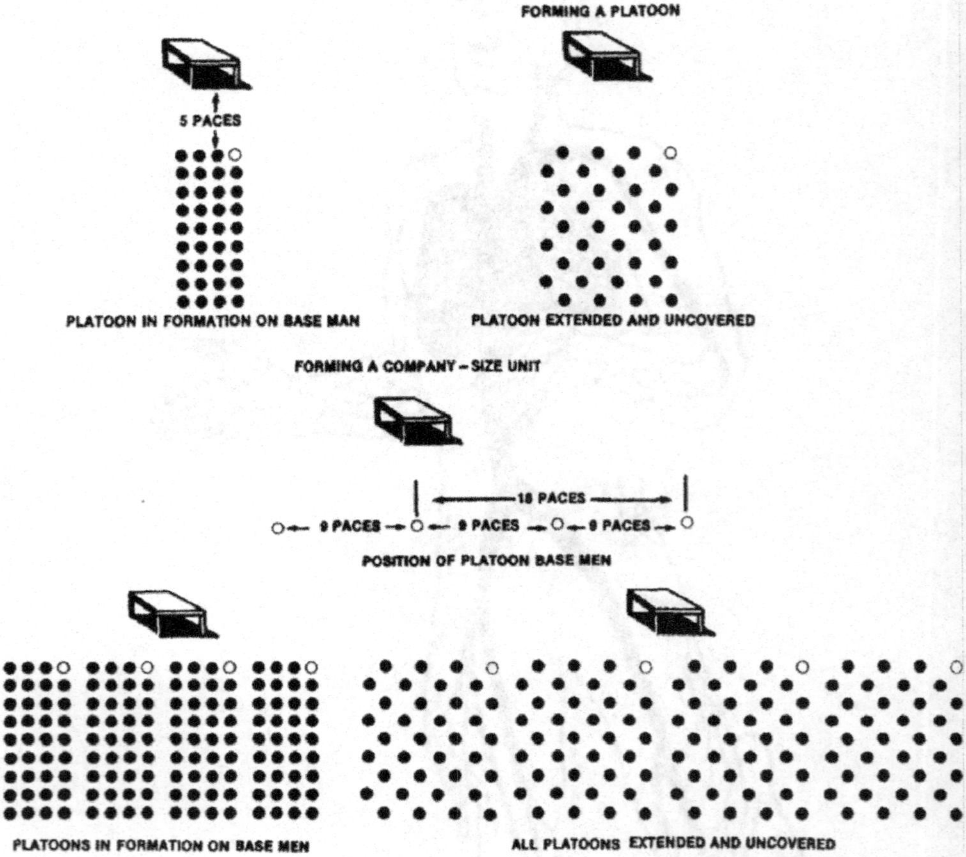

Figure A-l. Forming the Extended Rectangular Formation.

Figure A-2.Circle Formation.

Jungle Operations

Department of Defense

Pentagon
Publishing

FIELD MANUAL
NO. 90-5

* FM 90-5

HEADQUARTERS
DEPARTMENT OF THE ARMY
Washington, DC, 16 August 1982

JUNGLE OPERATIONS

CONTENTS

*This manual supersedes FM 31-35, 26 September 1969.

DISTRIBUTION RESTRICTION: Approved for public release; distribution is unlimited

The words "he," "him," "his," "man," and "men," when used in this publication, represent both the masculine and feminine genders unless otherwise specifically stated.

You may recommend changes to this manual to improve it. Key your comments to the page and line of text in which the changes are recommended. Give reasons for each comment to help understand and evaluate it. Send comments to: Commandant, United States Army Infantry School, ATTN: ATSH-B-ID, Fort Benning, Georgia 31905.

CHAPTER 1

The Jungle Environment

Section I. GENERAL

This chapter introduces jungle environments—where they are found and what they are like. Later chapters build on this information, providing guidance on fighting and living in the jungle.

Field Marshal Slim's words reflect the image of the jungle most armies carry into jungle warfare. At first, the jungle seems to be very hostile, but the hostility wanes as troops learn more about the jungle environment.

Jungles, in their various forms, are common in tropical areas of the world—mainly Southeast Asia, Africa, and Latin America.

"To our men. . . the jungle was a strange, fearsome place; moving and fighting in it were a nightmare. We were too ready to classify jungle as 'impenetratable' . . . To us it appeared only as an obstacle to movement; to the Japanese it was a welcome means of concealed maneuver and suprise . . . The Japanese reaped the deserved reward . . . we paid the penalty."

-Field Marshall Slim, Victor in Burma, World War II (Concerning the dark, early days of the Burma Campaign)

CONTENTS

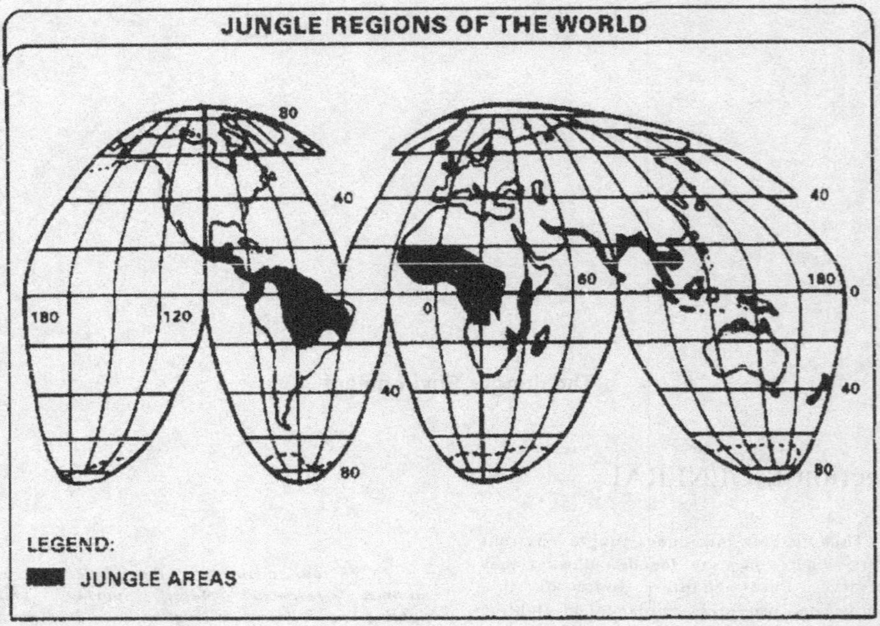

Section II. CLIMATE AND WEATHER.

The climate in jungles varies with location. Close to the equator, all seasons are nearly alike, with rains throughout the year; farther from the equator, especially in India and Southeast Asia, jungles have distinct wet (monsoon) and dry seasons. Both zones have high temperatures (averaging 78 to 95+ degrees Fahrenheit), heavy rainfall (as much as 1,000 centimeters [400+ inches] annually), and high humidity (90 percent) throughout the year.

Severe weather also has an impact on tactical operations in the jungle. The specific effects of weather on operations are discussed throughout this manual.

Jungle climates (high temperatures, high humidity, heavy rain) seriously affect:

- Men
- Clothing
- Equipment
- Weapons
- Vehicles
- Maintenance operations
- Training
- Tactics

Section III. TERRAIN AND VEGETATION

TYPES OF JUNGLES

The jungle environment includes densely forested areas, grasslands, cultivated areas, and swamps. Jungles are classified as primary or secondary jungles based on the terrain and vegetation.

PRIMARY JUNGLES

These are tropical forests. Depending on the type of trees growing in these forests, primary jungles are classified either as tropical rain forests or as deciduous forests.

Tropical Rain Forests. These consist mostly of large trees whose branches spread and lock together to form canopies. These canopies, which can exist at two or three different levels, may form as low as 10 meters from the ground. The canopies prevent sunlight from reaching the ground, causing a lack of undergrowth on the jungle floor. Extensive above-ground root systems and hanging vines are common. These conditions, combined with a wet and soggy surface, make vehicular traffic difficult. Foot movement is easier in tropical rain forests than in other types ofjungle. Except where felled trees or construction make a gap in the canopy of the rain forest, observation from the air is nearly impossible. Ground observation is generally limited to about 50 meters (55 yards).

Deciduous Forests. These are found in semitropical zones where there are both wet and dry seasons. In the wet season, trees are fully leaved; in the dry season, much of the foliage dies. Trees are generally less dense in deciduous forests than in rain forests. This allows more rain and sunlight to filter to the ground, producing thick undergrowth. In the wet season, with the trees in full leaf, observation both from the air and on the

ground is limited. Movement is more difficult than in the rain forest. In the dry season, however, both observation and trafficability improve.

SECONDARY JUNGLES

These are found at the edge of the rain forest and the deciduous forest, and in areas where jungles have been cleared and abandoned. Secondary jungles appear when the ground has been repeatedly exposed to sunlight. These areas are typically overgrown with weeds, grasses, thorns, ferns, canes, and shrubs. Foot movement is extremely slow and difficult. Vegetation may reach to a height of 2 meters. This will limit observation to the front to only a few meters.

COMMON JUNGLE FEATURES

SWAMPS

These are common to all low jungle areas where there is water and poor drainage. There are two basic types of swamps—mangrove and palm.

Mangrove Swamps. These are found in coastal areas wherever tides influence water flow. The mangrove is a shrub-like tree which grows 1 to 5 meters high. These trees have tangled root systems, both above and below the water level, which restrict movement to foot or small boats. Observation in mangrove swamps, both on the ground and from the air, is poor. Concealment is excellent.

Palm Swamps. These exist in both salt and fresh water areas. Like movement in the mangrove swamps, movement through palm swamps is mostly restricted to foot (sometimes small boats). Vehicular traffic is nearly impossible except after extensive road construction by engineers. Observation and fields-of-fire are very limited. Concealment from both air and ground observation is excellent.

SAVANNA

This is a broad, open jungle grassland in which trees are scarce. The thick grass is broad-bladed and grows 1 to 5 meters high. Movement in the savanna is generally easier than in other types of jungle areas, especially for vehicles. The sharp-edged, dense grass and extreme heat make foot movement a slow and tiring process. Depending on the height of the grass, ground observation may vary from poor to good. Concealment from air observation is poor for both troops and vehicles.

BAMBOO

This grows in clumps of varying size in jungles throughout the tropics. Large stands of bamboo are excellent obstacles for wheeled or tracked vehicles. Troop movement through bamboo is slow, exhausting, and noisy. Troops should bypass bamboo stands if possible.

CULTIVATED AREAS

These exist in jungles throughout the tropics and range from large, well-planned and well-managed farms and plantations to small tracts cultivated by individual farmers. There are three general types of cultivated areas—rice paddies, plantations, and small farms.

Rice Paddies. These are flat, flooded fields in which rice is grown. Flooding of the fields is controlled by a network of dikes and irrigation ditches which make movement by vehicles difficult even when the fields are dry. Concealment is poor in rice paddies. Cover is limited to the dikes, and then only from ground fire. Observation and fields of fire are excellent. Foot movement is poor when the fields are wet because soldiers must

wade through water about 1/2 meter (2 feet) deep and soft mud. When the fields are dry, foot movement becomes easier. The dikes, about 2 to 3 meters tall, are the only obstacles.

Plantations. These are large farms or estates where tree crops, such as rubber and coconut, are grown. They are usually carefully planned and free of undergrowth (like a well-tended park). Movement through plantations is generally easy. Observation along the rows of trees is generally good. Concealment and cover can be found behind the trees, but soldiers moving down the cultivated rows are exposed.

Small Farms. These exist throughout the tropics. These small cultivated areas are usually hastily planned. After 1 or 2 years' use, they usually are abandoned, leaving behind a small open area which turns into secondary jungle. Movement through these areas may be difficult due to fallen trees and scrub brush.

Generally, observation and fields-of-fire are less restricted in cultivated areas than in uncultivated jungles. However, much of the natural cover and concealment are removed by cultivation, and troops will be more exposed in these areas.

CHAPTER 2

Life in the Jungle

Section I. GENERAL

Soldiers must understand that the environment affects everyone. The degree to which soldiers are trained to live and fight in harsh environments will determine their unit's success or failure.

"Jungle fighting is not new to US soldiers, nor does the enemy have a monopoly on jungle know-how. US units adapted well to jungle fighting, and when we operated against the North Vietnamese Army along the Cambodian border we found that they had as much difficulty operating in the area as we did. The prisoners we captured were, as a rule, undernourished, emaciated, and sick with malaria. They stated that almost everyone in their unit had malaria, and many had died from it."

Report, 25th Infantry Division, Republic of Vietnam

There is very little to fear from the jungle environment. Fear itself can be an enemy. Soldiers must be taught to control their fear of the jungle. A man overcome with fear is of little value in any situation. Soldiers in a jungle must learn that the most important thing is to keep their heads and calmly think out any situation.

Many of the stories written about out-of-the-way jungle places were written by writers who went there in search of adventure rather than facts. Practically without exception, these authors exaggerated or invented many of the thrilling experiences they relate. These thrillers are often a product of the author's imagination and are not facts.

Most Americans, especially those raised in cities, are far removed from their pioneer ancestors, and have lost the knack of taking care of themselves under all conditions. It would be foolish to say that, without proper training, they would be in no danger if lost in the jungles of Southeast Asia, South America, or some Pacific island. On the other hand, they would be in just as much danger if lost in the mountains of western Pennsylvania or in other undeveloped regions of our own country. The only difference would be that a man is less likely to panic when he is lost in his homeland than when he is lost abroad.

CONTENTS

Section II. JUNGLE HAZARDS

EFFECT OF CLIMATE

The discomforts of tropical climates are often exaggerated, but it is true that the heat is more persistent. In regions where the air contains a lot of moisture, the effect of the heat may seem worse than the same temperature in a dry climate. Many people experienced in jungle operations feel that the heat and discomfort in some US cities in the summertime are worse than the climate in the jungle.

Strange as it may seem, there may be more suffering from cold in the tropics than from the heat. Of course, very low temperatures do not occur, but chilly days and nights are common. In some jungles, in winter months, the nights are cold enough to require a wool blanket or poncho liner for sleeping.

Rainfall in many parts of the tropics is much greater than that in most areas of the temperate zones. Tropical downpours usually are followed by clear skies, and in most places the rains are predictable at certain times of the day. Except in those areas where rainfall may be continuous during the rainy season, there are not many days when the sun does not shine part of the time.

People who live in the tropics usually plan their activities so that they are able to stay under shelter during the rainy and hotter portions of the day. After becoming used to it, most tropical dwellers prefer the constant climate of the torrid zones to the frequent weather changes in colder climates.

INSECTS

Malaria-carrying mosquitoes are probably the most harmful of the tropical insects. Soldiers can contract malaria if proper precautions are not taken.

Precautions against malaria include:

- Taking Dapsone and chloroquine-primaquine
- Using insect repellent
- Wearing clothing that covers as much of the body as possible
- Using nets or screens at every opportunity
- Avoiding the worst-infested areas when possible

PROTECTION AGAINST MALARIA

Mosquitoes are most prevalent early at night and just before dawn. Soldiers must be especially cautious at these times. Malaria is more common in populated areas than in uninhabited jungle, so soldiers must also be especially cautious when operating around villages. Mud packs applied to mosquito bites offer some relief from itching.

Wasps and bees may be common in some places, but they will rarely attack unless their nests are disturbed. When a nest is disturbed, the troops must leave the area and reassemble at the last rally point. In case of stings, mud packs are helpful. In some areas, there are tiny bees, called sweatbees, which may collect on exposed parts ofthe body during dry weather, especially if the body is sweating freely. They are annoying but stingless and will leave when sweating has completely stopped, or they may be scraped off with the hand.

The larger centipedes and scorpions can inflict stings which are painful but not fatal. They like dark places, so it is always advisable to shake out blankets before sleeping at night, and to make sure before dressing that they are not hidden in clothing or shoes. Spiders are commonly found in the jungle. Their bites may be painful, but are rarely serious. Ants can be dangerous to injured men lying on the ground and unable to move. Wounded soldiers should be placed in an area free of ants.

In Southeast Asian jungles, the rice-borer moth ofthe lowlands collects around lights in great numbers during certain seasons. It is a small, plain-colored moth with a pair of tiny black spots on the wings. It should never be brushed off roughly, as the small barbed hairs of its body may be ground into the skin. This causes a sore, much like a burn, that often takes weeks to heal.

LEECHES

Leeches are common in many jungle areas, particularly throughout most of the Southwest Pacific, Southeast Asia, and the Malay Peninsula. They are found in swampy areas, streams, and moist jungle country. They are not poisonous, but their bites may become infected if not cared for properly. The small wound that they cause may provide a point of entry for the germs which cause tropical ulcers or "jungle sores." Soldiers operating in the jungle should watch for leeches on the body and brush them off before they have had time to bite. When they have taken hold, they should not be pulled off forcibly because part of the leech may remain in the skin. Leeches will release themselves if touched with insect repellent, a moist piece of tobacco, the burning end of a cigarette, a coal from a fire, or a few drops of alcohol.

Straps wrapped around the lower part of the legs ("leech straps") will prevent leeches from crawling up the legs and into the crotch area. Trousers should be securely tucked into the boots.

LEECH STRAPS

SNAKES

A soldier in the jungle probably will see very few snakes. When he does see one, the snake most likely will be making every effort to escape.

If a soldier should accidently step on a snake or otherwise disturb a snake, it will probably attempt to bite. The chances of this happening to soldiers traveling along trails or waterways are remote if soldiers are alert and careful. Most jungle areas pose less of a snakebite danger than do the uninhabited areas of New Mexico, Florida, or Texas. This does not mean that soldiers should be careless about the possibility of snakebites, but ordinary precautions against them are enough. Soldiers should be particularly watchful when clearing ground.

Treat all snakebites as poisonous.

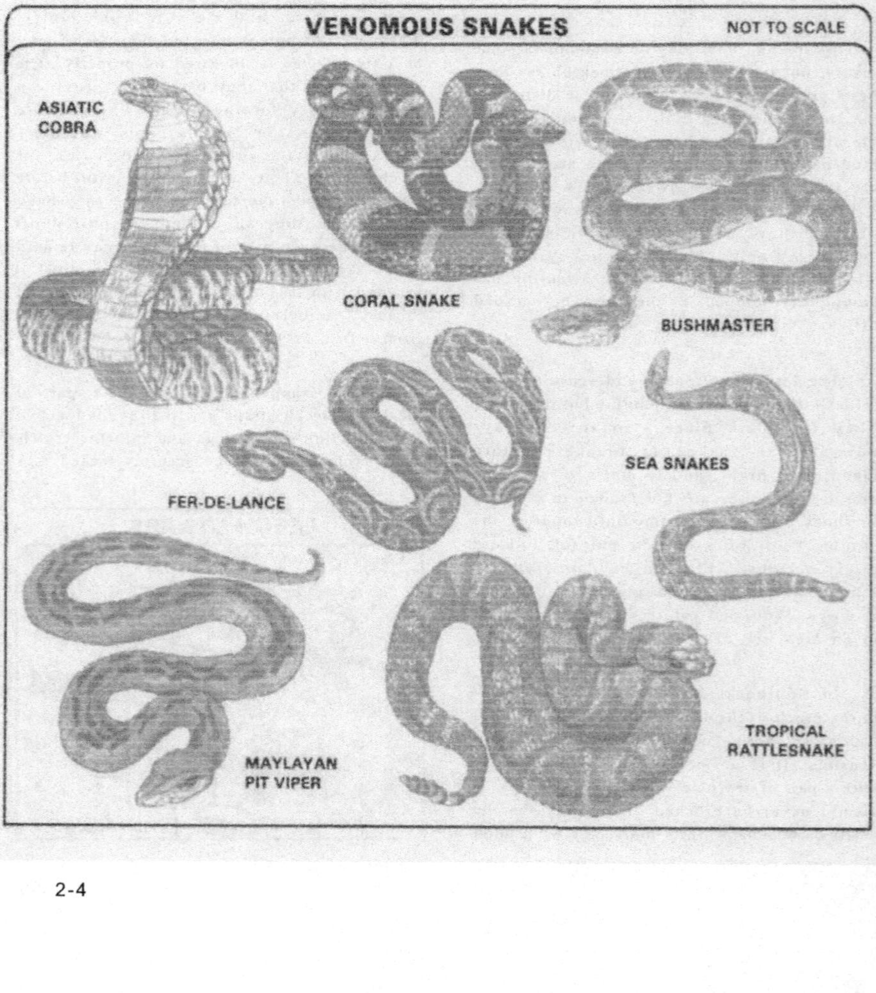

SNAKEBITE TREATMENT

Follow these steps if bitten:

- Remain calm, but act swiftly, and chances of survival are good. (Less than one percent of properly treated snakebites are fatal. Without treatment, the fatality rate is 10 to 15 percent.)
- Immobilize the affected part in a position below the level of the heart.
- Place a lightly constricting band 5 to 10 centimeters (2 to 4 inches) closer to the heart than the site of the bite. Reapply the constricting band ahead of the swelling if it moves up the arm or leg. The constricting band should be placed tightly enough to halt the flow of blood in surface vessels, but not so tight as to stop the pulse.
- Do not attempt to cut open the bite or suck out venom.
- Seek medical help. If possible, the snake's head with 5 to 10 centimeters (2 to 4 inches) of its body attached should be taken to the medics for identification. Identification insures use of the proper antivenom.

CROCODILES AND CAYMANS

Crocodiles and caymans are meat-eating reptiles which live in tropical areas. "Crocodile-infested rivers and swamps" is a catch-phrase often found in stories about the tropics. Asian jungles certainly have their share of crocodiles, but there are few authenticated cases of crocodiles actually attacking humans. Caymans, found in South and Central America, are not likely to attack unless provoked.

CROCODILES AND CAYMANS

WILD ANIMAL

In Africa, where lions, leopards, and other flesh-eating animals abound, they are protected from hunters by local laws and live on large preserves. In areas where the beasts are not protected, they are shy and seldom seen. When encountered, they will attempt to escape. All large animals can be dangerous if cornered or suddenly startled at close quarters. This is especially true of females with young. In the jungles of Sumatra, Bali, Borneo, Southeast Asia, and Burma there are tigers, leopards, elephants, and buffalo. Latin America's jungles have the jaguar. Ordinarily, these will not attack a man unless they are cornered or wounded.

Certain jungle animals, such as water buffalo and elephants, have been domesticated by the local people. Soldiers should also avoid these animals. They may appear tame, but this tameness extends only to people the animals are familar with.

POISONOUS VEGETATION

Another area of danger is that of poisonous plants and trees. For example, nettles, particularly tree nettles, are one of the dangerous items of vegetation. These nettles have a severe stinging that will quickly educate the victim to recognize the plant. There are ringas trees in Malaysia which affect some people in much the same way as poison oak. The poison ivy and poison sumac of the continental US can cause many of the same type troubles that may be experienced in the jungle. The danger from poisonous plants in the woods of the US eastern seaboard is similar to that of the tropics. Thorny thickets, such as rattan, should be avoided as one would avoid a blackberry patch.

Some of the dangers associated with poisonous vegetation can be avoided by keeping sleeves down and wearing gloves when practical.

HEALTH AND HYGIENE

The climate in tropical areas and the absence of sanitation facilities increase the chance that soldiers may contract a disease. Disease is fought with good sanitation practices and preventive medicine. In past wars, diseases accounted for a significantly high percentage of casualties.

Before going into a jungle area, leaders must:

- **Make sure immunizations are current.**
- **Get soldiers in top physical shape.**
- **Instruct soldiers in personal hygiene.**

Upon arrival in the jungle area, leaders must:

- **Allow time to adjust (acclimate) to the new environment.**
- **Never limit the amount of water soldiers drink. (It is very important to replace the fluids lost through sweating.)**
- **Instruct soldiers on the sources of disease. Insects cause malaria, yellow fever, and scrub typhus. Typhoid, dysentery, cholera, and hepatitis are caused by dirty food and contaminated water.**

WATERBORNE DISEASES

Water is vital in the jungle and is usually easy to find. However, water from natural sources should be considered contaminated. Water purification procedures must be taught to all soldiers. Germs of serious diseases, like dysentery, are found in impure water. Other waterborne diseases, such as blood fluke, are caused by exposure of an open sore to impure water.

Soldiers can prevent waterborne diseases by:

- **Obtaining drinking water from approved engineer water points.**
- **Using rainwater; however, rainwater should be collected after it has been raining at least 15 to 30 minutes. This lessens the chances of impurity being washed from the jungle canopy into the water container. Even then the water should be purified.**
- **Insuring that all drinking water is purified.**
- **Not swimming or bathing in untreated water.**
- **Keeping the body fully clothed when crossing water obstacles.**

FUNGUS DISEASES

These diseases are caused by poor personal health practices. The jungle environment promotes fungus and bacterial diseases of the skin and warm water immersion skin diseases. Bacteria and fungi are tiny plants which multiply fast under the hot, moist conditions of the jungle. Sweat-soaked skin invites fungus attack. The following are common skin diseases that are caused by long periods of wetness of the skin:

Warm Water Immersion Foot. This disease occurs usually where there are many creeks, streams, and canals to cross, with dry ground in between. The bottoms of the feet become white, wrinkled, and tender. Walking becomes painful.

Chafing. This disease occurs when soldiers must often wade through water up to their waists, and the trousers stay wet for hours. The crotch area becomes red and painful to even the lightest touch.

Most skin diseases are treated by letting the skin dry.

To prevent these diseases, soldiers should:

- **Bathe often, and air- or sun-dry the body as often as possible.**
- **Wear clean, dry, loose-fitting clothing whenever possible.**
- **Not sleep in wet, dirty clothing. Soldiers should carry one dry set of clothes just for sleeping. Dirty clothing, even if wet, is put on again in the morning. This practice not only fights fungus, bacterial, and warm water immersion diseases but also prevents chills and allows soldiers to rest better.**
- **Not wear underwear during wet weather. Underwear dries slower than jungle fatigues, and causes severe chafing.**
- **Take off boots and massage feet as often as possible.**
- **Dust feet, socks, and boots with foot powder at every chance.**
- **Always carry several pairs of socks and change them frequently.**
- **Keep hair cut short.**

HEAT INJURIES

These result from high temperatures, high humidity, lack of air circulation, and physical exertion. All soldiers must be trained to prevent heat disorders.

HEAT INJURIES			
TYPE	**CAUSE**	**SYMPTOMS**	**TREATMENT**
Dehydration	Dehydration is caused by the loss of too much water. About two-thirds of the human body is water. When water is not replaced as it is lost, the body becomes dried out—dehydrated.	The symptoms are sluggishness and listlessness.	The treatment is to give the victim plenty of water.
Heat Exhaustion	Heat exhaustion is caused by the loss of too much water and salt.	The symptoms are: Dizziness. Nausea. Headache. Cramps. Rapid, weak pulse. Cool, wet skin.	The treatment consists of: Moving the victim to a cool, shaded place for rest. Loosening the clothing. Elevating the feet to improve circulation. Giving the victim cool salt water (two salt tablets dissolved in a canteen of water). Natural sea water should not be used.
Heat Cramps	Heat cramps are caused by the loss of too much salt.	The symptom is painful muscle cramps which are relieved as soon as salt is replaced.	The treatment is the same as for heat exhaustion.
Heatstrokes	Heatstroke (sunstroke) is caused by a breakdown in the body's heat control mechanism. The most likely victims are those who are not acclimated to the jungle, or those who have recently had bad cases of diarrhea. Heatstroke can kill if not treated quickly.	The symptoms are: Hot, red, dry skin (most important sign). No sweating (when sweating would be expected). Very high temperature (105 to 110 degrees). Rapid pulse. Spots before eyes. Headache, nausea, dizziness, mental confusion. Sudden collapse.	Treatment consists of: Cooling the victim immediately. This is achieved by putting him in a creek or stream; pouring canteens of water over him; fanning him; and using ice, if available. Giving him cool salt water (prepared as stated earlier) if he is conscious. Rubbing his arms and legs rapidly. Evacuating him to medical aid as soon as possible.

Heat injuries are prevented by:

- Drinking planty of water.
- Using extra salt with food and water.
- Slowing down movement.

NOTE: For more details, see FM 21-10 for field hygiene and sanitation, and FM 21-11 for first aid for soldiers.

NATIVES

Like all other regions of the world, the jungle also has its native inhabitants. Soldiers should be aware that some of these native tribes can be hostile if not treated properly.

There may be occasions, however, when hostile tribes attack without provacation. If they attack, a small force should be able to disperse them.

To prevent a conflict, leaders should insure that their soldiers:

- Respect the natives' privacy and personal property
- Observe the local customs and taboos
- Do not enter a native house without being invited
- Do not pick fruits or cut trees without permission of their owners
- Treat the natives as friends

Section III. JUNGLE SURVIVAL

FOOD

Food of some type is always available in the jungle—in fact, there is hardly a place in the world where food cannot be secured from plants and animals. All animals, birds, reptiles, and many kinds of insects of the jungle are edible. Some animals, such as toads and salamanders, have glands on the skin which should be removed before their meat is eaten. Fruits, flowers, buds, leaves, bark, and often tubers (fleshy plant roots) may be eaten. Fruits eaten by birds and monkeys usually may be eaten by man.

Meats that can be found in most jungles include:

- Wild fowl
- Wild cattle
- Wild pig
- Freshwater fish *
- Saltwater fish
- Fresh-water crawfish

The following types of fruits and nuts are common in jungle areas:

- Bananas
- Coconuts
- Oranges and lemons
- Navele nuts
- Breadfruit
- Wild raspberries
- Nakarika
- Papaya
- Mangoes

Vegetables found in most jungles include:

- Taro *
- Yam *
- Yucca *
- Hearts of palm trees

*These items must be cooked before eating.

There are various means of preparing and preserving food found in the jungle. Fish, for example, can be cleaned and wrapped in wild banana leaves. This bundle is then tied with string made from bark, placed on a hastily constructed wood griddle, and roasted thoroughly until done. Another method is to roast the bundle of fish underneath a pile of red-hot stones.

Other meats can be roasted in a hollow section of bamboo, about 60 centimeters (2 feet) long. Meat cooked in this manner will not spoil for three or four days if left inside the bamboo stick and sealed.

Yams, taros, yuccas, and wild bananas can be cooked in coals. They taste somewhat like potatoes. Palm hearts can make a refreshing salad, and papaya a delicious dessert.

SHELTER

Jungle shelters are used to protect personnel and equipment from the harsh elements of the jungle. Shelters are necessary while sleeping, planning operations, and protecting sensitive equipment.

> **When selecting shelter, leaders should:**
>
> - **Choose high ground, away from swamps and dry river beds**
> - **Avoid trails, game tracks, or villages**
>
> (See app E for more on shelters.)

NAVIGATION

Navigation in thick jungle areas is difficult even for the most experienced navigators. Soldiers navigating in the jungle must use various aids. The compass is an obvious aid, but a soldier would never be able to move very fast in the jungle if he had to constantly move along a magnetic azimuth. Movement along a terrain feature, such as a ridgeline, is easier but can be extremely dangerous when establishing a pattern of consistency. A soldier must trust the compass, map, and pace count. A soldier should not keep his eyes riveted on the compass; however, it should be used as a check.

The shadows caused by the sun are an easily observed and accurate aid to direction. Allowances must be made for the gradual displacement of the shadows as the sun moves across the sky.

Other aids to maintaining direction include prominent objects, the course of rivers, prevailing winds, the stars, and the moon. *(See app B for more on navigation.)*

OBSERVATION

All movements of animals and men are marked by tracks and signs. Soldiers must learn to read signs left in soft ground, in streambeds, on roads and trails, and near watering places and salt licks. Animals seldom move without a reason; a few fresh tracks supply information about their maker, his direction, and probable intentions.

Animals avoid man. The animals, their tracks, and their behavior can reveal whether or not men are in the area. Jungle fighters can listen to the cries of animals and learn to recognize their alarm calls.

The ability to track and to recognize signs in the jungle are valuable skills. Throughout the soldier's time in the jungle, he should practice these skills. *(See app B for more on tracking.)*

CLOTHING AND EQUIPMENT

Before deploying for jungle operations, troops are issued special uniforms and equipment. Some of these items are:

JUNGLE FATIGUES

These fatigues are lighter and faster drying than standard fatigues. To provide the best ventilation, the uniform should fit loosely. It should never be starched.

JUNGLE BOOTS

These boots are lighter and faster drying than all-leather boots. Their cleated soles will maintain footing on steep, slippery slopes. The ventilating insoles should be washed in warm, soapy water when the situation allows.

INSECT (MOSQUITO) BAR

The insect (mosquito) bar or net should be used any time soldiers sleep in the jungle. Even if conditions do not allow a shelter, the bar can be hung inside the fighting position or from trees or brush. No part of the body should touch the insect net when it is hung, because mosquitoes can bite through the netting. The bar should be tucked or laid loosely, not staked down. Although this piece of equipment is very light, it can be bulky if not folded properly. It should be folded inside the poncho as tightly as possible.

CHAPTER 3

Preparation and Training to Deploy to Jungle Areas

Section I. GENERAL

Chapters 1 and 2 describe the jungle environment. Since many soldiers are unaccustomed to such an environment, they must make preparations before conducting jungle operations.

This chapter lists the factors to be considered when preparing for jungle operations and presents training tips for conducting jungle training.

Section II. FACTORS TO BE CONSIDERED WHEN PREPARING FOR JUNGLE OPERATIONS

When a unit is alerted for training or actual combat operations in a jungle environment, the commander must first consider the following:

- Where will the unit be training or operating?
- What are the climatic and terrain conditions of the objective area?
- How much time does the unit have to prepare?
- What available training areas have climate and terrain resembling the objective area?
- What type operations are to be conducted—conventional or counter-guerrilla?

CONTENTS

- Will the unit be taking its own equipment?
- Does any of the unit's equipment require modification (including camouflage painting)?
- What special equipment does the unit require?
- Does the unit have any jungle warfare instructors, soldiers with jungle experience, or linguists?
- What training assistance is available? Outside instructors? Training aids?
- Does higher headquarters have special standing operating procedures (SOP) for jungle war?
- Are all soldiers physically fit?
- What information is available about the enemy?
- What information is available about local civilians and allied forces in the objective area?

Once these questions have been answered, the commander can develop a program to prepare his unit to operate in the jungle.

The idea that a unit is technically and tactically proficient is only a small part of a unit's preparation. Emphasis should be placed on the mental, physical, and psychological aspects of operating in a jungle environment. The key to overcoming these problems lies with the unit chain of command in their efforts to develop a "will to win" as well as "will to train to win."

Section III. TRAINING TIPS

Units committed to jungle operations may have to fight as soon as they arrive in the operational area. Commanders must make the best use of the preparation time available. Measures which commanders should consider include:

Making use of time in garrison. Certain jungle subjects can be taught using classroom instruction. This training should begin as soon as possible, so that time in jungle training areas can be devoted to more advanced techniques.

Making use of local training areas. Although these training areas may not resemble jungle terrain, some jungle techniques can be introduced in them. This will provide a training base which can be expanded when the unit deploys to its jungle training or operational areas. In addition, physical training should begin in the local training area as early as possible before deployment.

Integrating individual training into unit training exercises. Rather than devote field training time to the individual skills required to live in the jungle, these skills should be introduced early in classes, and then practiced during unit training exercises.

The following lists can be used as a guide to subjects that should be covered:

INDIVIDUAL TRAINING

Common Subjects:	Staff and Leader Subjects:
Jungle environment and acclimation	Jungle maintenance and supply techniques
Living in the jungle	Jungle terrain appreciation
Survival, evasion, and escape	Enemy order of battle and tactics
Camouflage and concealment	Airmobile techniques
Tracking	
Operational area orientation	**Specialist Subjects:**
Jungle navigation	Language
Equipment recognition	Demolitions
Enemy orientation	Field expedient antennas
Physical conditioning	Medevac techniques

UNIT TRAINING

Movement

Obstacles and Barriers

Scouting, Surveillance, Patrolling, and Tracking

Air Defense

Adjustment and Conduct of Fires

Immediate Action Drill

Communications

Jungle Operations:

- Ambush/counterambush
- Raid
- Attack
- Defense
- Infiltration
- Airmobile operations
- Waterborne operations
- River crossing
- Road clearing

FIRE TEAM AND CREWMEMBERS

Identification and Marking of Mines and Booby-traps

Working with Helicopters

Weapons Training

ACCLIMATION

The first priority in preparation for jungle warfare is acclimation (getting accustomed to jungle climate). Troops who are not conditioned properly will not perform jungle warfare tasks reliably. Different people become acclimated to hot weather at different rates, but the following methods can be used in most units.

Exercise is the best method for acclimation, because troops in good physical condition will adapt easily to new climates. A 7- to 14-day conditioning period should be sufficient for most soldiers. Exercises should be moderately strenuous at the beginning, and become more demanding each day. Troops from warm climates will adapt faster than troops deploying from colder climates. Physical training in heated gymnasiums prior to deployment will also ease the acclimation process.

Leaders must be alert for symptoms of heat disorders during the acclimation period.

SURVIVAL, EVASION, AND ESCAPE

Convincing a soldier that he will survive alone in the jungle will go a long way in building his self-confidence. FM 21-76 contains details on survival, evasion, and escape training.

SWIMMING

Swimming is also a vital skill for the jungle fighter. Falling into a jungle pool or river can be a terrible experience, especially for a nonswimmer. All troops should be "drown proofed" as shown in FM 21-20. Units should identify their strong swimmers for lifeguard training and other more difficult swimming tasks.

CAMOUFLAGE AND CONCEALMENT

Training to conceal soldiers and equipment from ground and air observation is equally important to combat, combat support, and combat service support units. Proper use of camouflage will help to make up for an enemy's superior knowledge of the jungle area. Appendix E contains information about jungle camouflage and concealment techniques.

JUNGLE LIVING

Following a short period of classroom instruction, soldiers should experience jungle living conditions in the field. This training can be incorporated into other unit training.

Subjects which should be stressed include:

- Heat disorders
- Survival
- First aid
- Health, hygiene, and field sanitation
- Proper wearing of clothing
- Use of equipment in a jungle environment
- Prevention and treatment of snakebites and insect bites

During this period of training, use of garrison facilities should be kept to a minimum. Supplies should be brought to the field rather than the unit returning to the rear for them. Soldiers should learn to live without unnecessary personal comforts.

Land navigation should be practiced using jungle movement techniques. *(See app B.)*

TARGET AREA ORIENTATION

Classes on the host country should stress those facts which apply to operations.

Subjects could include:

- Terrain appreciation
- Climate
- Population and culture
- Language (phrase books may be issued)
- Road, railroad, and canal system
- Standards of conduct for US Army personnel
- Allied armed forces
- Reasons for US involvement

JUNGLE TACTICS

Chapter 5 describes tactics common to jungle fighting. These tactics should be taught first to leaders down to squad level. The leaders then train their own units. Stress should be placed on small unit tactics and operations with Army aviation. Since night operations, especially ambushes, are common in jungle fighting, units should emphasize night training.

MOVEMENT

Units should train in tactical marches.

Training should emphasize:

- Breaching of obstacles
- Scouting, patrolling, and tracking
- Off-road movement over rugged terrain
- Air defense and counterambush drills

WEAPONS TRAINING

Most jungle fighting takes place at close range. Soldiers should be trained in "quick fire," as outlined in *chapter 8, FM 23-9* Advanced training should be conducted on a "jungle range." On this range, soldiers move down a trail and engage pop-up and moving targets which appear suddenly at close range. Targets are operated by an assistant on signals from a lane grader. Targets should be exposed for 3 to 6 seconds. Only 2 or 3 rounds should be fired at each target. At a later stage, boobytraps and obstacles can be emplaced on the trail.

INTELLIGENCE

Soldiers should be trained in specific intelligence subjects.

These include:

- **Enemy organization and tactics**
- **Equipment recognition**
- **National markings**
- **Sound or signature recognition**

This last subject is particularly important in jungle operations, because soldiers will more often hear weapons firing than see them. If captured enemy weapons and equipment are available, they should be used as aids in this training.

MAINTENANCE AND LOGISTICS

Chapter 6 describes some of the jungle's effects on equipment. Operators need to learn techniques to keep their equipment operational.

Subjects to be covered include:

- **Effects of climate on equipment**
- **Jungle operational techniques**
- **Preventive maintenance**
- **Recovery and repair techniques**

Staff members and leaders should receive familiarization training on these techniques in order to supervise the operators. In addition, staff and leaders should learn those special supply requirements and procedures in the operational area. They should also be familiar with the capabilities of those logistical units supporting the force.

CHAPTER 4

The Threat in Jungle Areas

Section I. GENERAL

The jungle is an environment which stretches in a broad belt around the tropical areas of the world. Each of these areas has its own military, political, and economic conditions. As a result, it is impossible to describe one threat which applies to all jungle areas. Potential enemies which US forces might face in the jungle run the spectrum from lightly armed guerrillas all the way to conventional forces.

This chapter describes the main features of guerrilla and conventional forces as they are found in the jungle, and briefly outlines the types of potential threat forces in various jungle regions.

Section II. GUERRILLA FORCES.

WHAT GUERRILLAS ARE

Guerrillas are irregular forces. They normally constitute the military faction of a political resistance or a subversive movement. These forces engage in unconventional operations in order to undermine the power of an established government or to take political control away from other factions. Their goal is normally to establish a new government, often according to a radical political philosophy.

The basic guerrilla organization is a three- to five-man cell. These cells are capable of independent action. They also can be brought together for larger operations and dispersed later. Guerrillas are organized into cells for two reasons. One is for security. The fewer the people who can identify members of a guerrilla force, the better the chances are that it will survive. The second reason is for support. Guerrillas must live off the land to a large degree, and small cells are easier to support in this manner.

CONTENTS

HOW GUERRILLAS FIGHT

Guerrillas are usually weaker than conventional forces in terms of total resources. For that reason, guerrillas will not attempt to overwhelm large units of their opponents in combat. They will instead try to inflict as much damage as possible in lightning actions, withdrawing before the opposing forces can react. Guerrillas are most effective when they strike widely separated targets over a long period of time. This type of action will confuse, demoralize, and frustrate their opponents.

Typical missions which guerrillas conduct to accomplish their goals include:

- **Destroying or damaging vital installations, equipment, or supplies**
- **Capturing supplies, equipment, or key governmental or military personnel**
- **Diverting government forces from other operations**
- **Creating confusion and weakening government morale**

These missions are not normally accomplished by the use of conventional attacks and defenses. Instead, guerrillas rely on speed, surprise, and security. Guerrilla operations include raids, ambushes, mining and boobytrapping, and sniping.

Targets are selected by the guerrilla based on an analysis of how much the elimination of the target will disrupt the government, what the effect on the populace will be, the risk of being killed or captured, and the amount of weapons or supplies which can be seized. This analysis calls for timely intelligence, which is gained by active patrolling.

The retention of the initiative is the key to success in guerrilla operations. Guerrillas rely on their ability to strike where they are least expected, at points where the government forces are least prepared. If the guerrillas lose the initiative, and are forced to react to the operations of conventional forces, their effectiveness is greatly reduced.

Guerrillas are not normally organized or equipped for stand-and-fight type defensive operations. They prefer to defend themselves by moving, by dispersing into small groups, or by diverting the opponent's attention while they withdraw. Whenever possible, these operations are accomplished by offensive operations against the opponent's flank or rear. If the government forces persist in their attack, the guerrillas are prepared to disengage to keep their freedom of action. If forced to disperse into small groups, the guerrilla forces become less effective until they regroup to resume offensive operations.

One of the most important needs of guerrilla forces is support. This support can come from a number of sources. Food, for example, can be stolen or supplied by political sympathizers. Weapons can be gathered from raids on government installations. A foreign power may provide secret training, and shipments of food, weapons, ammunition, and equipment. If the guerrillas can be cut off from these sources of support, they will be much less effective.

To protect their operations, jungle guerrillas will normally establish bases from which they can operate. These bases will be in remote areas. The bases will be secured by a combination of guerrilla outposts and by a grapevine intelligence network established by political sympathizers. Although they may be difficult to find, there will normally be concealed routes into the bases, from which the guerrillas have access to their targets and sources of support.

GUERRILLA STRENGTHS AND WEAKNESSES

Guerrillas operate most effectively in countries where the people are discontented with government policies. If the people are apathetic or passively hostile to their government, the guerrillas will seek to develop this feeling into a popular base of support. If no such feeling exists among the people, it will be much harder for guerrillas to set up operations.

Some common guerrilla strengths include:	Some common guerrilla weaknesses include:
■ Highly motivated leadership ■ Strict, swift discipline among the guerrillas and sympathizers ■ Strong belief in a political, religious, or social cause ■ Capability to raise or lower the level of intensity from subversion to open warfare	■ Mental and physical stress, caused by long periods of isolation in an unstable environment ■ Fear of criminal prosecution by the government, or of reprisals against friends and family ■ Feeling of numerical and technological inferiority to counterguerrilla forces ■ Uncertain public base of support ■ Requirements to secure supply lines, transport means, and storage facilities or caches

Section III. CONVENTIONAL FORCES

TYPES OF CONVENTIONAL FORCES IN THE JUNGLE

Conventional forces committed to jungle operations can perform any one of a number of missions. The lowest level of involvement is the use of conventional forces to advise and assist native guerrilla or paramilitary forces, teaching them either how to fight or how to operate sophisticated equipment. A higher level of involvement is the use of conventional forces as a military cadre in units which are composed of native forces. Finally, the highest level of involvement is the operation of conventional forces in a conventional role, fighting major battles in the jungle.

Conventional jungle enemies may come from a number of places. It is possible that US forces committed to jungle operations will fight native conventional forces. It is also possible that US forces will fight conventional forces brought in from a sponsoring hostile power. In either case, most potential jungle enemies are infantry forces, supported with artillery, mortars, and armored vehicles, organized along the lines of Soviet forces. These forces may also have a capability to conduct tactical air (TACAIR) operations and nuclear, biological, chemical (NBC) warfare. They may be equipped with weapons and equipment that are a generation or two older than those found in more modern armies.

HOW CONVENTIONAL FORCES FIGHT IN THE JUNGLE

The way in which a potential conventional threat army fights in the jungle depends on the terrain, the combat experience of that army, and the degree to which it models itself after the forces of a sponsoring power.

In general, however, jungle enemies can be expected to follow these tactical principles:

- **Maintain the offensive; defend only to gain time.**
- **Embrace the enemy; stay close to reduce the effects of his firepower.**
- **Infiltrate at every opportunity.**
- **Operate during periods of limited visibility.**
- **Use surprise tactics; raids, ambushes, and patrols.**

A jungle enemy can be expected to be skilled in the art of camouflage, the chief means he has to evade his opponent's firepower. Camouflage will be for him a way of life. He will probably use bunkers and tunnels as protective survival measures. To slow opposing forces, he may use obstacles, mines, and boobytraps. He will move on covered and concealed routes, using darkness to conceal most of his operations. He will probably depend heavily on streams and rivers to provide concealed routes of movement and drinking water.

He can be expected to remove all intelligence indicators from the battlefield. He will go to great lengths to remove his dead, wounded, weapons, and even expended cartridges from the battlefield. He will try to leave no information relating to order of battle, strength, dispositions, or intentions.

"We captured numerous enemy documents which either condemned or commended certain units for the police of the battle field"

—Report, 25th Infantry Division, Kontum Province, Republic of Vietnam

Jungle enemies have also used deception means, such as explosive bullets and firecrackers, to mislead US units as to the size and disposition of the forces opposing them. Communications deception and jamming have also been used by jungle enemies against opponents.

Since the US Army is noted for employing an abundance of firepower, jungle enemies in the past have preferred to engage US units at extremely close range. At times, it is impossible for US commanders to use their supporting indirect fires without taking friendly casualties. The specific effects that this technique has on offensive and defensive operations will be discussed later. In general, however, the US ground commander must operate in such a manner that all his fire support can always be used effectively.

Jungle enemies can be expected to train hard to use the jungle to their advantage. In the offense, for example, they use the thick foliage to infiltrate positions and eliminate command posts (CP), key weapons, and vital facilities. These operations are conducted to take away their opponent's advantages in command and control, fire support, and logistical means. The intent is to put their forces on a more equal footing with their opponents. This situation can be exploited by a force with superior knowledge of the terrain.

When forced to defend, these forces will quite often prepare elaborate defensive positions, well camouflaged and concealed. In addition, defenders may use snipers, boobytraps, and ambushes to delay, create a sense of confusion and insecurity, and cause the attacker to surrender the initiative.

It is also possible that US forces committed to jungle operations will fight Warsaw Pact forces, probably members of airborne divisions. Although these troops will have newer and more sophisticated weapons than some of the troops native to jungle areas, *they probably will not be familiar with the local terrain and may not be well trained in jungle operations.*

STRENGTHS AND WEAKNESSES OF CONVENTIONAL JUNGLE ENEMIES

Much of a conventional jungle enemy's effectiveness depends on familiarity with the terrain. In general, this means that armies native to a battlefield area will be more effective than forces from outside. Even if these outside forces have a greater amount of firepower than the native forces, the lack of terrain familiarity may limit their ability to use that firepower.

Weaknesses of potential conventional jungle enemies will probably include:

- Larger units, much more difficult to
- Bigger targets for close-air support or artillery
- More difficulty in evading detection
- Less information from local sympathizers

Strengths of potential conventional jungle enemies will probably include:

- Adequate firepower for conventional attacks and defense
- Knowledge of the terrain and area
- Well-trained and disciplined soldiers
- Independence from local support

HOW THE THREAT DEFENDS

The threat defense is a temporary measure, adopted only when necessary. This does not imply, however, that the threat defense consists of half measures or that he is unskilled in defense techniques. Jungle enemies will use every trick possible to survive against massive amounts of firepower. His defense will be cleverly and carefully prepared.

A typical jungle enemy defensive position consists of a complex series of earth and timber bunkers, spider holes, and tunnels. These are positioned to achieve mutual support. Bunkers are built low to make them more difficult to see and engage by fire. They are well camouflaged—even the fields-of-fire may be cut from the waist down, so that they will be unnoticeable to a standing man. Weapons positions are planned to provide interlocking fires—lethal even during limited visibility. Boobytraps and obstacles are integrated into the defense to slow, demoralize, and confuse the attacker.

The jungle threat's concept of the defense is to trap the attacker by allowing him to move into prepared fields-of-fire. Fire is opened at extremely close range, sometimes at 50 meters or less. This is done for two reasons—first, to bring fires to bear from all sides, and, second, to force the attacker to remove himself before he can call for supporting fires.

While the key part of the jungle threat's defense is automatic weapons positions in bunkers, the enemy will also put snipers in the trees. In this way, the attacker cannot devote his full attention to the bunkers, because he must also deal with the snipers.

If the attacker is too strong, the jungle threat will attempt to withdraw over routes that have been planned and scouted to make the withdrawal as rapid as possible. Stay-behind ambushes, snipers, mines, and obstacles are used to slow the attacker.

HOW THE CONVENTIONAL ENEMY ATTACKS

Attack is the preferred form of combat for potential jungle enemies. Because most of these forces expect to have a disadvantage in firepower and technology when fighting US forces, most of them have developed special techniques to help make up the difference through surprise. They may, for example, probe a defensive position until the defender reveals the location of his key weapons. These weapons are then eliminated by infiltrators before the main attack. They may use firecrackers to create a diversion, drawing the defender's fire and deceiving him as to the size of the attacking force. They may infiltrate the defense to eliminate command posts, mortars, or artillery units.

"Decoy the Americans from one direction by smoke, firing, or shouting. Then attack him from an unexpected direction."

—Captured Japanese Document, World War II

Threat units will avoid attacking prepared defenses when possible. They prefer to attack a weak point, using the jungle, weather, and their own special training as much as possible. Sapper squads are specially trained to infiltrate minefields and obstacles in order to neutralize key positions or create a gap in the defense. The enemy may also isolate a position, so that their opponents will be tied down in trying to relieve it, or they may conduct raids to disrupt operations and lower the defender's morale. Darkness, poor weather, and rough terrain will be used to conceal these operations.

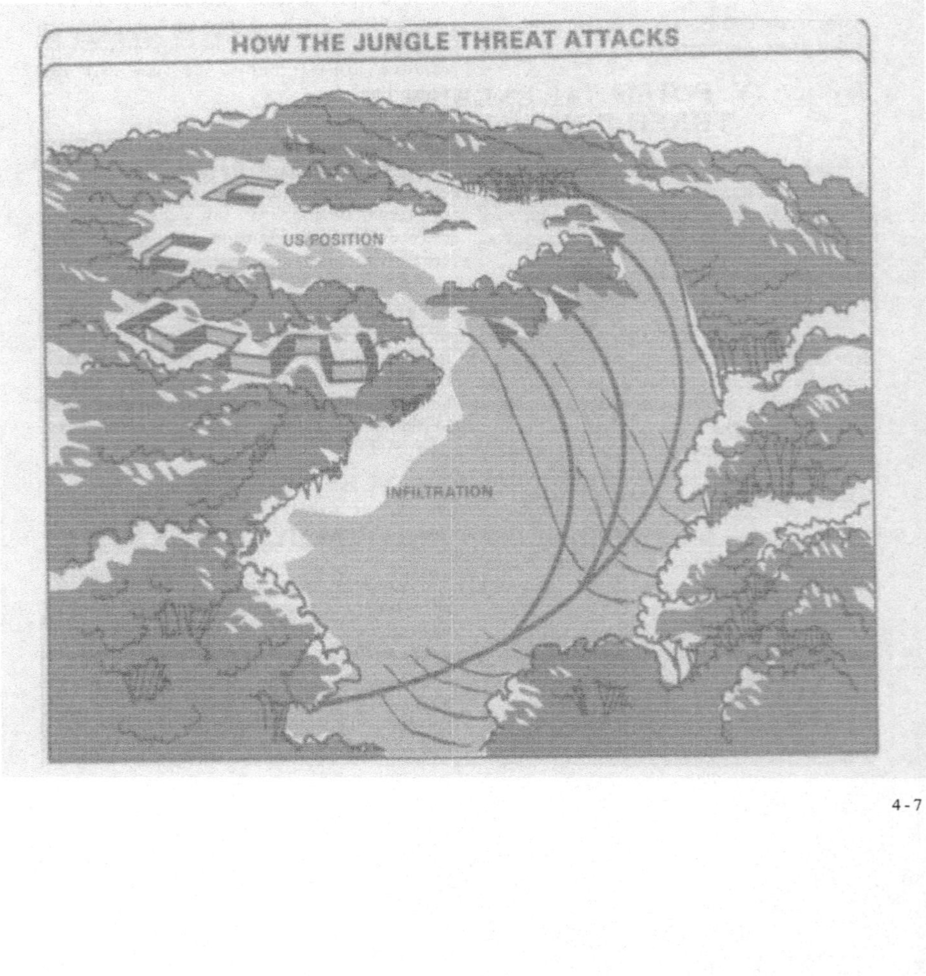

A commander should never assume that any jungle area is impassable to a well-trained jungle enemy. Experience has shown that such enemies are very adept at using extremely difficult terrain effectively as avenues of approach.

"Use fog and rain to catch the Americans off guard. Make an assault suddenly, from positions which the Americans believe unapproachable, such as cliffs, rivers, and jungles."

—Captured Japanese Document, World War II

Although the jungle enemy attacks swiftly, his attacks are planned in minute detail. If he is allowed to attack according to plan, the jungle enemy is an effective force. If the defender can interrupt even a minor part of the plan, the enemy will have difficulty in adjusting, and the attack will probably fail. This aspect of enemy operations places a high premium on the struggle for the initiative at all levels.

If the defending force can be pushed out of its positions and forced to retreat, the jungle enemy will probably make every effort to maintain contact through pursuit. He will harass the rear guard, at the same time sending forces to outrun and cut off the retreating force. He will then try to destroy the retreating force by ambush or encirclement.

Section IV. POTENTIAL ENEMIES IN THREE JUNGLE REGIONS

It is impossible to describe one jungle threat which applies to all areas of the world. There are, however, certain characteristics of potential threat forces that are peculiar to specific jungle regions.

LATIN AMERICA

The most likely threat the US forces may face in Latin American jungles are insurgent movements. These movements aim at the overthrow of a wealthy ruling class to install a new regime. The US is often viewed by the insurgents as an ally of the government, and US facilities and institutions are often targets for these movements.

The military faction of these insurgent movements consists of guerrilla forces similar to those described earlier. They are organized into small cells, are lightly armed, and are capable of concentrating for acts against major facilities and then dispersing after the operation. Although their ultimate objective will often be the establishment of control over the urban areas, they may use the jungle to provide a concealed and secure base of operations. The support of the local people is very important to their survival.

At the time US forces are committed to fight in Latin American areas, guerrilla forces are likely to be augmented with military aid and personnel from other sponsoring countries in the region. These forces may perform any one of a number of roles: advisors, guerrilla cadre, or limited conventional combat. Logistical and intelligence support may also come from these forces.

SUBSAHARAN AFRICA

The conflicts in this region since World War II have been waged by insurgent groups against perceived vestiges of colonialism or imperialism. Most of these colonialist and imperialist institutions are connected in the minds of the insurgents with the Western European powers. As a result, the instability in some areas of this region has provided a tempting target for provocation. To make matters more complicated, many of the conflicting factions are also struggling among themselves, due to political or ancient tribal differences. This in turn creates even more regional turmoil, and an even greater vulnerability for exploitation.

Conflicting factions in Subsaharan Africa consist primarily of guerrilla groups. These guerrillas, however, are often more heavily armed than Latin American guerrillas for two reasons. First, these groups have mortars, artillery, and recoilless weapons from national army formations which have been defeated or disbanded. Second, external powers have backed their favorite factions by supplying arms, ammunition, and equipment. For the most part, these guerrillas subsist by acquiring food and supplies from the countryside.

Foreign involvement in these guerrilla movements has consisted of advisors and cadre from sponsoring nations. Should US forces ever fight in this region, it is likely that they will encounter troops foreign to the nation. In addition, there is also a possibility that Warsaw Pact troops, primarily airborne or tactical aviation units, would be committed to such a region to fight US troops.

SOUTHEAST ASIA

In many respects, the potential threat array in Southeast Asia is the most complicated of any jungle region. There are active guerrilla movements in most Southeast Asian countries as well as tribal and cultural conflicts. There is a good possibility of foreign support or intervention.

The unique development in this region has been the rise of a regional power. Since the end of US involvement in Southeast Asia, this power has developed a potent conventional force, using equipment captured from the US and its allies or supplied by communist countries. More than any other potential threat native in a jungle region, it possesses the ability for sustained conventional operations against any US forces which might be deployed in the area. Its capabilities span the range from clandestine guerrilla operations to large-scale conventional attacks, supported by tanks, motorized units, artillery, and aviation.

Because there are already strong forces in this region, the probability of involvement of large numbers of world power forces is not great. There is a good possibility, however, that US troops committed in these areas might encounter weapons and equipment supplied by a world power. They might also encounter advisors from world powers that instruct and aid the native forces in the use of sophisticated equipment.

Finally, of all the regions discussed thus far, the chemical warfare threat will probably be greatest for US forces conducting operations in Southeast Asia.

Section V. WEAPONS USED BY POTENTIAL JUNGLE ENEMIES

Although potential jungle enemy forces vary widely from region to region, there are certain types of weapons which are commonly found in jungle countries. US forces should become familiar with these basic types of weapons in order to be able to recognize them on the jungle battlefield. They should also have a basic knowledge of the weapons' characteristics and know where the weapons are found in typical communist forces organizations.

Although guerrilla forces do not have the same type of organizational structure as conventional forces, they too will probably carry many of these weapons.

RIFLE PLATOON

The threat rifle platoon normally has three rifle squads of 6 to 10 men each. Weapons found in the platoon will include small arms, one to three light machineguns, and one to three grenade launchers. Typical weapons include:

AKM ASSAULT RIFLE

Recognition features are the pistol grip, the curved-box magazine, and the "underslung" barrel without a bipod. The AKM fires the M-1943 7.62-mm cartridge, also used in the SKS, RPD, and RPK. The AKM also comes with a metal folding stock.

CHARACTERISTICS OF AKM

Caliber and ammo	7.62 mm, M-1943
Cyclic rate of fire	600 rpm
Practical rate of fire:	
auto	100 rpm
semi	40 rpm
Magazine capacity	30 rounds
Effective range	400 m

AK-74 ASSAULT RIFLE

A new, small caliber version of the tried-and-proven AKM which takes advantage of modern ammunition technology. Two models are being produced: the standard stock AK-74 and the folding stock AKS-74.

CHARACTERISTICS OF AK-74

Caliber	5.45 mm
Ammunition	high-velocity ball (somewhat similar to US M16 rd)
Mode of fire	automatic and semiautomatic
Cyclic rate of fire	650 rpm
Maximum effective range	350 m

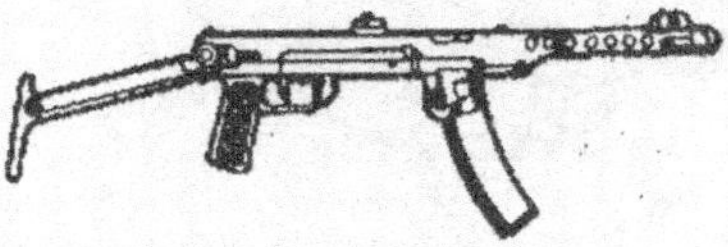

PPS SUBMACHINEGUN

The PPS is a fully automatic weapon. It has a hinged stock which folds up and forward. It also has a compensator welded on front of the barrel jacket. It fires the 7.62-mm M-1930 "P" ammunition from a curved-box magazine.

CHARACTERISTICS OF PPS

Cyclic rate of fire	650 rpm
Practical rate of fire	100 rpm
Magazine capacity	35 rounds
Effective range	200 m

RIFLE PLATOON CONTINUED

SKS CARBINE

This light, shoulder-fired weapon is recognized by a characteristic folding bayonet, a sporting rifle appearance, and triangular portion of the magazine which extends through the lower side of the stock, just forward of the trigger guard.

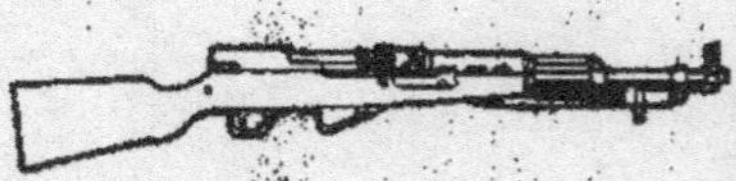

CHARACTERISTICS OF SKS

Caliber and ammo	7.62 mm, M-1943
Operation	gas, semiautomatic
Magazine capacity	10 rounds
Effective range	400 m

M-1890/1930 SNIPER RIFLE

The M-1890/1930 sniper rifle is a Soviet model fielded during World War II. It is still a standard weapon in many satellite armies. The M-1890/1930 mounts a telescopic sight. This sight is similar to the sight used on US hunting rifles.

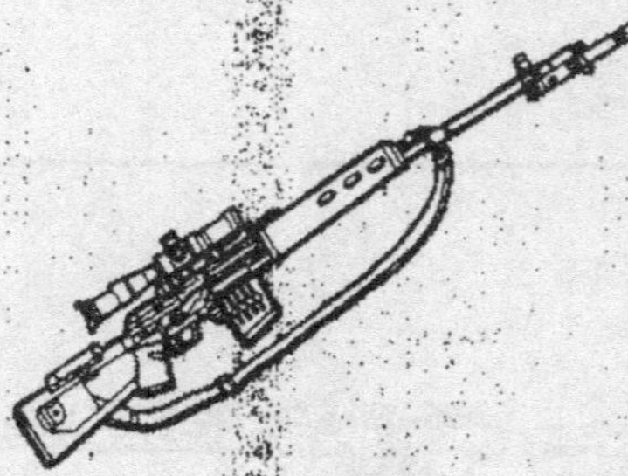

CHARACTERISTICS OF M-1890/1930

Caliber and ammo	7.62 mm, M-1908
Operation	turning bolt
Magazine capacity	5 rounds
Effective range	800 m
Maximum range	1,300 m

RPG-2 GRENADE LAUNCHER

The RPG-2 recoilless antitank rocket is fired from the shoulder. Recognition features are the warhead of the projectile which is larger than the launching tube, and the launcher itself, which is long, slender, and has a pistol grip handle and two sights.

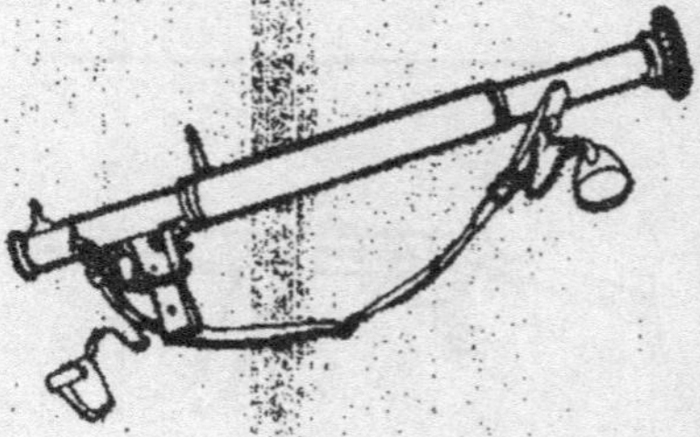

CHARACTERISTICS OF RPG-2

Caliber of tube	40 mm (1.58 in)
Caliber of projectile	82 mm (3.2 in)
Effective range	100 m
Rate of fire	4 to 6 rpm

RIFLE PLATOON CONTINUED

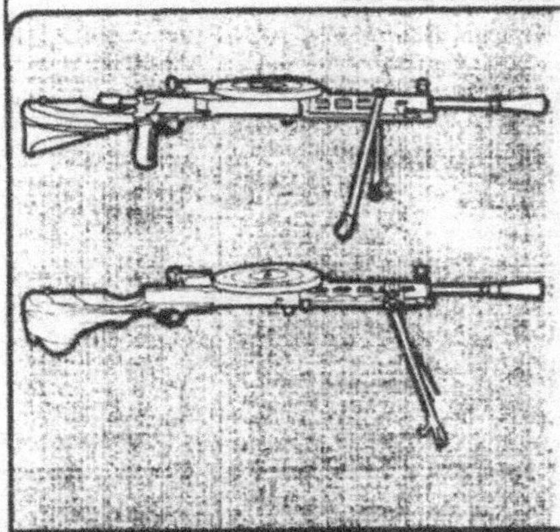

DPM AND DP LIGHT MACHINEGUNS

These machineguns can be recognized by the pan-type drum magazine on top of the receiver, slotted barrel casing, wooden stock with cheek rest, the fixed bipod, and flash suppressor. The DPM has a pistol grip and an operating-rod-spring housing projecting to the rear of the receiver. The DP has no pistol grip, and the operating-rod-spring housing does not project to the rear of the receiver.

CHARACTERISTICS OF DPM/DP LMG

Caliber and ammo	7.62 mm, M-1908
Cyclic rate of fire	550 rpm
Practical rate of fire	80 rpm
Magazine capacity	47 rounds
Effective range	800 m

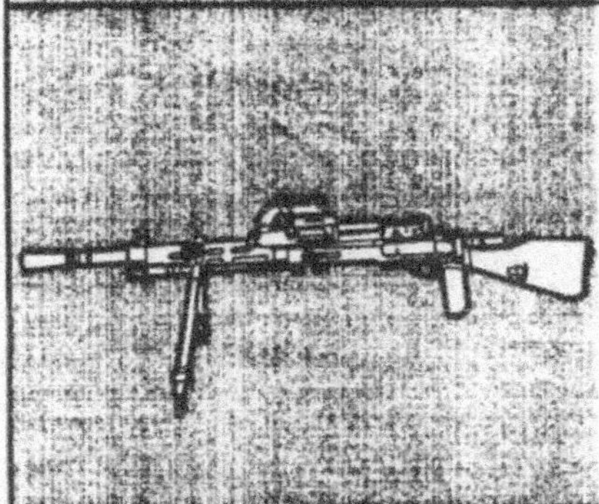

RP-46 LIGHT MACHINEGUN

The RP-46 was developed from the DP series. It has a detachable belt-feed mechanism that can be replaced by a DP type drum-feed mechanism. It is distinguishable from the DP and DPM by its carrying handle and heavier, thicker barrel.

CHARACTERISTICS OF RP-46

Caliber and ammo	7.62 mm, M-1980
Cyclic rate of fire	600 rpm
Practical rate of fire	230 to 250 rpm
Magazine capacity	200- or 250-round belt
Effective range	800 m

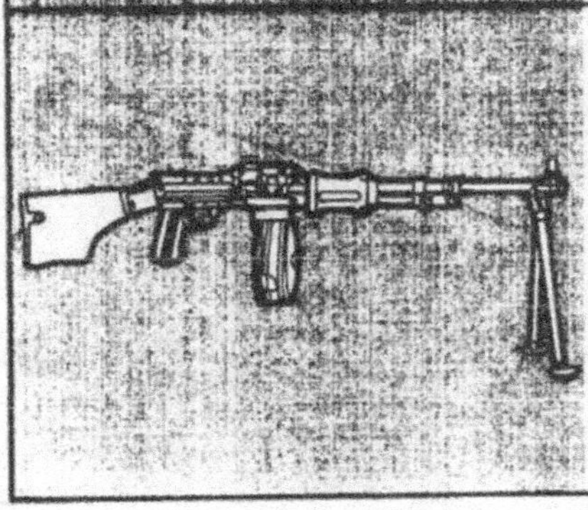

RPD LIGHT MACHINEGUN

The RPD uses the same ammunition as the AK-47 rifle. It has a drum (which houses a metallic link belt), a pistol grip, a permanently attached bipod, and an upper hand-guard which is flared at the ends. The RPD can fire full automatic only and has a chrome-plated barrel.

CHARACTERISTICS OF RPD

Caliber and ammo	7.62 mm, M-1943
Cyclic rate of fire	650 rpm
Practical rate of fire	150 rpm
Magazine capacity	100-round belt in drum
Effective range	800 m

RIFLE COMPANY

The rifle company usually has three rifle platoons and a mortar platoon with two or three 60-mm mortars.

60-MM MORTAR

CHARACTERISTICS OF 60-mm MORTAR

Length of tube	.724 m (28.5 in)
Weight	20.4 kg (45.0 lbs)
Elevation	40 to 85 degrees
Traverse	14 degrees
Maximum range	1,530 m

RIFLE BATTALION

The rifle battalion normally has three rifle companies. It may also have a heavy machinegun company, with 6 to 9 heavy machineguns, and a mortar company with 6 to 12 82-mm mortars.

DshK HEAVY MACHINEGUN

The DshK is used in a ground role, and on tanks and assault guns as an antiaircraft weapon. It has radial cooling fins on the barrel. In the ground role it has a wheeled mount and a shield.

CHARACTERISTICS OF DshK

Caliber	12.7 mm
Cyclic rate of fire	540 to 600 rpm
Practical rate of fire	125 rpm
Feed	50-round belt
Effective range; ground targets	1,500 m
Effective range; air targets	1,000 m

RIFLE BATTALION CONTINUED

82-MM MORTAR

There are three versions of the 82-mm mortar, but the M-1937 is by far the most common. All fire the same ammunition and can also use US 81-mm mortar rounds. (The US 81-mm mortar cannot fire the 82-mm ammunition.) The mortar is disassembled into three-pack loads for transport.

CHARACTERISTICS OF 82-MM MORTAR

Weight, travel position	56 kg (123 lbs)
Barrel length	1.22m (48.03 in)
Elevation	45 to 85 degrees
Traverse	6 degrees
Maximum range	3,040 m
Rate of fire	15 to 25 rpm
Projectile types	HE and smoke
Projectile weight:	
HE	3.3 kg (7.3 lbs)
Smoke	3.67 kg (8.0 lbs)

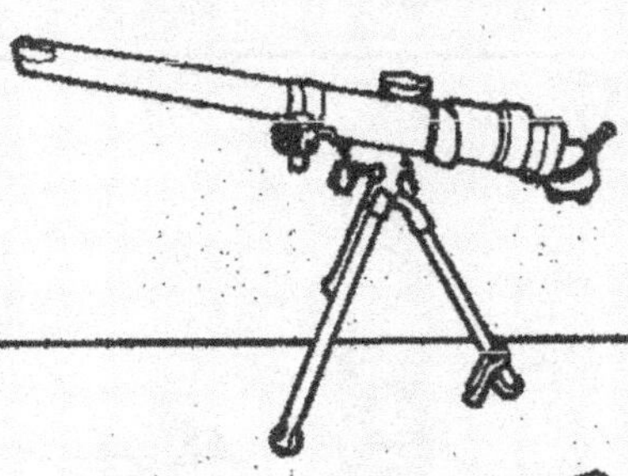

82-MM RECOILLESS RIFLE/TYPE 65

CHARACTERISTICS OF 82-MM RCLR

Weight	28.2 kg (62 lbs)
Caliber	82 mm
Effective range	450 m
Rate of fire	6 rpm
Ammunition	HEAT

75-MM RCLR

The 75-mm RCLR type 52 is a copy of the old US M20. The type 56 is a further development with the same performance as the type 52, but with a different mounting.

CHARACTERISTICS OF 75-MM RCLR

Length of tube:	
type 52	2.250 m (56.92 in)
type 56	2.280 m (58.16 in)
Weight, firing position:	
type 52	85 kg (187 lbs)
type 56	87 kg (190 lbs)
Rate of fire	10 rpm
Maximum range	6,675 m
Effective range	640 m

DIVISION AND HIGHER

Support from division and higher may include fire from 120-mm mortars, 76-mm guns, 122-mm howitzers, or 122-mm rockets.

120-MM MORTAR M-1938/1943

The 120-mm mortar in firing position looks like the 82-mm mortar but is much larger. It can be trigger-fired or drop-fired. An easily attached transport limber is provided for towing the mortar with a truck or armored personnel carrier. In addition to high explosive rounds, this mortar also fires smoke and incendiary rounds.

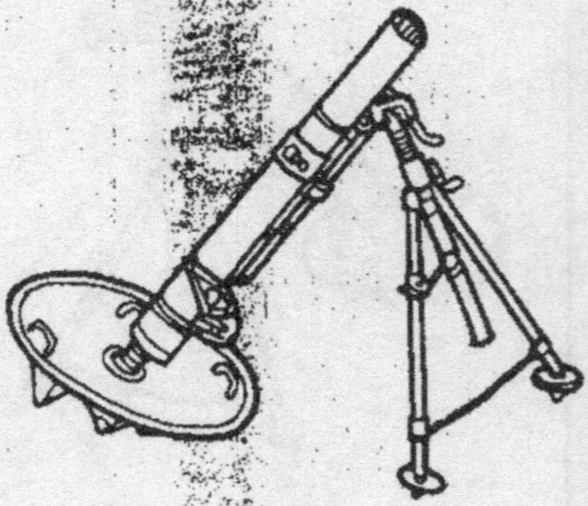

CHARACTERISTICS OF 120-MM MORTAR

Length of tube	1.854 m (51.66 in)
Weight in firing position	275 kg (605 lbs)
Elevation	65 to 80 degrees
Traverse	8 degrees
Maximum rate of fire	15 rpm
Maximum range	5,700 m
Weight of projectile: HE	15.4 kg (33.9lbs)

122-MM HOWITZER M-1938 (M-30)

This is a standard divisional artillery piece. The recoil mechanism is in a cradle below the tube, and the recuperator is above the tube.

CHARACTERISTICS OF 122-MM HOWITZER

Length of tube	2.8 m (9.2 ft)
Weight, firing position	2,500 kg (5,500 lbs)
Elevation	-3 to 63.5 degrees
Traverse	49 degrees
Rate of fire	5 to 6 rpm
Maximum range	11,800 m
Weight of projectile: HE	21.8 kg (48.0 lbs)

DIVISION AND HIGHER CONTINUED

122-MM SINGLE TUBE ROCKET LAUNCHER

The Soviet 122-mm rocket can be fired from a single tube launcher, consisting of a shortened BM-21 tube, mounted on a light tripod. The single-fired rocket has less range, but has the same warhead as the version fired from the multiple rocket launcher. The single-fired 122-mm rocket was used extensively in Vietnam; however, currently the BM-21 multiple rocket launcher is the most prevalent.

CHARACTERISTICS OF 122-MM ROCKET LAUNCHER

Length of tube	2.50 m (98 in)
Weight, tube/tripod	21.8/27.7 kg (48.0/60.9 lb)
Height, firing position	1.00 m (39.5 in)
Elevation	-4.5 to 42 degrees
Traverse	14 degrees
Maximum range	11,000 m

THREAT VEHICLES

TYPE 59

The type 59 tank is a Chinese version of the Soviet T-54. The major differences between the type 59 and the T-54 appear to be the type 59's lack of a gun stabilizer and infrared equipment, and the omission of power traverse. Both the gunner and loader have hand-traverse mechanisms. This results in a slow rate of engagement and difficulty in target identification on anything but a flat fire position.

PT-76

The PT-76 is the standard Soviet reconnaissance tank. The suspension of the PT-76 consists of six large, evenly spaced road wheels with no track return rollers. The drive sprocket is at the rear. The turret is circular and has flat sloping sides. The gun tube has a multibaffle or double-baffle muzzle brake. The PT-76 is amphibious. In the water, it is propelled by hydrojets which shoot water through ports in the rear of the tank. The gun fires the same ammunition as the 76-mm divisional gun M-1942 (ZIS-3).

CHAPTER 5

Tactical Operations

Section I. GENERAL

This chapter deals with tactical operations in the jungle. Before reading this material, soldiers should be familiar with the appropriate level of basic tactics in:

FM 7-8, *The Infantry Platoon and Squad (Infantry, Airborne, Air Assault, Ranger).*

FM 7-10, *The Infantry Rifle Company.*

FM 7-20, *The Infantry Battalion (Infantry, Airborne, Air Assault, Ranger).*

FM 90-4, *Airmobile Operations.*

FM 100-5, *Operations.*

Although jungle operations are conducted according to the basic guides contained in these manuals, there are special techniques which help to insure success in the jungle. These techniques result from the restricted maneuver, slow tempo, close combat, and limited visibility commonly found in the jungle.

Combat in the jungle is characterized by long periods of developing the situation and looking for the enemy; and short periods of violent, and sometimes unexpected, combat.

To meet these conditions, units must have:

- **Aggressive intelligence-gathering procedures**
- **Disciplined soldiers**
- **Solid SOPS proven in training and updated on a continuous basis**
- **Aggressive and tough-minded leadership**

These four points must be emphasized when a unit is engaged in jungle operations. The need for discipline is evident when one considers the extended periods of looking, often fruitlessly, for the enemy. When contact is made, maximum advantage can only be achieved through aggressive and violent action predicated upon solid SOPs. Aggressive leadership at the small-unit (squad and platoon) level is the one element that ties together the discipline and the training.

CHARACTERISTICS OF THE JUNGLE BATTLEFIELD

The thick foliage and rugged terrain of most jungles limit fields of fire and speed of movement.

CONTENTS

The following limitations may restrict fire and movement:

- Lack of line-of-sight and clearance may prevent visual contact between units, interlocking fires, and the use of tube-launched, optically-tracked, wire-guided missile (TOW) or Dragon missiles.
- Tree limbs may block mortars, flame weapons, 40-mm grenades, and hand grenades.
- Machineguns may not be able to attain grazing fire.
- Adjustment of indirect fire support is difficult due to limited visibility and may have to be accomplished by sound.
- Noise conditions differ in the jungle. There are large numbers of animals in jungle areas, and their noise (or lack of it) can give an indication of something out of the norm.
- Sounds in the jungle do not carry as far as on the conventional battlefield due to the amount of jungle foliage. The result is that noises are closer than first believed.
- Movement through jungle areas is also very difficult because:

  Heat, thick vegetation, and rugged terrain will tire troops rapidly, especially those carrying heavy weapons or radios.

  A lack of roads will hinder resupply and evacuation.

These terrain characteristics make jungle fighting different from fighting on more open terrain. To be effective jungle fighters, soldiers must learn to use these characteristics to their advantage. Potential jungle enemies train to exploit the jungle; so must the US Army.

CHARACTERISTICS OF JUNGLE OPERATIONS

The aspects of terrain and enemy discussed above result in fewer set-piece battles. Rather than conventional attacks conducted against conventional defenses, jungle battles are more often ambushes, raids, and meeting engagements. Battles are not fought for high ground as frequently as conventional battles. *Orientation is on the enemy rather than on the terrain.* Hills in the jungle are often too thickly vegetated to permit observation and fire, and therefore do not always qualify as key terrain. In the jungle, roads, rivers and streams, fording sites, and landing zones are more likely to be key terrain features.

The frequency of ambushes, raids, and meeting engagements makes it very important that units in the jungle practice immediate action drills. In the jungle firefight, the side which initiates contact and gains fire superiority in the first few seconds will normally have a decisive advantage.

CONTROL

Command and control are difficult in the jungle. The thick foliage allows leaders to see and control only a portion of their units.

To cope with this problem, commanders and leaders must:

- Plan their operations carefully
- Issue mission type orders
- Insure that each soldier understands his part of the mission

In addition, the thick jungle foliage and heavy monsoon rains often weaken radio signals, making communications difficult. To reduce the effects of the problem, use of the helicopter as a command and control vehicle is recommended. In that the heavy monsoon rains may not allow helicopters to always fly, an alternate means of command and control must be planned for.

FLEXIBILITY

While an appreciation of battlefield characteristics, jungle enemies, and characteristics of jungle operations is useful, flexibility is important to any leader involved in jungle operations. Successful operations require an extraordinary command adaptability—sometimes, a departure from orthodox thinking in favor of new and often untried procedures. Soldiers must learn to live *with* the jungle and adapt to its initially apparent disadvantages. Having done this, the unit can concentrate on the use of concealment, covered movement, and surprise.

SECURITY AND INTELLIGENCE

Commanders must stress effective security measures and aggressive intelligence-gathering techniques to prevent being surprised. The key is to give the front-line soldier an appreciation of the things to look for. Food remnants and feces can indicate how long ago an enemy unit occupied an area. Captured documents, equipment, and weapons may provide order of battle information and an idea of the enemy's logistical situation. Even an ammunition crate may yield a lot number and packing date. From this an intelligence specialist may be able to trace the enemy unit's place in the order of battle.

In the past, US forces operating in jungle warfare have generally been augmented by native scouts, attached down to platoon level. These scouts were auxiliaries, paid by the unit they supported from a fund established by higher headquarters for that purpose. Scouts familiar with the terrain and the enemy can be an extremely valuable asset. Local security regulations should provide guidance as to what friendly information can be given to scouts.

Surveillance, target acquisition, and night observation (STANO) devices, especially infrared, starlight scopes, and unattended ground sensors, are quite effective in gathering information about troop movements in the jungle. Radars and photography are not as effective because of the concealment of the foliage.

The local populace is one of the most valuable intelligence sources. Whether hostile, friendly, or indifferent, the people can provide information which, when processed, will help complete the intelligence picture.

JUNGLE INTELLIGENCE

Security prevents the enemy from gaining intelligence on US units. Active security measures, such as patrolling and the use of observation posts (OP), helps prevent US units from being ambushed or attacked by surprise. These measures do not lessen the need for passive security. Camouflage and noise and light discipline conceal US forces from enemy observation. To prevent being tracked by the enemy, bivouacs and trails must be policed. Odor discipline is also a security measure. The enemy can follow such odors as heat tabs, cigarette smoke, deodorant, and C rations.

TROOP-LEADING PROCEDURES FOR JUNGLE OPERATIONS

STANDING OPERATING PROCEDURES

A unit's jungle operations SOP should include actions which the unit does on a routine basis or actions that are earned out essentially the same way each time they are done. Examples of such actions include organizing for combat, resupply, bivouac and shelter preparation, movement techniques, and battle drill. Use of SOPs will save planning time.

PLANNING USE OF TIME

A unit planning for jungle combat follows the same planning sequence as in any other type of combat operation. In planning the use of available time, leaders must consider that many tasks in the jungle take more time than the same tasks in other environments. More time must be allowed for movement and security. This means that units may have to begin movements earlier in order to accomplish their missions within a specified time. This may leave less time for planning and preparation.

AIRMOVEMENT PREPARATION

If the unit is to be inserted by helicopter, the commander should go through the preparatory steps outlined in *FM 90-4*.

INSPECTION

Prior to beginning a misssion, unit leaders should inspect their troops to insure that:

- **They have all their needed equipment**
- **They have no unnecessary equipment**
- **Weapons are cleaned, lubricated, and zeroed**
- **Equipment and weapons are in working order**
- **Everyone understands his job and the unit's SOP**

CONTROL OF EQUIPMENT

Those troops who carry extra equipment should not be allowed to discard it. Captured US equipment has been used by jungle enemies in the past. In Southeast Asia and in the Pacific, recovered equipment was a major source of enemy supply.

SUPERVISION

Supervision must continue throughout the conduct of the operation. As the troops become tired after long periods of marching or digging in, they will tend to get lax and ignore good security habits. This is an especially common trend if they have not been in contact for a few days. Tight supervision is a must to insure that security patrols and OPs are dispatched and doing their jobs; that troops remain alert; and that fire, noise, and light discipline are not relaxed.

JUNGLE MOVEMENT

PLANNING AND ROUTE SELECTION

Before conducting a move in the jungle, leaders should make a map and aerial photograph reconnaissance. This reconnaissance will indicate possible danger areas, obstacles, and roads or clearings suitable for resupply.

In planning the route, leaders should consider the following:

- **Lines of drift, such as ridgelines, are easy to guide on because they avoid streams and gullies and because they are usually less vegetated.**
- **Danger areas, such as streambeds and draws, are usually more thickly vegetated. They offer excellent concealment, but travel along them is slow and difficult.**
- **Roads and trails should be avoided. Although they are easy to move on, they offer little concealment. These are the areas most likely to be under enemy observation. They are easy to ambush and are very likely to be mined or boobytrapped.**

MOVEMENT TECHNIQUES

Units moving in the jungle should normally use the jungle movement technique, but may use traveling overwatch and bounding overwatch when necessary. The file formation should be avoided in all but the most thickly vegetated areas.

To effectively use the jungle movement technique, certain key factors must be understood. They include the following:

- **Only the platoon should employ this movement technique.**
- **The lead fire team of the lead squad is always in a wedge (modified).**
- **The support elements may move with the headquarters element or be attached to a squad(s) depending upon likely threats.**
- **Each squad maintains an azimuth and pace.**
- **Immediate action drill (SOP) is essential.**
- **This technique is most effective during daylight movement.**

This movement technique is basically characterized as a formation of multiple columns which are mutually supporting.

Advantages of the jungle movement technique:

- Centralized control.
- Rapid deployment to maneuver or reinforce.
- Ease of movement (three routes).
- 360-degree security during movement and at halts.
- Multiple navigational aids (three azimuths and pace counts).
- Flexibility of adjustment during movement (danger areas, choke points).

Disadvantages of the jungle movement technique:

- Possibility of loss of contact at major obstacles due to multiple routes.
- Vulnerability to effectiveness of indirect fire weapons.

JUNGLE MOVEMENT TECHNIQUE FORMATION

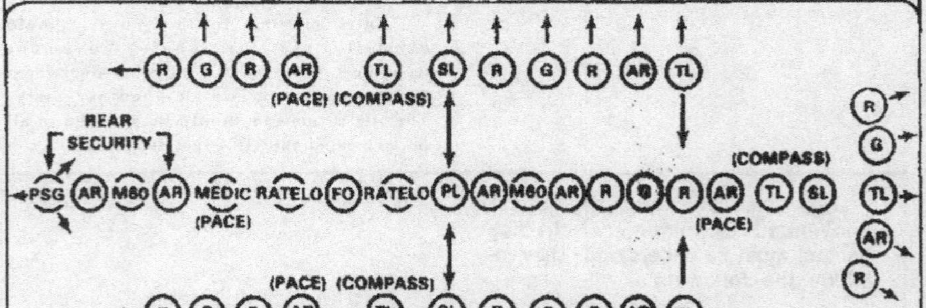

Note the positions of the platoon leader and adjacent squad leaders during movement. It is easily visualized how this formation can "balloon" its flanks at halts to form a hasty perimeter. Also, note the locations of the machinegun teams that allow for rapid attachment. The formation shown depicts a 45-man platoon; however, it can be effectively used with lesser numbers.

In traveling overwatch, the lead element performs the mission of point security, with troops from the rest of the unit performing rear and flank security. When contact is imminent, the unit moves into bounding overwatch. Bounds, as terrain allows, are normally 50 meters or less.

SECURITY

The thick foliage makes ambush a constant danger. Point, flank, and rear security teams will help keep a force from being ambushed. These teams must be far enough away from the main body that if they make contact the whole force will not be engaged. They should not be so far away, however, that they cannot be supported. These security elements must be alert to signs of the enemy, and should carry as light a load as possible so they are able to maneuver. Security duties should be rotated often to avoid fatigue. Scout dogs may also be used with the security element. These dogs often detect the enemy before he is detected by humans. The jungle heat is hard on them, however, and they must be rested frequently.

If contact is broken between elements or individuals, the rear element should remain in position. Those in front should return to establish contact.

HALTS

Units should plan halts on terrain which lends itself to all-round defense. During short halts, soldiers drop to one knee and face outward, their weapons at the ready. If the halt occurs at a trail crossing, security elements are sent out along the trail. The security element remains in place until the unit clears the crossing. During longer halts, units establish a perimeter defense. They run security patrols around their positions, and employ Claymore mines and early warning devices. Before an overnight halt, units should stop while there is still enough daylight to establish a secure perimeter defense, prepare ambushes, and dispatch patrols as necessary. If halted units are separated, connecting patrols should be run periodically to detect enemy infiltration.

Section II. RECONNAISSANCE, SURVEILLANCE, AND SECURITY OPERATIONS

RECONNAISSANCE

Reconnaissance operations are always important in jungle warfare. Many offensive operations in the jungle take on the aspects of a reconnaissance operation during their early stages. This is because the success of offense in the jungle depends on ability to find the enemy. The excellent concealment found in the jungle enables the enemy to operate unobserved both by day and night. Extensive patrolling is necessary to obtain information on his locations, strength, and disposition.

Reconnaissance is a responsibility of all leaders during jungle operations. Units with the capability to conduct reconnaissance should conduct frequent short patrols during the conduct of normal missions. These patrols should be coordinated with higher and adjacent units. In some situations, reconnaissance may become the primary objective of a major jungle operation.

Planning for a jungle reconnaissance should be thorough and well coordinated. Coordination with higher and adjacent headquarters will help insure maximum results from each patrol and eliminate duplication of effort. Radio is the primary means used to control reconnaissance operations. Each reconnaissance patrol must be prepared to make contact, develop the situation, and report to its controlling headquarters.

GROUND RECONNAISSANCE

In the jungle, these operations are accomplished by means of OPs and long- or short-range reconnaissance patrols. The value of OPs is somewhat reduced in the jungle because of the limited visibility. OPs are most effective when used along trails, roads, and streams to detect enemy movement. (Although OPs in other areas may not provide much useful information on the enemy, they still are effective in providing early warning.) Reconnaissance patrols in the jungle are normally squad-size. These patrols move in a manner to take advantage of natural concealment, and avoid becoming engaged with the enemy. The use of helicopters increases the depth behind enemy lines that such patrols can be employed. For further details on reconnaissance patrols, *see section VI.*

VISUAL RECONNAISSANCE

From the air, key terrain features can often be identified and the enemy detected in areas where there are gaps in the jungle canopy. Aerial photographs are important sources of information because photograph interpretation can disclose hidden enemy camps not visible to the air observer. Photographs can also be used to locate helicopter landing zones. Decoy reconnaissance flights can be used to confuse or deceive the enemy about upcoming operations.

RECONNAISSANCE-IN-FORCE

This method may be used to gather intelligence which cannot be gained by any other reconnaissance means. Examples of such intelligence include the enemy commander's plan for committing his reserves, or the trails used by a guerrilla force.

The commander conducting the reconnaissance-in-force will normally organize his unit into a number of reconnoitering forces. These reconnoitering forces conduct movements to contact, hasty or deliberate attacks, raids, reconnaissance, or patrols. Once the unit makes contact with the enemy, the commander must react on the intelligence gained. He must be prepared to exploit success or, if necessary, extricate the forces.

The size of the unit that conducts a reconnaissance-in-force depends on the nature of the intelligence to be gained and the chance that the reconnoitering force will have to fight on unfavorable terms. For example, if a battalion commander wants to find out how an enemy commander will commit his reserve, he may conduct a reconnaissance-in-force with his companies conducting limited objective attacks. If, on the other hand, a commander wants to find the routes used by a number of small guerrilla groups, the reconnaissance-in-force mission may be assigned to a company, which in turn will have its platoons conduct movements to contact.

RECONNAISSANCE-BY-FIRE

When using this method, the force fires on suspected enemy positions to cause the enemy to disclose his position by moving or returning fire. Reconnaissance-by-fire risks the loss of surprise. Its most effective use in the jungle is to find the flanks or gaps in enemy lines. Reconnaissance-by-fire from attack helicopters will often reveal the location of well-concealed enemy troops. Likewise, a reconnaissance-by-fire from armored vehicles firing into a wood line, either while moving crosscountry or along a road, can neutralize an enemy ambush. When using reconnaissance-by-fire, commanders must consider the difficulties of ammunition resupply in the jungle.

SURVEILLANCE

Surveillance operations in the jungle include using all techniques for establishing a continuous, thorough watch of the battlefield. This watch must be established both over large jungle areas and at selected key points such as trails, streams, and clearings. Surveillance operations are usually planned to support other missions.

GROUND SURVEILLANCE RADARS

These radars are best employed in those jungle areas where vegetation and terrain do not restrict line of sight. Night observation devices are also useful in such areas during periods of darkness. Unattended ground sensors, which are not affected by poor line of sight, are very useful in watching specific key areas. These electronic devices are affected by poor weather and are difficult to move in thickly forested areas. As a result, the use of manned OPs in jungle areas should always be planned.

AIR SURVEILLANCE

Surveillance of jungle areas from the air is most effective when pilots are familiar with ground operations and can recognize changes from normal patterns. Repeated flights by the same crews will attain this level of familiarity. In addition to visual surveillance, photographic coverage of an area can assist the surveillance effort. Side-looking airborne radar (SLAR) is not very effective in thick foliage, but can be used for surveillance along roads, trails, or streams. Likewise, infrared detection devices are limited by fog, clouds, rain, and vegetation. Airborne personnel detector devices (sniffers) were developed during the Vietnam war to detect human odors. These devices are extremely effective in detecting base camps of nonmechanized forces, but are limited by fog, rain, and windy conditions.

SECURITY OPERATIONS

Security must be a primary part of all jungle operations; therefore, specific security measures are covered as they apply to other operations throughout this chapter. Compared with operations in other types of terrain, security measures in the jungle must be intensified because of the poor observation and difficulties of control and movement. Operations must be slower than normal, and security forces must be closer to the units secured in order to provide adequate security in the jungle. Because it provides all-round security, the perimeter defense will be the defensive technique used most often by units operating independently in the jungle.

The "stand-to" is an important security technique in jungle fighting. When a unit stands-to, all of its soldiers don their fighting loads and occupy their fighting positions. The unit is 100 percent alert and ready to fight an attacking enemy. Stand-to procedures differ from unit to unit, but common stand-to times are before first light, before last light, before helicopter resupply, and before movement.

Section III. OFFENSIVE OPERATIONS.

SPECIAL FACTORS

The purpose and fundamentals of the offense as outlined in field manuals for other environments generally apply as well to offensive operations in the jungle.

> **There are, however, factors which require the use of special offensive techniques:**
>
> - **Thick foliage makes it difficult for leaders to control their soldiers or to detect the enemy**
> - **Fire support is difficult to observe and adjust**
> - **Momentum and speed are difficult to maintain**

FUNDAMENTALS OF THE JUNGLE OFFENSE

When considering the use of special offensive techniques, commanders must remember that some offensive fundamentals acquire a new significance in the jungle.

Probably the most important and most difficult of these fundamentals is the requirement to *see the battlefield.* Above all else, the attacker must know the battlefield. As a result, he relies heavily on security patrols, information provided by air and ground reconnaissance, and proper movement techniques.

Key to effective operations in jungle warfare is the fundamental of using weapon systems to their best advantage. In addition to organic weapons, the ground commander must closely coordinate the employment of the supporting weapons available to him. The dense foliage found in some jungles may prevent heavy weapons from moving directly with the infantry. In those cases, TACAIR support and helicopter weapons must make up the difference. TOWs and Dragons, on the other hand, are of limited use in most jungle environments. The soldiers that man these weapons may be more effective as security forces or as reinforcements for maneuver elements. The primary jungle weapons are individual infantry small arms, supported by machineguns and mortars.

To *concentrate overwhelming combat power* against enemy weakness in jungle operations, the attacker must be able to bring up other elements quickly to support an element that is engaged. In determining how far he can separate his subordinate units, the commander must consider the factors of mission, enemy, terrain and weather, and troops and time available (METT). The separation may be expressed in time or in distance. The ability to provide mutual support must not be overlooked. Gaps between units should be covered by scouts and connecting patrols. Targets should be planned along the unit's route so that supporting fires can be responsive. The use of helicopters will permit even more rapid concentration of forces and provide additional firepower *(chap 6).*

The jungle also increases the difficulty of efforts to *provide continuous support.* The key is constant planning, coordination, and maximum use of helicopters. Fires must be planned along the attack route so that they can be delivered in the shortest amount of time. Procedures for calling attack helicopters must be standardized and rehearsed. Likewise, combat service support must be timely and responsive. Ammunition and water must be loaded on pallets in the trains so that they can be brought forward as soon as needed.

CONDUCT OF THE JUNGLE OFFENSE

Because it is so hard to gather intelligence, jungle offensive tactics must be characterized by continuous reconnaissance. A unit attacking without timely information on the location of the enemy may subject its elements to enemy ambush without being able to support them. In such a situation, they may be defeated in detail.

"It is a situation that too frequently occurs in the Vietnam fighting. The forward element, losing men and becoming pinned down, compromises the position of all others. What has started out as an attack loses all form and deteriorates into a costly rescue act."

-S.L.A. Marshall, BIRD

Successful jungle attacks usually combine dispersion and concentration. For example, a rifle company may move out in a dispersed formation so that it can find the enemy. Once contact is made, its platoons close on the enemy from all directions. In this way, they move to support each other and destroy the enemy.

Operations are enemy-oriented, not terrain-oriented. Wherever the enemy is found, that is where he should be destroyed. If he is allowed to escape, he will only have to be found again, with all the risks involved.

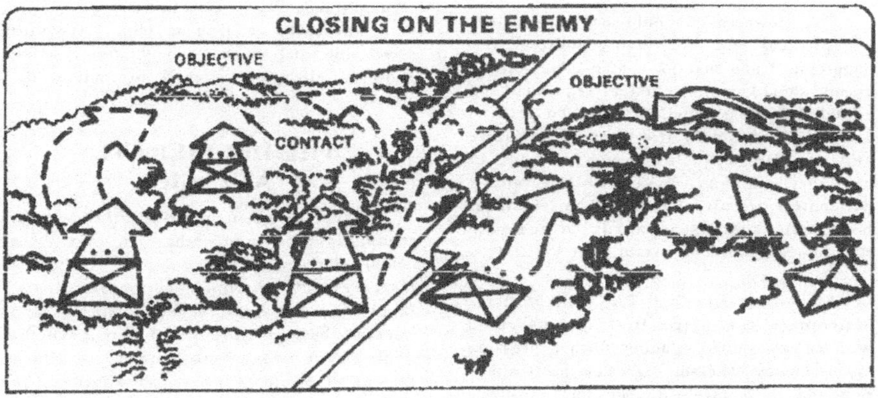

MOVEMENT TO CONTACT AND HASTY ATTACK

These two forms of combat are discussed together since they account for the majority of combat actions in the jungle. The successful follow-on action to movement to contact is a violently executed hasty attack. This action capitalizes on the advantage of surprise and the confusion prevalent in the jungle. Movement to contact in the jungle differs little from the general concept described in FM 7-20. The decision to employ single or multiple columns may depend solely on terrain and foliage considerations.

For companies and battalions, multiple columns are a sound movement practice because this formation provides more firepower to the front and because it is easier to deploy troops from two or three columns than from one file. In order to speed up deployment, units should develop and rehearse SOPs and immediate action drills. Troops should try to maintain a distance of five to seven paces between men, but must maintain visual contact. All-round defense and security measures must be maintained throughout movement.

Once contact with the enemy is made, the unit's first action is to build up a large volume of fire. The commander then assesses the situation and deploys his unit to overrun the enemy's positions while they are still suppressed. In this way, he seizes the initiative. There should be no delay in the troops' movement from the march formation into assault formation. Security elements protect the rear and prevent the enemy's counterattack. These forces may be used later to exploit a success, but should not be committed until the commander understands the situation.

The slowness of jungle maneuver makes a rapid call for supporting indirect fire important. Upon making contact, fires on the enemy should be immediately requested and adjusted from planned targets. To receive effective and timely fire support, accurate and continuous land navigation is necessary. Means for controlling attack helicopters are also important; this includes both radio and visual means. Adjustment techniques should be established by SOP.

Supporting fires and TACALR or attack helicopters can place fires on suspected withdrawal routes, placing further pressure on the enemy. The success of the hasty attack depends to a large degree on the unit's vigorous execution of unit SOPs and the leadership of the squad and platoon leaders.

As the situation is developed and an enemy position is located, a violent assault should be made over the enemy's position. Soldiers stay on the alert for hidden enemy positions, snipers in the trees, and tunnels through which the enemy might move to attack the attacker's rear. This thorough technique will also provide enough information of the enemy and security to permit the commander to use his reserve force for exploitation, if needed.

The assault should be made using fire and maneuver. Soldiers should cover each other, moving by crawls and short rushes. Fire should be well-aimed shots and short bursts of automatic fire.

In such a fast-moving situation, it might be possible, for example, for a platoon to receive a fragmentary order (FRAGO), move to a pickup zone (PZ), and conduct an air assault to an objective. In this case, the planning might take place on the PZ or even in the vicinity of the objective rally point (ORP).

After the objective is seized, it must be secured immediately with a hasty perimeter, OPs, and early warning devices. This is to detect and repel an enemy counterattack or to allow the attacker to prepare to continue the attack.

THE DELIBERATE ATTACK

Based on information gained from reconnaissance and other sources, the commander may formulate a plan to attack a larger objective using a deliberate attack. Jungle terrain favors reduced distances and intervals between troops and units, and the deployment from movement to attack formation as far forward as possible. *In thickly vegetated terrain, the use of some of the same control techniques used in a night attack may be required.*

Fire support is as essential in the jungle as in other types of terrain. Unsupported troops are likely to incur heavy casualties when attacking jungle positions, especially considering the difficulties of employing organic weapons. Targets must be pinpointed by reconnaissance, and fires must be adjusted within very close range of attacking troops. During the assault, these supporting fires must continue until shifted by the assaulting commander. They are then adjusted onto targets which will assist the progress of attacking forces by blocking enemy counterattacks or withdrawal. Due to poor observation, indirect fire may have to be adjusted by sound *(see app I)*.

Assaulting troops move over the objective using aggressive fire and movement to overcome enemy resistance. Assaulting troops again must be alert to snipers, mines and boobytraps, hidden positions, and tunnels which would permit the enemy to maneuver into the rear of attacking forces. Assaulting platoons and squads move in a single direction, with fires concentrated on enemy positions as they are located. Attacking elements must adjust their progress using base elements and phase lines. Smoke may be used to screen the flanks of the penetration from enemy observation and reduce his ability to deliver effective fires. Once an initial penetration is secured, it is exploited until the objective is taken.

After the objective is overrun, it must be secured immediately with a hasty perimeter, OPs, and early warning devices to detect and repel an enemy counterattack or to allow the attacker to prepare to continue the attack *(see IV)*.

INFILTRATION

Jungle areas are ideal for infiltration. Dense vegetation and rugged terrain limit the enemy's ability to detect movement. As a technique to move through the enemy's positions, infiltration can be used with other offensive maneuvers to gain an advantage in the jungle. Although jungle infiltrations are normally conducted on foot, under certain circumstances helicopters or watercraft may be used.

Infiltrations are normally difficult to control. Chances for success are better if troops are well trained, well briefed, and well rehearsed. Roads, trails, and streams should be avoided because they will normally be under enemy surveillance. Movement by stealth is normally slow and exhausting. Phase lines (PL), infiltration routes, and adequate communications must be used to control the operation and to coordinate fires with movement.

CONTROL OF AN INFILTRATION

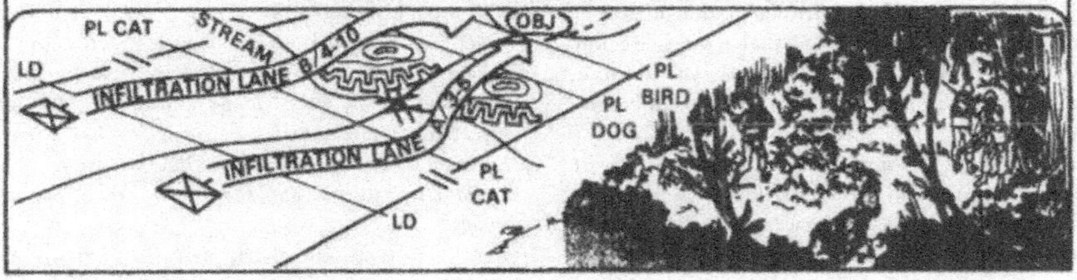

EXPLOITATION AND PURSUIT

Local successes should be exploited as soon as possible to cut off the retreat of isolated enemy forces. Airmobile troops are most effective to block enemy retreat in the jungle. They can also disrupt and harass enemy reserve, logistical, and command operations. Artillery, TACAIR support, and attack helicopters may also be used to block escape routes. Tanks may be used along trails or roads or in less dense areas if properly secured with infantry. During exploitations, rear areas must be secured against the actions of bypassed or infiltrating enemy.

Pursuit operations in the jungle should be conducted to maintain contact with the enemy. The precautions required to secure against ambush can slow pursuit operations considerably. Because attacking troops become more fatigued than defending troops, pursuit operations should be conducted using troops from the reserve. As is the case with most jungle offensive operations, airmobile forces, air cavalry, and attack helicopters can be used very effectively in pursuit operations.

ATTACK AGAINST A FORTIFIED POSITION

As mentioned earlier, jungle enemies have often used strongly fortified defensive positions to protect themselves from the effects of US firepower. If it is necessary to attack such a position, troops will probably encounter bunkers, barbed wire, mines, and boobytraps. The enemy will often have to be burned or blasted out of such positions. These operations will require attacks on a narrow front, great amounts of firepower, and limited objective attacks.

Fortified enemy positions in dense jungle are often so well concealed that troops are not aware of their presence until they have physically encountered them. In these cases, the best course of action is usually to adjust forward dispositions enough to allow use of supporting indirect fires, to deploy additional forces to block possible withdrawal routes, and then to maneuver under the cover of supporting fires to defeat the fortifications in detail.

Maximum use of combat intelligence is required when attacking fortified areas. Aerial photographs, electronic intelligence, interrogation of prisoners of war (PW), and aggressive patrolling are all means of gaining the required information. Patrols also keep the enemy off balance and limit the enemy's ability to patrol.

In addition, combat engineer, and special weapons (such as flame) and equipment should be attached to the maneuver forces to assist in destroying the fortifications.

Destruction of the enemy in those types of positions takes a lot of time and effort. Other means should be used as much as possible. Here is a technique that proved successful in the past:

"The use of CS riot control gas could not be overlooked... On one occasion, a battalion made a night attack with gas masks following an aerial CS attack. A helicopter made several low passes on the windward side of the area and dispersed about 250 CS grenades. This was followed by 20 minutes of artillery fire, about half of which was VT fuze fired into the enemy positions. Behind a walking barrage of artillery fire, one company assaulted, and, once inside the objective, flareships lit up the area. Eighteen enemy were killed, while no casualties were suffered by the friendly troops."

—Report, 25th Infantry Division, Duang Nhgai Province, Republic of Vietnam

Section IV. DEFENSIVE OPERATIONS

SPECIAL FACTORS

The purpose and fundamentals of the defense as outlined in field manuals for other environments also apply to defensive operations in the jungle.

There are, however, certain factors which require the use of special techniques:

- **Thick foliage makes it difficult to detect the approach of an attacking enemy**
- **Slowness of jungle movement makes it difficult to react to an enemy threat**
- **Limited visibility between defensive positions**
- **Limited fields of fire**
- **Psychological impact of fighting in a strange environment**

FUNDAMENTALS OF THE JUNGLE DEFENSE

As in the offense, jungle defensive operations are based on the same fundamentals used in other area operations. Some of these fundamentals acquire a special significance in the jungle.

To succeed in the jungle defense, a commander must *understand the enemy and see the battlefield.* The enemy will probably be expert in using the environment to his advantage, and the defender must understand enemy techniques. To counter the threat of infiltration, the defender must employ all-round defense and all surveillance means available. No amount of electronic means can eliminate the need for frequent patrolling.

The defender must *exploit every advantage* that he has, particularly the abundant concealment provided by the foliage and the weather. A force which remains concealed may disrupt an enemy's attack by using surprise fire from hidden locations. Though not as common as in other types of terrain, features which lend themselves to the defense, such as rivers, gorges, and ridges, should be used if they dominate likely avenues of approach. The defender must appreciate the defensive characteristics of the terrain and environment.

The defender must *maximize the effectiveness of key weapons,* which in the jungle are infantry small arms, mortars, and artillery. The poor trafficability also increases the importance of attack helicopters and TACAIR.

In planning to *concentrate combat power at critical times and places,* the defender must first plan the massing of small-arms fire. Since fields-of-fire will be limited, positions must be placed close together for mutual support. To move troops and weapons rapidly to supplementary or alternate positions, it maybe necessary to cut paths through the bush. Units should rehearse these maneuvers.

As in the offense, the problem of *providing continuous support* is to a large degree solved by effective communications and the use of helicopters.

PLAN OF THE JUNGLE DEFENSE

Planning for the jungle defense should provide for a covering force area, a main battle area, and a rear area. Forces in each area must be provided fire support. Obstacles are planned to improve the natural defensive strength of the terrain. Plans are also formulated for counterattacks. The following factors should be considered when planning for the jungle defense:

Day and night in jungle regions are each roughly 12 hours long. Nights, especially under jungle canopies, are extremely dark. Defensive preparations should begin at least 2 hours before nightfall.

The heat and humidity will fatigue troops rapidly.

Tropical rain will flood positions unless they are adequately drained. During the rainy season, defensive positions should be dug on high ground, if possible.

Because jungle terrain favors infiltration, the use of starlight scopes, OPs, and early warning devices is very important.

After the commander organizes the ground and secures the area to be defended, he then positions his Dragons, if they can be used effectively, and machineguns and clears fields of fire. Leaders must insure that troops do not cut too much vegetation. In order to be mutually supporting, positions will be closer together than on other types of terrain. In addition to attaining mutual support, this helps prevent enemy infiltration. Mines and obstacles should be emplaced where they are covered by friendly fires. These should be located beyond hand grenade throwing range of the defensive positions.

PLANNING THE DEFENSE IN JUNGLE TERRAIN

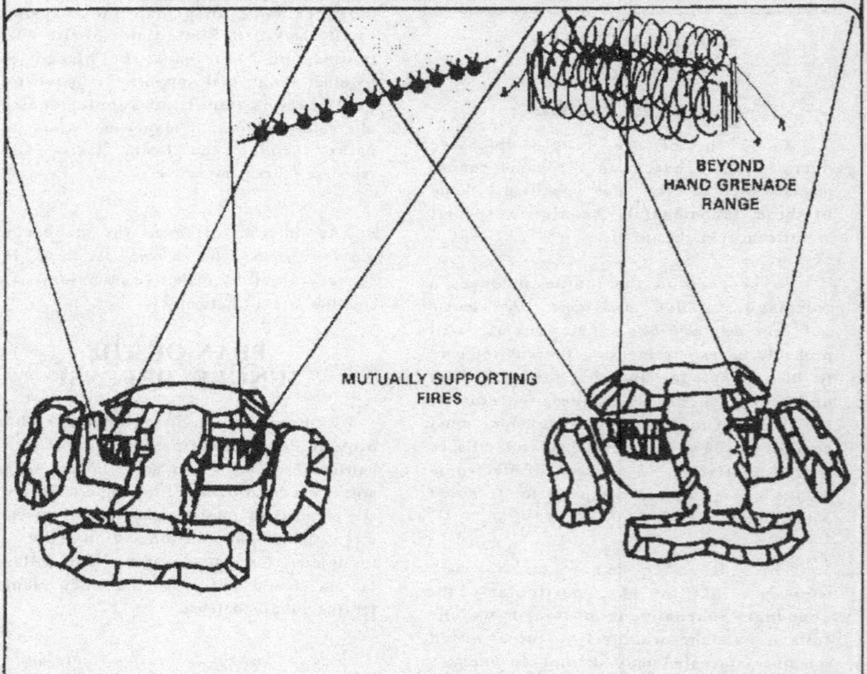

SECURITY

Since enemy tactics, jungle terrain, and bad weather favor attacks conducted by stealth, security should be the leader's first concern. OPs, early warning devices, ambushes, and patrols are all measures which will prevent a unit from being surprised. Patrols must be planned according to an irregular schedule. A system of recognition signals must be used to prevent the engagement of friendly units.

NOTE: If mechanical ambushes are used, units should stop patrolling in that area, and should provide some means, such as communications wire, to guide OP personnel and prevent them from straying into the mechanical ambush.

DEFENSIVE FORMATIONS

The basic defensive technique in the jungle is the perimeter defense. Two other very effective defensive techniques for jungle operations are the triangle and "Y" formations *(see app J)*. Whether a unit is operating independently or as part of a larger defensive position, it must be prepared to defend itself against an attack from any direction. Initially, these formations will be formed by platoons or by companies. Larger units should position their companies in depth to provide all-round defense. Alert and aggressive patrols and OPs will defeat enemy attempts to infiltrate between positions. Later, if time permits, platoon and company formations can be connected with fighting positions and trenches. Even then, however, companies and platoons must be prepared for all-round defense.

PRIORITY OF WORK

Units in the defense must pay particular attention to their priority of work. Since more security measures must be taken than normal, fewer troops will be available to prepare defensive positions at any one time. Positions should be prepared and camouflaged as in any other situation. Overhead cover should be prepared using strong wood and sandbags. Claymore mines and trip flares should be emplaced in front of the defensive positions. Fields of fire should be cut low, leaving enough foliage so as not to reveal the location of the defensive position *(app E)*.

To counter enemy reconnaissance efforts, units should shift the positions of machineguns after dark. After a few days, the entire unit's position should be changed. If a unit remains in position for a longer period of time, it is more likely to be reconnoitered by the enemy and subject to attack.

COMMAND AND CONTROL

Command and control are extremely difficult in the jungle defense. Commanders must place great emphasis on planning, coordination, and small-unit leadership. Decentralized control is important to insure that subordinate units can react to multiple threats. Aggressive leadership at the small-unit level is necessary in fighting off isolated assaults at close range. Alternate communications means must be established wherever possible. An example might be a communications system using wire as the primary means, radio as the secondary means, and pyrotechnics for certain prearranged signals.

Defensive targets for artillery and mortars should be planned on stream and trail junctions, and any other likely enemy avenues of approach. Artillery and mortar fire should also be used to cover the many areas of dead space found in jungle terrain. It is also a good idea to confirm the location of the defense on the ground by using artillery marking missions integrated into registration missions. Signals for the employment of TACAIR and attack helicopters, both day and night, and for medevac and resupply helicopters must also be planned.

CONDUCT OF THE JUNGLE DEFENSE

There are occasions when a unit will have to establish a defense with minimum planning time. This normally occurs when an attack is stalled, at dusk when the unit is still in contact with the enemy and no night attack is planned, or when an intermediate objective must be secured before continuing the attack. These situations are more dangerous in the jungle than in other areas because of the dense foliage and the closeness of the enemy.

The normal course of action in these cases is to establish a perimeter defense. Dragons, if they can be used effectively, and machineguns are positioned immediately where they have the best fields of fire. As soon as possible, OPs and other local security measures are established. Frontages are smaller than in other types of terrain, especially at night, to guard against enemy infiltration. Indirect fires are registered and fighting positions are dug as soon as possible.

Once these actions have been completed, steps are taken to improve the defense. A primary consideration in improving a defense is to expand the perimeter to gain "working room. " This may require limited attacks, massed artillery and mortar fire, or close-in machinegun fire to force the enemy to withdraw. If the position will be occupied for a long time, it should be made as strong as possible.

This may be done as follows:

- **A small reserve is formed as soon as possible to react to enemy threats**
- **Local security is pushed forward**
- **Counterattack plans are developed**
- **Wire communications are established and pyrotechnic signals planned**
- **Machineguns are employed singly in order to cover as many enemy approaches as possible**

If troops remain in a defensive position for a long time, they must not become complacent. Leaders must inspect weapons, positions, and the cleanliness of troops. They must also develop plans for alerts, feeding, maintenance, and bathing. These activities must be scheduled according to a random pattern so that the enemy cannot take advantage of a set routine.

An alert system must be established, so that a portion of the defensive force is always awake. Although the poor observation in jungles favors the enemy's attack at any time, the early hours of the morning afford him the greatest chance of surprise unless positive alert measures are taken. All troops should stand-to before dawn.

The jungle enemy will try to probe a position to locate the flanks of positions and key weapons. Soldiers must not give away their positions by premature firing. Claymores and hand grenades should be used to engage these probes. When probed, riflemen near machineguns should fire, not the machinegunners. Machinegunners must use their pistols for self-defense instead of their machineguns. When the enemy attacks, he will try to isolate friendly positions and destroy them one at a time. Well-planned, mutually supporting fires will prevent this.

COUNTERATTACK

If enemy forces penetrate a position, a counterattack is the best way to expel them. Troops in the area of the penetration must stay in their positions and continue to fire to support the counterattack. If they leave their positions while the enemy is being expelled, they increase the chance that they will be hit with friendly fires.

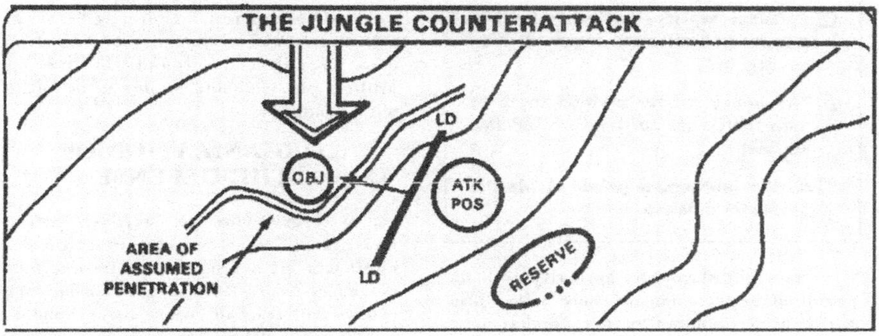

THE BATTALION COMBAT BASE

When engaged in tactical operations in the jungle, elements of the battalion will often establish a base for command and control and fire support resources, protected by a perimeter defense. These resources are called the battalion combat base.

The location of the perimeter defense to defend the battalion combat base will depend upon the:

- **Forces available to defend the combat base**
- **Ability to support subordinate units with indirect fire**
- **Defensibility of terrain**
- **Ability to communicate with subordinate units**

Prior to establishing the battalion combat base, the commander should conduct a reconnaissance to determine the defensibility of the terrain. He also plans forces required for the defense. While the defense must be capable of defeating the largest attack which the enemy is likely to conduct, it must use the minimum forces necessary. The combat support company, reinforced as necessary with an attached infantry platoon, is the largest force that is realistically available for preparing and defending the perimeter. To use a larger force would probably leave insufficient forces to fight the more important combat. Since some elements, such as the antitank platoon of the combat support company, may not be employed effectively in the jungle, they will often be available for use in the perimeter defense. The battalion commander will normally designate the combat support company or headquarters and headquarters company commander to be the battalion combat base commander and will have him take charge of the construction and execution of the perimeter defense.

Forces normally under control of the battalion combat base commander include:

- **The antitank platoon**
- **A Redeye section, if attached (both to man the perimeter and to provide anti-aircraft fire)**
- **The heavy mortar platoon (both to man the perimeter and provide fire support)**
- **A rifle platoon, if provided for the perimeter defense**

The scout platoon is normally used for patrolling or screening missions, rather than being used in manning the combat base perimeter.

During construction of the perimeter defense, it is vulnerable to enemy attack. Consequently, it is imperative to complete the perimeter defense as quickly as possible and to provide maximum security during construction.

The threat of infiltration attacks must be emphasized. The enemy may not be able to conduct large scale attacks on fortified positions, but he may be capable of disrupting operations by infiltrating one- or two-man teams through the perimeter to place explosive devices on command and control facilities, artillery pieces or mortars, or ammunition storage areas. This infiltration is often preceded by a deceptive attack or probe by ground forces. Troops in the perimeter must maintain constant security, using early warning systems and continuous patrolling. Starlight scopes, OPs, unattended ground sensors, and tripflares are also used. Wire obstacles should be used to keep infiltrators out of critical facilities.

A battalion combat base may have to remain in place for a long time. Continuous firing of mortars and landing of helicopters makes concealing its location very difficult. These two factors make it necessary to harden the perimeter defense. Overhead cover and sandbagged bunkers must be provided for all fighting positions. The tactical operations center (TOO and CP should have similar protection and may also be dug underground. Mortars and artillery pieces should be dug in or fortified with sandbags.

ORGANIZATION OF THE DEFENSE

A reserve for the defense may be constituted from attachments, such as engineers (if available), or from off-shift personnel from TOC and CP elements. This reserve will react to enemy attacks, and will reinforce the defense or counterattack. They must be rehearsed on signals and actions until they become proficient. Mortars are employed to provide close-in fire support. Artillery pieces can provide direct fire but probably will not be able to provide indirect fire support of the perimeter. Hence, the perimeter should be located within range of other artillery and mortar units for additional protection.

The battalion combat base commander assigns sectors to subordinate platoons, insuring that likely avenues of approach are dominated by Dragons, machineguns, and artillery in direct fire role, if possible. He then plans indirect fires. The commander specifies points at which adjoining platoons must coordinate. The platoon leader selects each position and designates the personnel to man it.

The platoon leader must insure that he has complete coverage throughout his sector to deal with not only a mass attack but also infiltration of small elements.

Once the positions are selected, the platoon leader insures that a priority of work is adhered to.

EXAMPLE OF PRIORITIES OF WORK

Work should be accomplished in the following steps, consistent with the tactical situation and the availability of resources.

Step 1: Air assault/ground assault seizes the site.

Immediate security established to Include OPs.

Area swept for boobytraps.

Mortars laid.

Step 11: Communications established.

CP set up.

Position of TOC dug in.

Selected TOC personnel displaced to perimeter defense.

Step III: Perimeter positions established.

Fields of fire cleared.

Reserve force established.

Wire to all positions.

Step IV: Barriers and obstacles placed around perimeter defense.

Early warning devices emplaced.

Security/ambush patrol plans established.

Final protective fire (FPF) fired in.

Step V: Positions sustained.

Positions are hardened with overhead cover.

All other positions improved.

More fields of fire cleared.

Landing zone enlarged.

Latrine, generators, and ammunition supply point (ASP) established.

NOTE: Throughout work, camouflage must be applied.

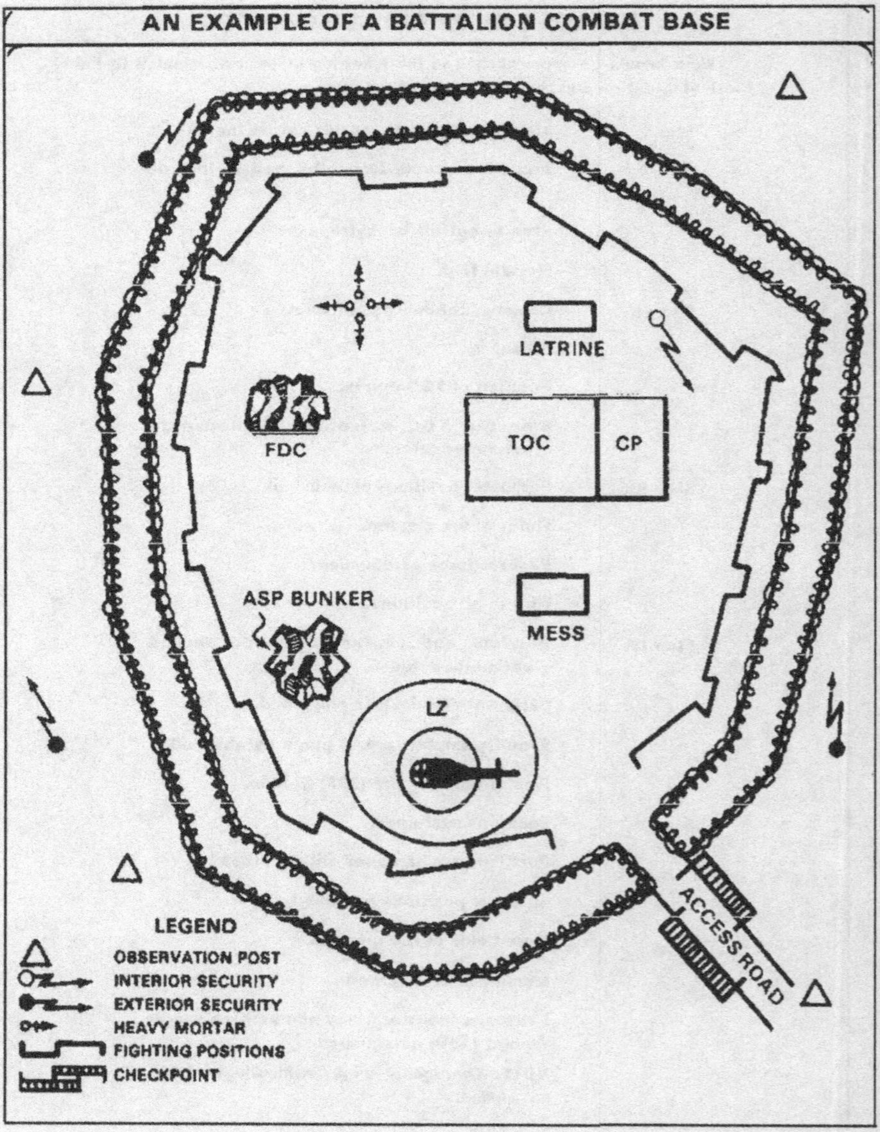
AN EXAMPLE OF A BATTALION COMBAT BASE
LATRINE
FDC
TOC
CP
ASP BUNKER
MESS
LZ
ACCESS ROAD
LEGEND
OBSERVATION POST
INTERIOR SECURITY
EXTERIOR SECURITY
HEAVY MORTAR
FIGHTING POSITIONS
CHECKPOINT

SECTION V. RETROGRADE

Retrograde operations are conducted in much the same manner as the defense, with the additional requirement to gain a mobility advantage over the enemy. This is done by taking measures to increase the mobility of the unit conducting the retrograde (reconnaissance and preparation of routes, use of helicopters, etc.), and by taking measures to decrease the mobility of the enemy (ambushes, artillery fires, mines, wire obstacles, etc.).

WITHDRAWL

Withdrawals may be conducted under enemy pressure or not under enemy pressure as explained in FM 7-8 and FM 7-10. Because of the cover and concealment provided by the jungle, a withdrawal under enemy pressure may be conducted using deception in much the same manner as a withdrawal not under enemy pressure. The ruggedness of the terrain and the strength of the attacking enemy are key factors in this type of operation. Routes, assembly areas, and new positions must be thoroughly reconnoitered. They may also be marked if such marking does not compromise security. Densely vegetated areas will require increased use of guides. Control in such areas is very difficult; therefore, leadership, planning, and rehearsal are crucial.

"J will never again tell my platoon to withdraw • especially in the jungle - without telling it where to go. I had a hell of a time getting them together. "

—Platoon Leader, New Guinea, World War II

Unattended ground sensors employed along likely enemy avenues of approach can be used to provide information on enemy movement and activities. This information can in turn be used to place fires on the enemy.

DELAY

The delay in sector is normally the most frequently used type of retrograde in the jungle. Ambushes may be used to halt the enemy's pursuit, and can aid the main body in breaking contact along the delay route. Mechanical ambushes, wire obstacles, and minefield will also delay the enemy. In the jungle, the delay is normally conducted in several phases: defense, withdrawal under pressure, breaking contact, and movement to and occupation of new positions where the defense starts again. As in the withdrawal, units should reconnoiter and clear routes to the rear.

Section VI. OTHER COMBAT OPERATIONS

Patrols, raids, and ambushes are nor-really used more often in the jungle than in more open terrain. The jungle permits small units to move undetected during reconnaissance patrols and achieve surprise when conducting raids and ambushes. These are small-unit operations. They depend heavily on the skill and stealth of infantry platoons and squads, and are demanding operations.

One jungle myth that was popular both during World War II and the Vietnam War was that "the night belongs to the enemy. " The enemy has no better night vision or stealth than do trained US soldiers. He will often use the night as a means to avoid US firepower, but with the proper use of patrols, raids, and ambushes, this problem can be eliminated.

"On any given night in Vietnam, American soldiers staged hundreds of ambushes, for the ambush is one of the oldest and most effective military means of hampering the enemy's nighttime exploits."

—J.A. Cash, Seven Firelights in Vietnam

Successful jungle patrols, raids, and ambushes result from detailed planning, intensive training, and constant rehearsal. Troops must be alert. A unit which has moved cross-country through the jungle until late in the afternoon will not be in a condition to succeed if it has to go out on ambush that night. Commanders must realize that such operations require time to prepare, train, and rest.

The basic techniques for patrols, raids, and ambushes can be found in *chapter 5, FM 7-8*. Specific techniques which are effective in jungle operations are described below.

PATROLS

A patrol is a detachment sent out by a larger unit to conduct a combat or reconnaissance operation. The operation itself is also called a patrol. The mission to conduct a patrol may be given to a fire team, squad, platoon, or company. The leader of the detachment conducting a patrol is referred to as the patrol leader.

CATEGORIES OF PATROLS

The planned action at the objective determines the patrol's category. There are two categories of patrols:

Combat (ambush, raid, or security) Patrol. This patrol provides security and harasses, destroys, or captures enemy troops, equipment, and installations. A combat patrol also collects and reports information, whether related to its mission or not.

Reconnaissance (area or zone) Patrol. This patrol collects information or confirms or disproves the accuracy of information previously gained.

Regardless of the category of the patrol, there are four key principles to successful patrolling. These are:

- **Detailed planning.**
- **Thorough reconnaissance.**
- **Positive control.**
- **All-round security.**

ORGANIZATION FOR A PATROL

The patrol leader decides what elements and teams are needed for his patrol, selects men or units for these elements and teams, and decides what weapons and equipment are needed. He should, however, use his unit's normal organization (squads and platoons) and chain of command (squad and platoon leaders) as much as possible to meet these needs. For example, a combat patrol may be organized like this: the company headquarters is the patrol headquarters; the 1st platoon is the assault element; the 2d platoon is the security element; and the 3d platoon and weapons platoon make up the support element.

GENERAL ORGANIZATION

A patrol generally consists of a patrol headquarters and the elements needed for the mission.

Patrol Headquarters. The headquarters (HQ) of a company-size patrol normally consists of the same number of men as a regular company headquarters. However, regardless of a patrol's size, its leader tailors the headquarters to meet mission needs. The patrol headquarters has the same responsibilities as any other command element.

Reconnaissance Patrol. In an area reconnaissance (recon), a patrol has a reconnaissance element and a security element. In a zone reconnaissance, a patrol has several reconnaissance elements. Each one provides its own security.

Combat Patrol. A combat patrol normally has an assault element, a security element, and a support element. At times, the support element may be omitted by combining it with the assault element.

In general, jungle terrain affords excellent concealment, provides some cover, and hinders enemy observation and movement. During rainy periods, the sound of movement is less obvious. Wet ground and wet vegetation also muffle noise. All of these factors favor the patrolling unit.

On the other hand, the difficulties of movement and control and the ease with which the enemy can infiltrate friendly units are disadvantages to units patrolling in the jungle. These factors can best be overcome by training and discipline.

Silence, in both voice and movement, is essential at all times during a jungle patrol. With practice, it is possible to move steadily, deliberately, and carefully through the jungle, parting the undergrowth instead of crashing through it or cutting through it with machetes. Troops should avoid walking on dry leaves, sticks, rotten wood, or anything that would make noise. Machetes should be used to cut trails only as a last resort. Talking should be done in a whisper, and arm-and-hand signals should be used whenever possible.

Trails should be avoided. Patrols should make every effort to hide signs of movement, especially when moving through untraveled territory or near enemy positions.

Some techniques which may be used include:

- **Requiring all troops to wear boots that have the same pattern on their soles.**
- **Requiring troops to carry only the mission essentials, and do not let them litter.**
- **Cautioning troops to avoid small saplings, when going up hill. The shaking of overhead branches can be seen and heard at a distance.**
- **Requiring troops to keep off trails. If necessary to monitor or guide on a trail, patrols should move parallel to the trail and not on it.**

Native scouts are valuable in patrolling because they are often very familiar with the terrain. Patrol leaders must realize, however, that a scout's function is only to show direction and provide information. He should never lead the patrol. The correct position of a scout is with the patrol leader, so that the leader can make decisions based on the scout's advice.

RAID

Raids in the jungle environment must be keyed to reliable intelligence. The actions of the raiding unit must be decisive and rapid in order to catch an elusive jungle enemy. A raid's success depends on good intelligence and a sound plan.

Jungles favor raid operations. The excellent concealment enables skilled raiding patrols to operate deep in enemy territory. Platoon-sized units are best suited to jungle raids. Supporting artillery fires should be planned, but due to difficulties of control, timing, and communications, jungle raids may be executed without artillery support. Surprise is a key ingredient of a successful raid.

Raids that require deep penetration into enemy-held areas are best executed by establishing a patrol base in the general area of the final objective. From there, reconnaissance patrols can be sent to scout enemy positions while the remainder of the force completes its preparations for the raid. Helicopters and watercraft are effective means of transporting a raiding force rapidly to the vicinity of its objective without depleting their physical strength in a difficult march.

AMBUSHES

The ambush is more important, more effective, and more frequently used in jungle fighting than in any other type of combat. Jungle terrain provides many opportunities for a well-concealed force to gain surprise. Surprise is essential for a successful ambush.

Destruction of enemy forces is the primary purpose of most ambushes, but other benefits result from a well-executed ambush program.

These benefits include:

- **Disruption of enemy operations, since troops become reluctant to move and fight in areas where ambushes are frequent.**
- **Capture of prisoners and equipment which may yield intelligence data.**
- **Capture of supplies, thus increasing combat effectiveness at the expense of the enemy. In some instances, this is the primary source of supplies for guerrilla forces.**

More than in any other type of terrain, jungle ambushes require high standards of discipline. Soldiers on an ambush must be prepared to remain in the same position for hours at a time, without being able to sleep, talk, or smoke. They must endure insects and resist the desire to make any quick moves to swat or brush the insects away. All these require extensive training to develop the patience and self-discipline required.

The location for an ambush should be chosen after a careful analysis of the terrain, using maps, aerial photographs, and personal reconnaissance. The site chosen must contribute to the surprise of the ambush. Many times the selection of a site for surprise alone will be more effective than attempting to ambush from a site which is in other respects tactically sound but at which the enemy is sure to be suspicious. Covered routes of approach and withdrawal, good fields of fire, and canalization of the enemy are characteristics of a good site. The site should always be reconnoitered and approached from the rear.

In no other operation is camouflage more important than in the ambush. Weapons should fire through screens of undisturbed, living foliage. Spoilage resulting from the preparation of positions must be removed from sight. There can be no unnecessary noise or movement. If reliefs are used, they should be scheduled so that only a few men move at any one time. One or two men moving are harder to detect than an entire relief moving at once.

DEFENSE AGAINST AMBUSH

Since ambushes are more frequent and effective in the jungle than in any other type of terrain, a unit moving through the jungle must take all possible measures to reduce its vulnerability to ambush. The most effective means of countering an ambush is to detect it before entering the kill zone. This, however, is not always possible.

Dismounted troops have an advantage over mounted troops in avoiding ambushes because they do not have to move on roads or trails. Commanders of dismounted units should make a map and aerial photograph reconnaissance to detect likely ambush sites and plan routes which avoid them. During movement, security to the front, rear, and flanks should be maintained at all times. Alert troops, good noise discipline, and well-rehearsed signals are other means which will reduce the chances of ambush. Accurate land navigation, continuous fire support planning, and counterambush drills are also important antiambush techniques.

Mounted troops are very vulnerable to jungle ambushes, especially where the foliage grows up to the edge of a road. Ambush of vehicular columns traditionally has been a primary tactic ofjungle enemies. As a result, traffic in jungle areas must be tightly controlled and kept to a minimum. All vehicles should have armed riders. Armored vehicles should escort convoys, and traffic information should be carefully guarded. Fire planning and route selection and reconnaissance are important for mounted troops as well as for dismounted troops *(app M, FM 7-8).*

PREPARATION OF TRUCKS FOR MOVEMENT

To prepare trucks for jungle convoys:

1. Lower windshields and cover them with tarps (this reduces danger of flying glass and stops glare).
2. Place tailgates halfdown to ease dismounting (troops can then shoot and dismount fast).
3. Put at least two layers of sandbags on the floors of the cabs and cargo areas to protect against mines.
4. Stack sandbags along the sides of the cargo beds to protect troops from small-arms fire from the flanks.
5. Stack and fasten sandbags around the engine compartments to protect the engines (do not block the flow of air to radiators).
6. Place troop seats in the center, facing out, so troops can react fast.
7. Install wire-cutting apparatus in front of trucks to prevent injury of troops from wire stretched across the road.
8. Remove the rear safety straps for fast dismounting.

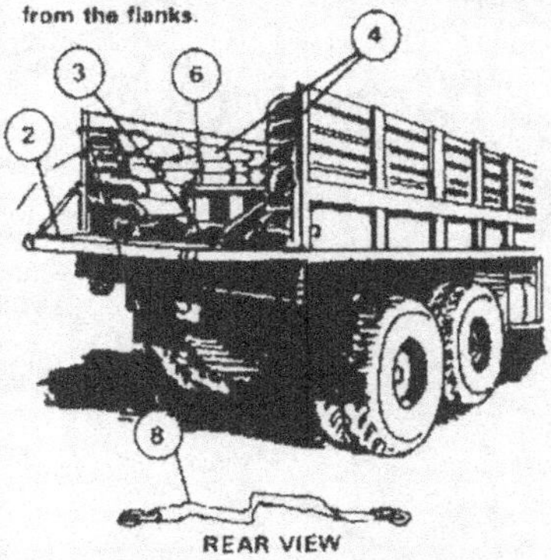

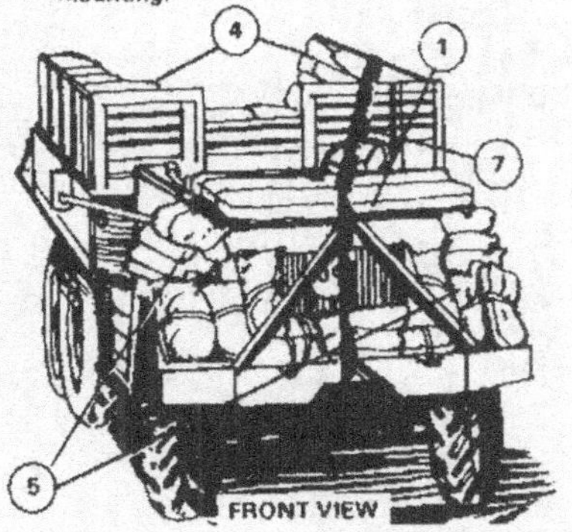

COUNTERAMBUSH MEASURES

Troops must also be trained in counter-ambush measures. The key is early detection followed by reflex-type counteraction, a high volume of return fire, and relentless pursuit. The most effective counterambush measures are well-rehearsed immediate action drills. Every soldier must know exactly what he is supposed to do.

Dismounted troops should react to an ambush immediately, firing into the ambushers without orders. Building and retaining fire superiority is the best initial defense against an ambush. If a patrol finds itself in an enemy ambush, it must get out of the kill zone immediately. It must take the following immediate actions:

Troops in the kill zone, without order or signal, immediately return fire, and quickly move out of the kill zone by the safest way. (There is no set way to do this; it must be each soldier's decision for his situation.) Smoke grenades can help conceal the troops in the kill zone.

Troops not in the kill zone fire to support the withdrawal of the troops in the kill zone.

If a dismounted patrol is ambushed, it should attempt to break contact and reorganize in the last designated rally point.

Mounted troops who are ambushed should attempt to drive rapidly out of the kill zone. Vehicles approaching the kill zone should stop so they do not enter it. Troops should then dismount and maneuver to destroy the ambush.

In any case, the rapid call for supporting artillery and mortar fire will help the ambushed force to gain fire superiority and will assist the maneuver to destroy the ambush.

CHAPTER 6

Helicopter, Armor, Mechanized Infantry, And Combat Support Operations

Section I. GENERAL

In the jungle, infantry battalions are the heart of both the offense and the defense. These battalions are normally supported by a mixture of helicopter, armor, artillery, air defense, engineer, and other units.

This chapter outlines those techniques which are most effective in supporting infantry engaged in jungle operations.

Section II. Helicopters

HELICOPTER ROLES

The helicopter is a combat system that is unaffected by the poor trafficability of jungle areas. As a result, it has become a vital part of US jungle operations. US forces have used helicopters in the jungle successfully for heliborne fire, reconnaissance, air assault, command and control, resupply, and medical evacuation. This section will discuss heliborne fire, reconnaissance, air assault, and command control in the jungle. *For resupply and medical evacuation, see chapter 7.*

In all of these operations, bad weather and enemy air defense are major considerations in the jungle as they would be anywhere else.

CONTENTS

Heliborne fire is provided by the attack helicopter unit.

The attack helicopter unit performs three basic missions:

- **Overwatch and security**
- **Engagement of targets located by ground elements**
- **Independent target engagements**

Attack helicopters in an overwatch and security role protect other assets, such as other helicopters or convoys. Attack helicopters in the heliborne fire role can augment the fires of units fighting on the ground. In jungle operations, they maybe the only additional means of fire available. Finally, attack helicopters may acquire and engage targets independent of any other operations.

HELICOPTER ARMAMENT

Although the Cobra, AH IS, armed with the TOW, has become the standard US attack helicopter, the nature of the typical jungle enemy and the terrain make the 2.75-inch rocket (HE) a more useful weapon against known point positions. The "Flechette" 2.75-inch rocket, 7.62-mm minigun and 40-mm grenade launcher, also on the attack helicopter, are extremely effective against dismounted troops.

COMMUNICATIONS

The most important consideration when using attack helicopters to augment the fire of ground troops is communications. Both radio and visual communications means must be established and maintained. Radio communications are used to pass target information from the ground commander to the helicopter. Visual communications are used to mark the location of friendly troops. Common daytime signals are a smoke grenade, a panel, and a mirror. A flashlight, a strobe light, and a chemical light stick are effective for signaling at night.

COMMUNICATION WITH ATTACK HELICOPTERS

While the helicopters are en route to the target area, a member of the attack helicopter team will contact the ground commander over the radio. At this time, the ground commander must transmit *target handoff information.*

TYPICAL EXCHANGE OF TARGET HANDOFF INFORMATION

IDENTIFICATION	Helicopter: —	"55T7 this is 24J6."
	Ground Commander: —	"24J6 this is 5517."
ENEMY INFORMATION	Ground Commander	"I am pinned down by an estimated enemy company. I am receiving AK-47, RPD, and RPG fires."
OWN SITUATION	Ground Commander:	"I am in a perimeter defense. There is no artillery impacting at this time. I will mark my location with smoke."
	Helicopter	"I'm about 30 seconds from your reported location. Pop a smoke grenade."
VISUAL IDENTIFICATION	Ground Commander: —	"WILCO."
	Helicopter: —	"I identify yellow smoke."
	Ground commander	"Roger, yellow smoke. Enemy is at azimuth 120 degrees from yellow smoke, estimated distance 200 meters."
COORDINATION	Helicopter:	"Roger, I'll make my run from northeast to southwest."

RECONNAISSANCE

Reconnaissance missions are normally given to air cavalry. Air cavalry troops and squadrons are equipped with attack helicopters, scout helicopters, and utility helicopters. The air cavalry troop is organized with a troop headquarters, aeroscout platoon, aeroweapons platoon, aeroreconnaissance platoon, and service platoon. The troop is task organized for specific reconnaissance missions.

The aeroscouts use terrain flying while looking for signs of the enemy. Attack helicopters from the aeroweapons platoon support them and engage targets as they are acquired. If a thickly vegetated area cannot be reconnoitered from the air, the aeroreconnaissance platoon can be inserted on the ground by the utility helicopters.

AIRMOBILE OPERATIONS

This employment of air cavalry is effective in the jungle. The information gathered can be rapidly processed into intelligence. This in turn can be used to plan timely commitment of troops into the area.

Airmobile operations in the jungle are the most rapid means of concentrating combat power at a critical time and place on the ground. The doctrine for airmobile operations is found in FM 90-4, and it is generally applicable to jungle operations.

Special considerations in jungle operations include:

- **Utility helicopters will not be able to lift the same size loads that they can in more temperate areas. Sometimes this may result in as few as 5 soldiers per sortie for UH-1H and 19 soldiers for UH-60.**
- **Radio and visual communications between ground and air are vital to insure proper coordination. The thick jungle foliage will often make this more difficult than in other types of terrain.**
- **There maybe few suitable landing zones (LZ). Most landing zones will be only large enough to support one or two helicopters at a time.**
- **Units may have to be resupplied totally by air.**

One of the most common uses of the helicopter in jungle operations is command and control. Command and control (C&C) helicopters are specially outfitted with radios which allow communications with elements on the ground, supporting artillery, and other aircraft. These helicopters also have seats in them arranged so that commanders, S3s, and fire support coordinators (FSCOORD) can talk to each other and perform their respective functions. Such helicopters are normally available to command groups down to and including infantry battalion level.

To use the C&C helicopter effectively, a commander should exploit its unique capabilities:

- **The helicopter's speed enables a commander to influence the action in widely separated areas.**
- **The helicopter's powerful radios and flight altitude permit the commander to communicate more effectively and reliably than he could on the ground.**
- **The unrestricted observation from the air allows the commander to see the terrain better, adjust fires more accurately, and control attack helicopters more effectively than he could on the ground.**
- **When used for a commander's reconnaissance, the C&C helicopter is a significant help in planning operations. Likewise, a battalion commander who takes his company commanders in the C&C helicopter to issue an operation order (OPORD) can use the terrain below as a reference— like a readymade sandtable.**

The C&C helicopter can be one of the jungle commander's greatest assets. It can also be abused. The keys to its effective use are a few common sense considerations.

The helicopter cannot become a substitute for command presence on the ground. (A commander cannot merely fly over a unit and be satisfied that he knows its situation. The view from the air often gives a distorted picture of conditions on the ground.)

The helicopter should not interfere with ground operations. (If a unit has to devote time to cutting an LZ every day for a visit from the commander, it will get little else done.)

The helicopter must not violate a ground unit's security. (A helicopter hovering overhead soon reveals the location of a unit on the ground.)

The helicopter must not intrude on a ground unit's internal radio nets. (The helicopter's radios are more powerful, and will interfere with the ground commander's ability to control his own unit.)

The helicopter is vulnerable to ground fire. (Although the nature of the air defense threat will be different in every conflict, no commander can expect to survive if he maintains a stationary position anytime in the vicinity of ground contact or orbits at high altitudes. The helicopter is extremely vulnerable to small-arms fire and heat-seeking missiles.)

Section III. ARMOR OPERATIONS IN THE JUNGLE

Up to this point, this chapter has covered infantry tactics in the jungle. Armor may also play a role on the jungle battlefield. Because the tank's mobility is greatly restricted on jungle terrain, its role is different in the jungle than that on more open terrain. Instead of rapid envelopments and deep penetrations, tanks primarily provide fire support for infantry in jungle fighting.

Tanks can perform the following missions in the jungle:

- **Support attacking infantry. In the meeting engagement, they move to the point of contact to provide direct fire support.**
- **Provide heavy fire support in the defense and conduct counterattacks.**
- **Support infantry in retrograde operations.**
- **Secure convoys and protect lines of communications and key facilities.**
- **Attack or defend against enemy armor.**

TACTICAL OPERATIONS

OFFENSIVE OPERATIONS

In the movement to contact, infantry will normally lead, supported by tanks from overwatch positions whenever the terrain permits. Terrain and vegetation will dictate distances between the forces, but the tanks should be close enough to the infantry to be able to move forward quickly to provide immediate fire support.

In a meeting engagement or attack, once the infantry makes contact, tanks move to positions from which they can support the infantry.

TANKS IN POSITION TO SUPPORT INFANTRY

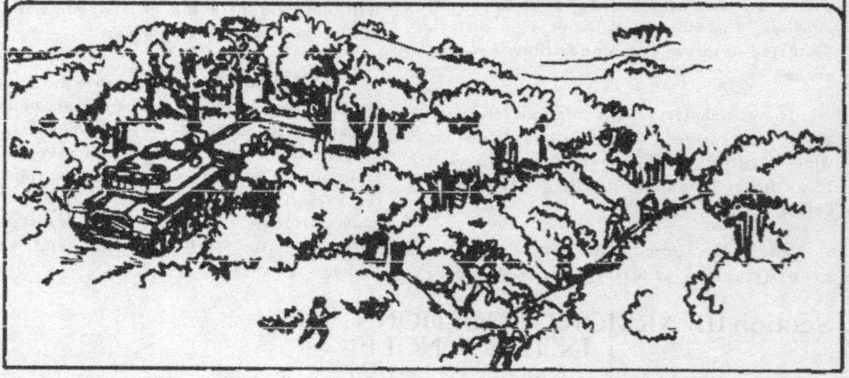

Tanks can fire their machineguns and main guns using high explosive antitank (HEAT) and antipersonnel rounds to destroy or suppress the enemy, clear jungle foliage, and destroy fortified positions.

Should the enemy have armor, the tanks can more effectively attack the enemy armor than can the TOW or Dragon antitank guided missile because the tank gun can better fire through the jungle foliage.

When terrain permits, infantry can fix the enemy, while tanks create a penetration to split the enemy defense. The tanks and infantry can then destroy the severed forces.

DEFENSIVE OPERATIONS

Tanks are used in the jungle defense as they are used in any other environment. They add greatly to close defensive firepower and serve as a mobile counterattack force.

In the position defense, tanks should be positioned to:

- **Block possible armor or motorized penetrations**
- **Move quickly to fill gaps in the defense**
- **Provide antiarmor fire**

Tanks in the jungle defense are vulnerable to infiltrators; consequently, all-round security and close coordination with infantry are essential.

Hurricane fencing should be erected as a screen in the defensive position to provide stand-off against enemy use of rocket-propelled grenades (RPG) and antitank guided missiles (ATGM).

RETROGRADE

Tanks can support retrograde operations in much the same way as in the defense. Contrary to general practice, in jungle retrograde operations tanks will often disengage first, covered by the infantry.

LIMITATIONS

RESTRICTIONS ON USE OF ARMOR

Vegetation. Densely forested areas, where tree trunks are close and heavy above-ground root systems exist, may make movement difficult. Thick stands of bamboo may slow or even stop tanks.

Topography. Many jungles exist in rugged mountainous areas which will impede tanks. Swamps, coastal river basins, and other areas intersected by waterways pose obstacles to tanks because of their soft soil and frequent deep channels.

Weather. Rainy seasons (monsoons) cause rivers and streams to rise and become unfoldable. Heavy rains may also cause damage to roads.

TECHNIQUES TO OVERCOME ARMOR RESTRICTIONS

All of these factors limit but do not exclude the use of tanks.

A commander can overcome those limitations by using these techniques:

- **Know the terrain. Know where tanks can travel. Avoid areas which are obviously impassable.**
- **Know the weather. What were recent conditions? What is expected? What effect will these conditions have on the use of tanks?**
- **Move dismounted infantry in front of tanks on unfamiliar and heavily vegetated terrain. They can check and verify conditions, act as guides, and provide security.**

TIPS FOR MOVEMENT IN THE JUNGLE

Jungle conditions vary greatly from place to place and season to season.

Here are some movement tips which apply to most jungle areas:

- **Red silt soils tend to break down quickly when wet. They may support a single tracked vehicle but may become untrafficable with heavy use.**
- **Inundated areas containing yellowish reeds and cloudy water usually have bottoms too soft to support tanks.**
- **Rice fields, in the dry season, are usually trafficable to tanks and personnel carriers. During the rainy season, they may be untrafficable. Fields with standing water in the wet season may have a bottom too soft to move on; those containing clear water and green vegetation usually are on firm ground and are trafficable. One technique for determining if rice fields are trafficable is "poling." Using this technique, troops precede tanks across the field they are to cross, sinking poles (1 to 1 1/2 inches in diameter, with flat ends) into the ground. If the poles cannot be sunk, the field is usually firm enough to cross.**
- **River and stream bottoms usually are untrafficable. The armored vehicle launched bridge (AVLB) can span 17 meters (57 feet) and is more than adequate for most stream crossing. Care must be taken to insure that the shoulders of the banks can support the AVLB while tanks cross. When the AVLB is not avilable, perforated steel planks can be used to provide a firm surface on which tanks can ford small streams.**

Section IV. MECHANIZED INFANTRY

Mechanized infantry units are able to provide a commander fighting in a jungle environment a greater flexibility than that afforded by nonmechanized infantry. Mechanized infantry may be employed as *a mounted force, a dismounted force, or airmobile infantry.*

CAPABILITIES

OFFENSIVE

In offensive operations, mechanized infantry may conduct the following offensive actions:

- Movements to contact
- Attacks-hasty or deliberate
- Reconnaissance-in-force

DEFENSIVE

In defensive operations, mechanized infantry may be used:

- In a strongpoint defense
- As part of a linear defense
- To establish a perimeter defense

OTHER OPERATIONS

Mechanized infantry units, especially when formed into tank-infantry company teams, may also be used in other operations such as:

- Route clearance and security
- Convoy escort
- Reserve of a higher headquarters (either for offensive or defensive operations)
- Raids

CONSIDERATIONS FOR EMPLOYMENT

As mechanized infantry may be used in different ways, the precise mission it is assigned must be weighed against its strengths and weaknesses in a jungle environment.

STRENGTHS	WEAKNESSES
■ **Quick reaction over distances.** The unit has its own transportation immediately available, and considerable firepower to accompany it. In savanna or on firm but lightly vegetated ground, mechanized infantry is able to move considerable distances in a relatively short period of time. ■ **Firepower.** A mechanized infantry force has greater firepower than other infantry units. The carrier's caliber .50 machinegun can be used to strengthen defenses or provide overmatching fires. ■ **Large load capacity.** The armored personnel carrier (A PC) provides mechanized infantry units larger capacity to carry food, water, and ammunition than other units. ■ **Communications.** A mechanized infantry unit has a communications advantage because of both the number and types of radios it uses.	■ **Trafficability.** When used as a mounted force, mechanized infantry units may be limited in the scope of their operations by the terrain in which they are employed. ■ **Noise.** The operating noise of the tracked vehicles limits the ability of mechanized infantry units to achieve surprise while moving. ■ **Logistics requirements.** Mechanized infantry units increase logistics requirements, because of increased petroleum, oils and lubricants (POL) consumption; greater repair parts requirements; more frequent recovery operations; and larger maintenance requirements.

Disposition of Ml 13 APC when troops are dismounted. When employing mechanized infantry as an airmobile force or as dismounted infantry, consideration must be given to the disposition of the M113 APCs. It is usually advisable for a driver to remain with each vehicle. In addition, some infantrymen may have to provide security for the carriers and thus will not be able to fight with the dismounted element.

MECHANIZED OFFENSIVE OPERATIONS IN A JUNGLE ENVIROMENT

The offensive operations that mechanized infantry units are most likely to conduct in a jungle environment are movement to contact and reconnaissance-in-force.

A movement to contact is used to gain or regain contact with the enemy, and to develop a situation. It serves as the first stage of operations against an enemy force which has not been located, and normally ends in a meeting engagement.

A reconnaissance-in-force is employed to force the enemy to react so that friendly elements can develop information about the enemy by discovering his location, disposition, and intent. It is used when other means of gaining information about the enemy are not available.

Hasty and deliberate attacks are conducted in much the same manner as they are in conventional terrain. The techniques, however, may differ due to the jungle environment, especially the strict application of control measures in controlling maneuver elements in the thick jungle vegetation. Also, an attack will normally be conducted on a more narrow front due to limited visibility.

MOUNTED MOVEMENT

Hasty and deliberate attacks are conducted in much the same manner as they are in conventional terrain. The techniques, however, may differ due to the jungle environment.

When moving in the jungle, either cross-country or along roads, it is best to move in multiple columns on as broad a front as possible. The lead elements should be preceded by an aerial route reconnaissance. Indirect fire support must be closely coordinated and instantly available, as contact is often violent and without warning. TACAIR and attack helicopters provide another responsive means to react to enemy contact.

DISMOUNTED MOVEMENT (APC ACCOMPANYING)

When in contact or contact is anticipated the troops dismount and fight on foot. The caliber .50 machinegun on the APC can provide fire support maneuvering elements.

RECONNAISSANCE BY FIRE

Movement in the jungle carries with it the prospect of ambush at any time. Reconnaissance by fire is a technique that can be used to decrease this danger. In this technique, lead elements engage suspected or likely enemy locations with automatic weapons as the unit moves. The enemy is then made to move or return fire. When considering a reconnaissance by fire, the commander should realize that it uses a lot of ammunition and reveals his location to the enemy. Reconnaissance by fire is not an excuse for indiscriminate fire by individuals.

MOVEMENT TO CONTACT

Mounted movement to contact in the jungle is characterized by deliberate use of traveling overwatch and bounding overwatch. The overwatch positions will be closer than in other areas, due to the vegetation and terrain. Contact with the enemy will usually be made at close range (within 200 meters); therefore, automatic weapons should be kept well forward. Frontal, flank, and rear security may be established mounted, but more often should be provided by dismounted elements. The security elements should be rotated frequently to prevent fatigue.

RECONNAISSANCE-IN-FORCE

When conducting a reconnaissance-in-force, a mechanized infantry unit may have lead elements conduct a mounted movement to contact. At likely ambush sites along the route, however, these elements must dismount and patrol. These patrols are normally conducted by a platoon's maneuver element. The earner teams, usually with two men in each, should overwatch the maneuver element. They can be called forward when required.

DELIBERATE ATTACKS

Deliberate attacks in the jungle are usually conducted dismounted, with Ml 13s supporting the attacking elements. Care must be exercised to insure that the maneuvering infantry does not mask the supporting fires during the assault. The assault can be preceded by artillery preparation and use of close air support.

MECHANIZED INFANTRY DEFENSIVE OPERATIONS IN A JUNGLE ENVIROMENT

Mechanized infantry units are a formidable force when defending. The most common types of defensive operations they might participate in are forming a defensive perimeter, establishing a strongpoint, or taking part in a position defense.

DEFENSIVE PERIMETERS

Defensive perimeters may be established at any time but are normally used during periods of limited visibility to increase the security of the force and to allow time for maintenance and rest. The perimeter is a hasty defense technique. It is usually only a temporary arrangement and is moved frequently. A perimeter is not oriented against a particular enemy force, but takes advantage of terrain to obtain the greatest security possible.

Establishment of a Perimeter. An M113 leaves a track that is difficult to hide and easy to follow. Therefore, a mechanized infantry perimeter is difficult to conceal. Consideration should be given to the placement of ambushes along the back track path leading into the perimeter. The perimeter is very similar to that used by dismounted infantry but it is adjusted to accommodate the tracked vehicles. It is best to have a quartering party precede the unit but this may not be possible in all cases. A unit SOP should be the basis for specifying the establishment of a defensive perimeter. M113s may be used to clear fields of fire where they are capable of doing so. (An APC is capable of clearing considerable brush in a short period ot time.) The M113s are then parked in a circle, wagon train style, facing out. Individual fighting positions are prepared in front of the parked vehicles. Dismounted observation posts (OP) and ambush patrols are sent out. These elements may use Claymore mines and early warning devices. The caliber .50 machineguns may be dismounted or remain on the Ml 13s but are manned at all times. When the soldiers are allowed to sleep, they should be in or close to fighting positions in order to be near overhead protection. Ramps should normally be raised, with access to the M113 through the troop door, in order to provide protection to the crew from indirect fire.

Defense of a Perimeter. An enemy ground attack against a perimeter defense maybe by a surprise assault or be preceded by preparatory fires. The unit should continually have all the automatic weapons manned and ready to fire, *but care must be taken to insure patrols and OPs are back within the perimeter before machinegun FPLs are fired.* M113s should not be moved during a night attack on a defensive position, except under emergency conditions, due to the danger of injuring prone friendly troops.

STRONGPOINT DEFENSE

A strongpoint is a defensive position which is fortified as extensively as time and materials permit. It is normally located on a terrain feature critical to the defense. Individual fighting positions should be prepared with overhead cover. The M113s should be placed in hull defilade positions, which are formed by natural terrain features or dug by a bulldozer. Each position should be deep enough to protect the vehicle. The caliber .50 machinegun may be dismounted or remain on the vehicle. If it remains on the vehicle, the position should allow the caliber .50 machinegun to attain grazing fire. If materials are on hand and time is available, a chain link fence may be emplaced 10 to 15 meters in front of each Ml13 position to cause premature detonation of antitank rounds. Each squad's position and OP should be linked with the platoon leader in a telephone hot-loop, with radio used as an alternate means of communication.

SUGGESTED DEFENSIVE TECHNIQUES FOR MECHANIZED INFANTRY UNITS

Each Ml13 can carry a large amount of equipment. The mechanized unit should take advantage of this by habitually carrying equipment to aid in preparing defensive positions.

This equipment may include:

- **Concertina and barbed wire**
- **Engineer stakes**
- **Pierced steel planking**
- **Sandbags**
- **Chain saws**
- **Pioneer tools**
- **Hurricane fencing (RPG/ATGM screen)**
- **Night vision devices**

M60 machineguns should be dismounted wherever possible and placed in prepared positions. The caliber .50 machinegun may be used either mounted or dismounted. It should be dismounted when the terrain prevents the gunner from getting a suitable mounted firing position. During mounted movement, the caliber .50 machinegun should be manned by the track commander and the M60 machinegun manned in the cargo hatch. *(See FM 7-7.)*

OTHER MECHANIZED OPERATIONS

When operating in the jungle, mechanized infantry units may be given missions which they would not normally perform in other types of terrain. Such missions require special planning and coordination before and during the conduct of the mission.

ROUTE CLEARANCE AND SECURITY

These operations are conducted when a route has been closed or unused. They are usually conducted so that supply activities may occur, but should not be confused with convoy escort operations, which require different tactics and techniques. Route security and clearance operations are oriented on a specific route and the surrounding areas, to insure that vehicular operations are not interrupted along that route.

Route clearance operations are conducted to eliminate the enemy along the road and to remove any explosives which may have been placed there. Whenever possible, route clearance is a combined arms effort involving as a minimum the use of armor, infantry, engineers, artillery, and Army aviation. Route clearance involves deliberate, detailed, and coordinated actions which are slow. The route must often be walked by mine-sweep teams, and the areas adjoining the route must be cleared by dismounted infantry. During route clearance operations, the mechanized infantry elements normally provide security forces for the combat engineer mine-sweep teams. The infantry is also used to clear the areas on either side of the road.

Route security missions are characterized by continuous activity to prevent the enemy from cutting the route or ambushing elements using it.

Patrolling is the key to route security, day and night. Patrols should be dispatched so there is no pattern. Patrols are usually squad-sized and are assigned specific areas of responsibility. They may at times move mounted but dismount where appropriate. These areas of responsibility should extend to 1 kilometer each side of a route, for the mission is to prevent enemy mining or ambushes. A reserve force should be centrally located to be able to react in any direction.

Bridges or large culverts along roads are prime locations for enemy demolitions and ambushes. They must be kept clear and intact. A good way to do this is to ambush near them at night, and patrol from them during the day. Night observation devices should be used in ambush operations.

CONVOY MOVEMENTS

Coordination must take place before and during a convoy movement. Each patrol must know the time of entry and time of exit of each convoy, the numbers and types of vehicles, and whether the convoy has an armed escort. Just prior to the convoy's approach, the security patrols intensify patrolling the areas immediately next to the road, and assume security positions along the known or suspected enemy avenues of approach. The convoy commander should have each patrol leader's callsign and frequency, and notify him when he is entering and leaving the area.

The elimination of the enemy is the responsibility of the unit through which the convoy is passing. If the convoy makes contact with the enemy, it should get out of the contact area as quickly as possible. The convoy commander will call the appropriate patrol leader or area commander and inform him of the contact. It is the area commander's responsibility to relieve the convoy, reinforce friendly elements, and coordinate indirect fire or air support. The convoy, or its escort, fires on ambush locations until they can get out of the kill zone.

Convoy Escort. The purpose of a mechanized infantry convoy escort is to move with and secure a convoy. The convoy commander establishes the route, checkpoints, fire coordination, and communications. He supervises rehearsal of actions on contact. He also identifies critical loads in the convoy and priorities for cross-loading. The convoy commander and escort should operate on a common radio frequency. Cross-loading spreads critical cargo in more than one vehicle. Convoys can use aerial reconnaissance aircraft as a part of the escort.

The mechanized infantry in their M113s and tanks, if available, both lead and follow the convoy. Wherever possible, the M113s will travel in pairs. If tanks are present, they normally lead the M113s. Communications among all convoy elements are essential. The convoy will travel according to the rate of the slowest vehicle.

The suppression and elimination of hostile fire is the responsibility of the escort commander. His first priority is to preserve the convoy and its contents. Upon being engaged by the enemy, the convoy will speed up to get out of the kill zone. Accompanying mechanized infantry and armor will assault by fire first, and maneuver as necessary. Any activity other than in the immediate area must be coordinated with the commander in whose area the ambush occurs, particularly if indirect fires or pursuit becomes necessary.

TANK-INFANTRY TEAMS

Tanks and mechanized infantry are often restricted by jungle terrain, but where they can be employed, they complement each other well. For this reason, mechanized infantry and armored units are often teamed. The two most common uses of tank-infantry teams in the jungle are for movements to contact and for convoy escorts.

Each must know how the other operates and the characteristics of all weapons systems. The tankers must always know where dismounted infantrymen are and the infantry must protect the tanks from close-in enemy attack.

VEHICLE RECOVERY

Mechanized infantry's primary advantage is its ability to move rapidly as compared to nonmechanized infantry. This often will be curtailed by jungle terrain. Every effort must be made to prevent M113s from becoming stuck or sustaining mechanical malfunctions that render them inoperative. The two most common occurrences are throwing a track and getting stuck in mud. The easiest way to prevent either is to precede each M 113 with a ground guide.

If an MI 13 becomes stuck, the primary recovery vehicle is the M578 vehicle track retriever (VTR). One is assigned to each company. It is equipped with a crane and boom, chains, and various other equipment, and it is usually located in the company trains. *It should be escorted to and from a recovery site as it is not a combat vehicle.* Every effort should be made to recover vehicles by using another MI 13 before the M578 is used, unless an M578 is immediately available. The M578 should be moved only when needed because it has less trafficability than the M113, weighs more, and is wider, which make movement in the jungle difficult for the M578.

Thrown Tracks. An M113 usually throws a track when the track is broken or detached from the drive sprockets. Repair time depends on a number of variables but usually varies from 45 minutes to 4 hours. The best preventive for thrown tracks is maintenance and good driving techniques. These involve keeping the track in good repair and under proper tension, the drive sprockets turned or changed, and supervising the drivers.

Mired M113s. Despite the MI 13s ability to move over most terrain, getting stuck in mud is not uncommon, especially when operating near rice fields or in areas of high rainfall. In these circumstances, ground guides are of little help. Infantry squads must develop expertise in expedient recovery methods. Each M113 should carry tow cables and also chains. Since M578s will not always be available, the squad must know how to recover the vehicle without VTR support. In such instances, other M113s pulling in tandem may be necessary. A tank is even better because it is much heavier than an M113, is more powerful, and generally has greater traction, except in mud.

LOAD PLANS FOR THE MI 13

Load plans are designed to standardize loading of equipment and supplies and evacuation of priority items if the MI 13 is rendered inoperable. The plans further serve to standardize operations within a unit. Load planning requires a reverse sequence—that is, items that are used infrequently and are less critical to the mission than others are loaded first. Items that are used on a daily basis and are essential to mission accomplishment must be readily available. When planning loads, consideration must also be given to maintenance. For example, radios should be easy to get to, and the engine compartment should not be blocked. The load plan diagram should also establish a numerical or alphabetical priority of equipment and supplies.

Section V. COMBAT SUPPORT

RIFLE COMPANY MORTAR SECTION

The rifle company's mortar section's mission is to provide close and continuous indirect fire support to the company. Where trafficable road networks are available, the mortar section can displace to perform its missions while mounted. It may, however, have problems in the jungle environment due to a limited number of suitable firing positions, as a result of heavy rains and restricted overhead clearance. The use of chain saws and demolitions may be necessary to obtain mask and overhead clearance. It also may be possible to fire from positions on or near roads and trails.

When the company operates in terrain where vehicles cannot go, several difficulties surface. Moving a dismounted mortar section in rugged jungle terrain with all of its weapons components and ammunition is difficult. The infantry platoon can be tasked to assist; however, this will severely hamper their ability to maneuver. If the mortar fire is absolutely essential to the company's scheme of maneuver, then consideration should be given to the employment of only one mortar. Also, consideration should be given to the displacement of mortars by helicopter.

Ideally, the rifle platoons will operate within range of the mortars. Resupply is provided by vehicle or helicopter. If the company is operating within range of a 4 .2-inch mortar platoon or an artillery battery, then consideration may be given to using the mortar section as infantry. This will allow the company more operating flexibility and considerably lighten the soldier's fighting load.

FIELD ARTILLERY

The mission of the field artillery is to provide close and continuous indirect fire support; destroy, neutralize, or suppress the enemy indirect fire; and coordinate all supporting fires. A jungle environment poses unique challenges to both maneuver and field artillery commanders by restricting movement, observation, fields of fire, communications, and target acquisition.

Jungle battles are characterized by limited visibility and close combat. The preponderance of fire support is employed danger-close, and must be carefully coordinated with supporting fire direction centers (FDC) and adjacent maneuver units. Adjustment is frequently conducted using creeping techniques and sometimes by sound.

The following factors must be considered when employing field artillery in a jungle environment.

MUNITIONS EFFECTS

Heavy vegetation degrades the effects of all types of munitions. More firepower or alternative fuze action is normally required to achieve effective results.

The following guidance is offered when requesting or selecting shell-fuze combinations:

- **Proximity (VT) fuzed rounds lose much of their effect in tree canopies or other thickly vegetated areas by exploding above or high in the treetops.**

- **Delay fuzes give better effects in heavy vegetation.**

- **Time fuzed rounds also tend to lose some effect, in the jungle by exploding in tree canopies, due to the fuze impact function. Time fuzes, however, can be extremely effective when used for close, defensive fires (300 to 900 meters) with high-explosive (HE) munitions.**

- **Superquick fuzing is useful in open, lightly vegetated, or secondary growth areas. Rounds may explode too soon in high, thick tree canopies, and like time fuzes, lose most of the shell's effect. In low canopy jungles, however, airbursts can increase casualty-causing effects. As the tree canopy is thinned out by bursting munitions, a switch to VT or time fuze maybe required if airbursts are the desired method of attack. Also, superquick fuzes should not be used in muddy or water covered terrain, such as swamps or paddies. The effects of superquick fuzed HE ammunition will be sharply reduced as blast and shell fragments will be directed upward.**

- **Delay fuze can penetrate thick canopies. The fuze triggers in the treetops, and detonates the round in the air at a lower level. However, when used against troops in inundated areas or areas with soft, boggy soil conditions and no canopy to trigger the fuze overhead, delay rounds will bury themselves in the ground, losing their antipersonnel effect.**

- **Improved conventional munitions (ICM) should not be used in thick foliage and water-filled rice paddies. The ICM submunitions may hang up in heavy foliage or fall undetonated to the ground, creating a hazardous area, or they may submerge in water, rendering them ineffective. Appropriately fuzed HE rounds should be substituted for ICM, as necessary.**

Rearward protection for individual fighting positions must be provided when manning a perimeter around an artillery unit (105-mm) which has beehive ammunition. Infantrymen must also be advised of the signals for firing antipersonnel (beehive) munitions and periodically see a demonstration of effects to insure understanding and compliance.

The heat and humidity of tropical areas speeds the deterioration of all types of ammunition. Further, ammunition packing materials and subcomponents are also subject to more rapid deterioration and require protection.

This protection should include:

- **Adequate shelter and ventilation for ammunition.**
- **Rotating stocks. The first-in should be the first fired.**

OBSERVATION

Ground observation is limited, and aerial observation may be more advantageous. All available air assets, proficient in observed fire procedures, to include USAF, should be used when priorities and level of risk/advantage are favorable. Limited visibility might also make it difficult for ground observers to locate themselves and other friendly units. Determining location by one of the indirect fire means may be necessary. Marking rounds may be fired. In severe cases adjustment is possible by sound.

POSITIONING

Positioning of artillery pieces may pose problems due to soft soil, lack of overhead clearance, and piece masking by trees and terrain. When operating in unfamiliar areas, the battery commander's reconnaissance is important. Construction of prefabricated firing platforms, in accordance with FM 5-15, may be necessary when operating in low-lying areas or near bodies of water. In an area where no suitable firing positions are known to exist, an engineer representative should accompany the battery reconnaissance party. The engineer can provide invaluable expertise when a firing position must be carved out of the jungle or selected on poor terrain.

MOBILITY

The mobility of field artillery units may be impaired by bad weather, lack of roads, or poor terrain trafficability conditions. If roads are used for movement, proper preparations must be made. Roadsides may lack adequate room for firing positions during road movements. Road movement requires ground security forces, supporting artillery, and helicopter or USAF cover to maintain clear routes and provide protection and fire support. If roads are too restrictive, artillery units must be well trained in airmobile operations as this may become their primary mode of transportation in a jungle environment. Engineer support may be required to clear and maintain roads and to prepare adequate LZs.

SURVEY

Line-of-sight limitations severely hamper survey operations. Target area survey may be virtually impossible. Radar, celestial observation, simultaneous observation, hasty survey, and observed firing charts should be used to help offset the probable lack of survey. Firing positions should receive the highest priority in survey planning.

COMMUNICATION

Radio relays and elevated antennas are often required to overcome line-of-sight communications restrictions. Cross-country wire can be laid by helicopter, but specially trained and equipped crews are required.

PLANNING AND COORDINATION

Jungle battles are normally very decentralized, but fire support planning and coordination must be centralized.

Since movement is difficult, fire support may be restricted by the inability to quickly move mortars or artillery so that they can provide continuous coverage. Advance planning and continuous situation monitoring are necessary to avoid such difficulties.

Observed fires are coordinated routinely. Unobserved fires require particular attention to protect friendly units because of the observer's difficulty in accurately locating himself and targets. All indirect fires must be coordinated by fire support officers at all levels. Fire support officers must constantly monitor the location of supported maneuver units to be capable of providing rapid clearance when required.

BATTERY DEFENSE

When the enemy situation is vague or significant enemy guerrilla activity is present, artillery units can be collocated with or provided security by elements of an infantry battalion. Additional security is needed because of the artillery's limited capability for self-defense in the face of a determined enemy attack. The artillery unit's capabilities will also enhance the maneuver unit's defense. When the infantry unit provides security, this permits the battery to fulfill its primary mission of fire support.

The star formation of howitzers is optimal for all-round defensive coverage. However, due to terrain restrictions, a lazy W or variation thereof may be utilized.

Additional security in the form of fire support must be planned for and provided by adjacent artillery and mortar units. This additional fire support should be planned on and around unit positions with both HE and illumination. Once in position, the battery's howitzers, FDC, and TOC form an inner perimeter; the infantry occupies the outer perimeter. Each howitzer will be assigned a sector for direct fire. Each howitzer section chief will make a range card. It will include data necessary to engage targets within his assigned sector.

Each howitzer should have beehive (105-mm only) and/or **HE**, with point detonating and time fuzes allocated for battery defense. Battery defense can be improved by collocating one 81-mm mortar with the battery. The mortar can provide close-in illumination and close indirect fire (within 70 meters of its own position). As soon as possible, howitzer positions should be parapeted, vital installations dug in, and fighting and sleeping positions prepared.

HOWITZER FORMATIONS

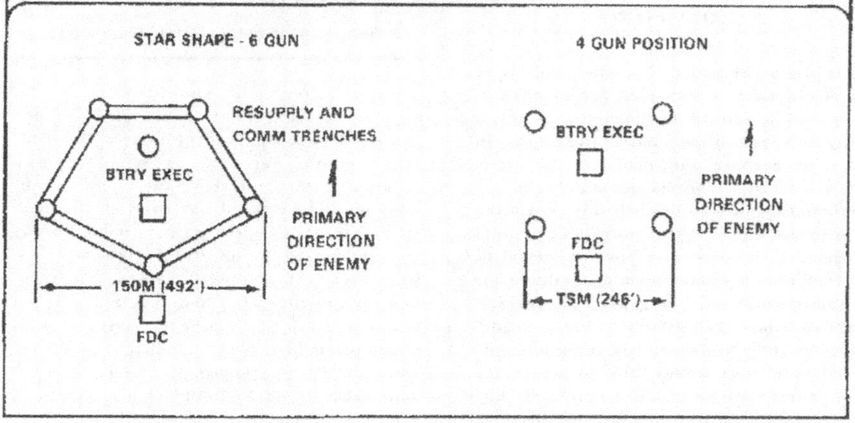

SPLIT BATTERY OR PLATOON OPERATIONS

In order to provide artillery coverage to separated maneuver elements, the battery may have to operate in a split battery or in platoon configurations. If split, the firing elements should be located within maximum range so that both elements can mass fires on the most important targets. Each firing position should be laid out so that each gun can support another gun's position with direct fire, and so located that other artillery units can provide supporting indirect defensive fires. If firing positions are too widely separated, mutual support and the massing of fires will not be possible. Fractional battery operations have a major disadvantage of reducing the number of battery personnel who can be used for battery defense. Augmentation with at least an infantry platoon at each location becomes essential. Separated firing elements should be collocated with a maneuver battalion's defensive position. The battery commander will probably go where the greatest threat exists or with the larger part of the battery. The executive officer will command the other elements of the battery. The communications assets of the battery will be distributed so that each element can monitor and transmit on at least two frequencies. For sustained fractional battery operations, the FDC section must be augmented with additional personnel. This can be done when battery personnel are cross-trained. Split techniques should be avoided where possible as they violate the critical principle of mass and result in drastic reductions in effectiveness.

STRONGPOINT

If it is anticipated that a battery will be in one position for an extended period of time, the position should be hardened to offer protection from direct and indirect fire. In order to prepare the position, adequate materials must be provided, and the battery commander must work closely with the engineers and the maneuver unit commander. If possible, a position should be selected near a usable road to facilitate the movement of heavy engineer equipment, supplies, and ammunition to the position. However, if the strongpoint is accessible only by air, the time needed to construct the strongpoint will be greatly increased. The initial construction efforts will be directed toward the construction of a defensible tactical position by nightfall of the first day.

The following must be completed first:

- **Fighting positions with overhead cover**
- **Howitzer positions with parapets and overhead cover for ammunition**
- **Tactical wire emplaced and Claymore mines positioned**
- **FDC and TOC positions dug in with overhead cover**
- **Range cards and defense plans made**

The firing unit should occupy the position in a formation which provides 6400-mil coverage and optimal battery defense. As the perimeter is expanded, any vegetation which presents site-to-crest problems must be cleared. As soon as the perimeter has been improved, construction efforts are centered on the howitzers. Howitzer positions are reinforced with trail logs, erosion control, powder pits, and revetment of the inside and top of the parapet, using logs or ammunition boxes and sandbags. Sandbags alone can be used to parapet the howitzer, in which case, 1,500 to 2,000 sandbags for each parapet are needed.

105-MM HOWITZER POSITION DIMENSIONS

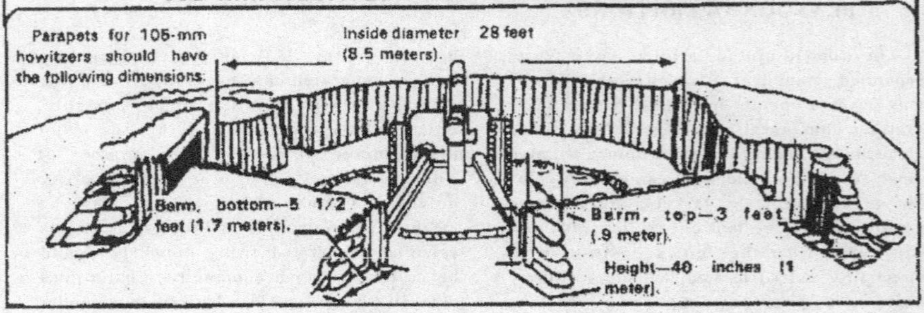

Larger caliber weapons will require special assistance from supporting engineers.

NAVAL GUNFIRE

Naval gunfire, along with close air support, is employed in amphibious operations prior to the landing of artillery units. After maneuver and artillery elements are established ashore, US Navy ships can provide reinforcing artillery support in jungle areas near the coast. Destroyers or smaller inshore fire support ships may be able to provide support along large jungle rivers.

When a unit is conducting operations within range of naval gunfire, a naval gunfire liaison officer will normally be located in the nearest fire support coordination center. He will arrange for ground and aerial observers for the maneuver units. If these are not available, forward observers can adjust the ship's fires. The call-for-fire and adjustment procedures for naval gunfire are basically the same as for artillery support. *(See FM 6-40 for details.)*

NOTE: Naval gunfire provides low trajectory ordnance. This might be a problem in heavily forested jungle areas because the projectile may strike tall trees, detonating prematurely over friendly troops.

TACTICAL AIR SUPPORT

The roles of TACAIR forces are the same in the jungle as elsewhere. In the jungle, where contacts are often unplanned, these aircraft offer another source of responsive combat support. Heavily armed fighter-bombers may be able to engage targets closer to friendly troops than can artillery.

FORWARD AIR CONTROLLERS

Airborne forward air controllers (FAC) are more effective than ground FACs in densely vegetated areas because they have a broader view of the battlefield. The airborne FAC usually can see the target area better for controlling aerial delivery of munitions. All means must be used to mark friendly and enemy locations—smoke, mirrors, and panels. Fires already being delivered can also be used to mark targets. As is true with other aerial systems, the air defense threat is a primary consideration when using airborne FACs.

MUNITIONS

The munitions used vary with the type vegetation and terrain. Unfinned napalm and cluster bomb units may lose much of their effect if they detonate in the thick canopy of primary jungle. Finned napalm, on the other hand, is more stable and accurate and may be used as close as 160 meters to friendly troops in more open areas. The "hard bombs" range in size from 250 to 3,000 pounds and can be used in primary jungle. The FAC will advise danger-close distances. Another weapon, the 20-mm cannon, can be brought in as close as 25 meters from troops under cover or 200 meters from exposed troops.

AIRCRAFT

The A-10 aircraft is particularly well suited to operate in the jungle. It flies slow enough so that the pilot can see his targets. It can also remain in the area for a long time. The A-10 can employ any of the Air Force's close air support munitions, and is fitted with the 30-mm cannon, which is especially well suited for ground support. Safety considerations for the 30-mm cannon are the same as for the 20-mm cannon.

When requesting TACAIR, the requester should specify the type terrain and vegetation it will be targeted on. This will help the Air Force decide what type of ordnance and fuzing to employ.

AIR DEFENSE AND AIR DEFENSE ARTILLERY

Jungles normally provide good concealment from the air. This may reduce air defense requirements. Thickly vegetated areas increase the effectiveness of passive air defense measures, such as camouflage and dispersion. Exposed areas, such as clearings, roads, or river crossings, and facilities where troops tend to concentrate, such as supply points and headquarters, will require air defense artillery (ADA) protection.

The jungle affects ADA mobility, maintenance, and communications in the same way as it affects other combat support resources. In addition, target acquisition is more difficult, and radars are less effective. ADA gunners may not be able to see enemy aircraft until they fly over their position. Reaction times will be short, and many engagements will take place after crossover.

PRIORITY FOR AIR DEFENSE

The jungle concealment afforded to maneuver forces decreases the likelihood that enemy air will attack them. Thus, it is more likely that air attacks will be directed against base complexes, airfields, and logistical facilities. These areas will normally receive first priority for air defense in jungle operations.

HAWKS

Hawk battalions will normally be assigned general support missions for the defense of critical assets, such as base complexes, airfields, port facilities, and lines of communications. As much as possible. Hawk coverage will be extended over the areas in which maneuver forces are operating. To guard against attack by an infiltrating enemy, Hawk elements may be located within base complexes. While this reduces the security problems, it may sacrifice some of the Hawk's early engagement capability. Radars may have to be mounted on towers or berms to be effective.

CHAPARRAL/VULCAN

Chaparral/Vulcan battalions in the jungle can be used in a similar manner to Hawk battalions. In addition, Vulcans can be used to protect convoys from both air and ground attack. Chaparrals may be prepositioned to protect traffic at critical points. When used in this manner, Chaparral/Vulcan units in the jungle are normally attached to the maneuver units they support. Chaparral/Vulcan positions which afford 360-degree coverage and long-range fields-of-fire will be hard to find. Chaparral/Vulcan units may have to clear trees and shrubs to have adequate firing positions.

REDEYE AND STINGER

Redeye and Stinger will cover maneuver units when they are crossing open areas, such as rice paddies or rivers. Redeye and Stinger gunners will also find it hard to find positions which offer 360-degree coverage and long-range fields of fire, and may also have to clear trees and shrubs for adequate firing positions. These positions should only be occupied long enough to fire because they are easily detected from the air. Redeyes and Stingers defending convoys on jungle roads should be positioned within and move with the convoys. Redeyes and Stingers in jungle operations are normally attached to the units they support.

ENGINEERS

Engineers "open up" the jungle. Road building and repair; installation, port, and airfield building; and river crossing operations are continuous tasks for the engineers.

Like most other type supporting units, engineers do their jobs much the same in the jungle as anywhere else. But, also like other type units, they have special considerations to deal with.

ROADS

Road construction in the jungle is usually affected by (1) poor drainage and heavy rainfall, and (2) poor subgrade foundation.

These problems can be dealt with by:

- **Avoiding low ground for road construction**
- **Laying long sections of pontoon bridging, corduroy, or chespaling road through low swampy ground**
- **making roads wider, which thins the overhead foliage, so the sun can get through to dry them out**
- **Using subgrade materials to support heavy traffic** (See TM 5-330 for details.)

AIRFLELDS

Construction of LZs, helipads, and airstrips in remote areas are also important engineer tasks in the jungle. To accomplish these tasks, engineers are inserted into the area to be cleared. Using demolitions and tools, they make an initial clearance of trees and underbrush. For large jobs, follow-on engineers with heavy equipment are brought in to finish the work.

THE LAYING OF A CHESPALING ROAD

Another problem is protecting the surface of these sites from erosion in the rainy season, and protecting aircraft from dust in the dry season. Steel matting, T17 membrane (a tough rubberized fabric), or Peneprime (oil surfacing) are all materials which will keep the dust down on jungle LZs.

PRIMARY SUPPORT

Engineers support the infantry in the jungle primarily by clearing the way for the movement of friendly forces (mobility) while impeding the movement of the enemy (countermobility). Both of these types of support enhance the third area of engineer support—survivability.

Some specific ways engineers can help in all three areas are:

- **Mobility. Improving or building lines of communications and helping move troops and supplies across barriers—particularly water obstacles.**
- **Countermobility. Creating obstacles to support the defense; removing them to assist in the offense.**
- **Survivability. Constructing strongpoints and assisting in the construction of other defensive positions.**

It is also essential that the infantry support the engineers. Engineers are extremely vulnerable during construction activities and need infantry protection to do their jobs.

COMBAT SUPPORT COMPANY IN A JUNGLE ENVIRONMENT

The infantry battalion has its own direct and indirect fire support, reconnaissance/security, and communications section. Their effective employment is critical to the commander's ability to mass his combat power against the enemy. All of the battalion's combat support elements except the communications platoon are assigned to the combat support company.

The combat support company commander has a unique role in that he:

- **Is a special staff officer, making recommendations on the employment of his elements to the battalion commander**
- **Retains command responsibility for the platoons and sections of the company even though he normally loses operational control in a tactical environment (He still supports them logistically and administratively unless the platoons or sections are attached to a rifle company.)**
- **May become the battalion combat base commander** (See battalion combat base, chapter 5.)
- **May be used as a maneuver element commander**

The battalion commander must determine how to use the combat support company elements through a thorough examination of the battalion's mission and the combat support company's state of training. Only then can he select the appropriate role for the combat support company commander and his unit.

If and when the combat support company commander becomes a maneuver element commander, the commander of headquarters and headquarters company must assume the responsibility of commanding the combat base. In addition, soldiers must be taken from the least essential jobs in the battalion trains to replace the combat support company soldiers. Normally, the scout platoon, the antitank platoon, and the battalion reserve element would be formed into a maneuver element.

FUNCTIONS OF THE SCOUT PLATOON

The mission of the scout platoon is to perform reconnaissance, provide limited security, and assist in controlling the battalion's movement. The scout platoon is normally employed as a unit under battalion control. However, under certain conditions it may be attached to another unit or operate as a separate element with an area of operations of its own. Generally, under all three of these conditions, it is the primary mission of the scout platoon to find the enemy. In a jungle environment it normally performs route, zone, and area reconnaissance. The scout platoon has a limited capability to conduct security operations, to provide early warning of enemy maneuver, and to deny the enemy information concerning the battalion disposition and movements. The primary security mission for the scout platoon is the screen. A screen is a series of OPs from which enemy movement can be observed. When vegetation allows good observation, OPs are located on high ground and patrols are only used for local security around the OP. When observation ranges are very short, the OPs may have to be moved to the lower areas. Patrols between OPs are required to detect enemy movement. The distances between OPs may be extremely long so they must be alert for signs of enemy infiltration.

FUNCTIONS OF THE ANTITANK PLATOON

The primary mission of the antitank platoon is to destroy enemy tanks. Its secondary missions are to engage other vehicles, crew-served weapons, fortified positions, and other point-type targets. Against an enemy with no tanks, or in terrain which does not permit the use of long-range antitank weapons, the antitank platoon may be used as a rifle platoon.

CHAPTER 7
Combat Service Support

Section I. GENERAL

Combat service support elements in a jungle environment retain the same basic missions and capabilities as in other environments. However, they must make adjustments due to terrain, weather, and vegetation.

Jungle operations subject personnel and equipment to effects not found in other environments. Trafficability and security problems, discussed in other chapters, often affect combat service support elements as much as maneuver forces.

The lack of an extensive all-weather transportation network in many jungle areas makes the mission of support units more difficult. Transportation difficulties may dictate that maneuver units be resupplied by air, pack animals, or human portage.

US Army forces in the jungle will normally operate at the end of a long line of communication. Cargo space must not be wasted on unessential items. Transportation priority must be given to essential materials, and support areas must not become crowded with comfort items.

Jungle combat operations are characterized by ambushes and infiltration. The security threat caused by infiltrators will require that lines of communication be patrolled frequently and convoys be escorted. Consequently, it is essential that combat service support is performed as far forward as the tactical situation permits. This not only improves response time, but reduces road movement and allows the combat service support elements to take advantage of the security offered by combat units.

The heat and humidity ofjungle areas also have harmful effects on men and equipment.

CONTENTS

Weapons tend to rust quickly, and must be cleaned and oiled more frequently than in most other areas.

Canvas items rot and rubber deteriorates much faster than in more temperate areas.

Battery life is shorter than is normal.

Electrical connections corrode quickly.

Lenses and dials become quickly fogged with internal moisture.

Troops drink more water, requiring greater water purification and transport means.

Section II. LOGISTICS

TRAINS

The variations ot terrain, weather, and vegetation in the jungle affect the organizing, positioning, and securing of combat service support.

EFFECTS OF THE JUNGLE ON OPERATING THE TRAINS

Organizing the trains is often simplified in the jungle. Because of the terrain, aerial resupply will usually be common practice. The responsiveness provided by aerial resupply requires fewer supplies stockpiled in the combat trains. Thus, combat trains consisting of medics, a maintenance element, and small amounts of C rations, ammunition, and lubricants may be the only requirement. The remainder of the logistical assets will be located in the field trains, to include: petroleum, oils, and lubricants (POL); most vehicles; stockpiled ammunition and other ordnance items; the aid station; rations; maintenance contact teams; and, when available, water purification facilities.

Trains should be far enough forward to respond quickly to troop needs, yet far enough to the rear to have minimum exposure to the enemy. The jungle offers excellent sites for concealment, defilade, and dispersion. Since most resupply is done by air, the combat trains may often be located with the field trains in the brigade trains area. Thus, they can be a greater distance to the rear than normal. All trains should be located near LZs. Field,

A JUNGLE LANDING ZONE

combat, and company trains should also be located close to a road, river, or trail which may be used as an alternate method of resupply. If none of these are available, LZs and supply routes may have to be constructed.

Securing the trains area is a major problem in the jungle. Infiltration of enemy units into rear areas is highly probable. In the defense, the trains should be located within a reserve unit defensive perimeter *(see TV, chap 5).* In the offense, the trains personnel normally have to provide their own security. If other forces are available—for example, the reserve—they are integrated into this security plan.

Frequent displacement of the trains is often required, either to evade enemy infiltration or to keep up with combat elements. The combat trains usually move with the combat elements when they displace. They can move independently of the forward combat elements if a unit is provided to secure them. The field trains will normally displace when the brigade trains displace.

RESUPPLYING UNITS

Airlift is key to jungle resupply operations to include movement of supplies to attacking or defending units. Returning aircraft should be used to evacuate casualties.

Since the combat trains in a jungle environment are smaller yet less mobile than in more open terrain, most of the class III (POL) and V (ammunition) will be stocked in the field trains. The helicopter can provide timely resupply when ammunition, POL, etc., are not stored in the combat trains. These supplies can often be moved more quickly by air (helicopter or airdropped by parachute) from the field trains than overland from combat trains. The normal method in the jungle is to deliver supplies directly to forward companies (unit distribution).

SECURITY OF SUPPLY ROUTES

A special security problem exists when supplies are not moved by air. Enemy ambushes and mines are a constant threat on ground supply routes. The following actions may be taken to minimize the threat to supply routes.

Commanders should locate fire support elements, command posts, and defensive positions near supply routes. Subordinate units can then conduct clearing and route security operations. Each unit should be given an area to clear and secure. Maneuver units should operate on the flanks of the route to guard against ambush.

CLEARING TEAM

Infantry, armor, and engineer elements can organize mine-clearing teams which methodically search for and clear mines, bodytraps, and roadblocks. Engineers with mine detectors and probes sweep the road. The infantry and tanks provide security against ambush. Any mines detected are marked and either deactivated or blown in place.

Working behind the mine-clearing teams, dump trucks fill holes in the road with gravel. Other engineer teams replace culverts, repair bridges and stream crossings, and clear vegetation back from the edge of the road.

The road-clearing process is repeated each morning before traffic starts to move. The enemy may replace removed mines during the night.

Security against ambush and attack is provided mainly by patrols to either flank of the route to discover enemy forces before they can get into position to launch an ambush. These patrols should range as far out as possible to reduce the danger of the main body being ambushed. Routes should be patrolled at irregular intervals throughout the day and night.

Clearing the vegetation back from the road's edge will also help to prevent ambush, but may be a very large undertaking. It also increases the enemy's difficulty of mining the road, and facilitates friendly aerial observation and airmobile operations along the road.

Airborne artillery observers, forward air controllers, attack helicopters, and air cavalry reconnaissance increase the security by flying reconnaissance and surveillance missions.

At night, counterambush patrols should be sent out. Scout dogs are very useful to these patrols in detecting the enemy. As additional security on the supply route, STANO devices should be emplaced to detect enemy movement at critical points along the route.

Bridges are vulnerable links in any supply route. They must be protected. A static security post may be required to secure the bridge against attack by guerrillas or infiltrators. Static security posts are best organized in a perimeter defense. The size of the defending force depends on the mission, the size and characteristics of the hostile force, and the importance of the bridge to the mission. Such posts in remote areas will be larger than those nearer supporting forces. The parent unit must be prepared to counterattack with reserves to assist each post.

It will usually be necessary to form armed convoys, escorted by armed vehicles from a military police, cavalry, or maneuver unit. At least one helicopter should fly ahead of the main body to give early warning of ambushes or evidence of hasty mining and provide additional communications capability. On-call fires should be planned on likely ambush sites. Convoys should not be scheduled at regular intervals.

If ambushed, the escort vehicles should immediately return fire. The cargo vehicles should be moved out of the kill zone as rapidly as possible. Attack helicopters and trailing combat vehicles can then maneuver to destroy the ambush.

CLASSES OF SUPPLY

CLASS I (SUBSISTENCE ITEMS)

Nonperishable canned, dried, or dehydrated items primarily should be used when the unit is not actively engaged in combat. This requires kitchen facilities, except for refrigeration. Rations (individual combat meal, long-range patrol, or small detachment rations) normally will be issued to units actively engaged in combat. The number of rations earned by the individual soldier should be determined by such factors as weight, how and in what quantities food can be brought forward, when resupply will be conducted, and the estimated duration of the operation. Hot meals should be served when the combat situation allows. These should be prepared in the field trains and delivered by helicopter. Feeding is usually accomplished during the day because of the danger and difficulty of movement at night and the possibility of enemy ambush or night attack.

CLASS II (INDIVIDUAL EQUIPMENT AND GENERAL SUPPLIES)

In a tropical environment, rapid deterioration is a primary consideration in class II supply. Issue clothing, particularly combat boots and socks, lasts a very short time. Normal wear and tear on clothing may require a resupply every 5 to 6 days. Class II resupply can be a major morale builder. Class II requirements should be estimated well in advance and special provisions should be made for timely resupply. At battalion level, limited emergency supplies of assorted combat boots, socks, uniforms, and similar items of short wear periods should be stocked in the field trains.

CLASS III (POL)

The supply of class III items does not initially present a great problem, as relatively few vehicles will be forward during the operation. However, aviation fuel will be required in large amounts. Units must establish class III distribution points for vehicles and aircraft, as required, in the field trains. Tank trucks will be used as far forward as possible. Collapsible fuel drums can be brought forward by helicopter. Units must take care to insure that diesel and motor gasoline fuel drums are available and marked. When circumstances permit the use of tracked vehicles, planning must include additional POL since tracked vehicles normally have a high fuel consumption rate when operating in rugged terrain. Additional lubricants for weapons will also be required due to the greater requirement for cleaning and protection.

CLASS IV (CONSTRUCTION MATERIALS)

Barrier material and special equipment is extremely bulky and heavy. The use of large amounts of it for defensive positions will create transportation problems. Heavy lift helicopters are the most practical means to move these items.

CLASS V (AMMUNITION)

Because of weight and bulk involved, resupply of ammunition and explosives presents a difficult problem. The nature of the terrain and climate in the jungle magnifies the problems in ammunition storage. Steep terrain should be avoided when selecting suitable sites for storage. Level, firm ground may become too soft in heavy rain. Rain and intense heat combine to speed the deterioration of containers and ammunition, especially if it is taken out of the packing material. To counteract these conditions, adequate dunnage, shelter, and ventilation must be used. Ammunition in the field trains should be left in packing materials until ready for use. Frequent checks should be made for deterioration of packing material.

When the ammunition is unpacked and taken to the field, it is more susceptible to rain, heat, and humidity. Therefore, as a general rule, ammunition which is not used within two weeks should be exchanged for new ammunition. While in the field, however, measures must be taken so that ammunition is never exposed needlessly to the weather. It should be carried in ammunition pouches or containers and protected when stored in positions.

CLASS VT (PERSONAL DEMAND ITEMS)

Such items as soap, cigarettes, candy, etc., are important morale builders. They should be supplied on a regular basis.

CLASS VII (MAJOR END ITEMS)

Storage areas must be designed to protect these items from the jungle environment. Weapons, vehicles, and other items in storage must be inspected more frequently than in other environments. Protective lubricants must be maintained on all surfaces that can rust.

Operational readiness of all stored equipment must be insured by frequent inspections and maintenance. Commanders should turn in those items for which they have no need in the jungle. When major end items are not in use, they are stored or maintained in rear areas.

CLASS VIII (MEDICAL SUPPLIES)

Because the heat and high humidity can cause deterioration of certain medical materials and medicines (for example, penicillin, hydrocortisone liquids, sulfa liquids), frequent resupply and replacement of these items may be required. There will also be a priority need for refrigeration equipment to store class VIII supplies. There will be an increased requirement for antiseptic material because the high heat and humidity increase the incidence of infection in wounds.

CLASS LX (REPAIR PARTS)

Equipment parts that deteriorate or wear out faster in the jungle environment must be determined. Prescribed load list (PLL) must reflect the increased turnover of these parts.

CLASS X (NONMILITARY ITEMS)

These items, for example, agricultural tools and equipment, are not required for the conduct ofconventional combat operations in the jungle; however, they are very important in the conduct of civil-military operations. These items should not be stocked unless a unit is actively participating in a civil-military operation.

OTHER SUPPLIES

Maps exposed to humid jungle air quickly deteriorate. A greater stockage of maps will be required than in other environments.

Water is critical in the jungle. Water must be obtained from the divisional water points using battalion transportation or purified by units themselves in the field. There is a greater demand for water purification tablets and water points. Units should also carry a greater number of lyster bags and other water containers. Close supervision should be exercised to insure that soldiers do not drink unpurified water.

The lack of suitable landing zones makes water resupply extremely difficult if the aircraft cannot land. Numerous techniques have been tried in dropping water through the jungle canopy, for example, free drop, parachute. However, they have met with only limited success and for the most part are impractical. The following water resupply technique has proven to be extremely successful. Water should be placed in either collapsible 5-gallon containers or in 2-quart containers. These containers should be placed inside lightweight, portable containers, that is, rucksacks or duffel bags, which a soldier can quickly move off the DZ. No more than 10 gallons of water (80 pounds) should be placed in each bag for lowering. The bags will be lowered from a helicopter by use of the equipment belay system.

Equipment needed for platoon-size element to supply water by helicopter, using collapsible containers and the Equipment Belay System:

Two 120-foot climbing ropes.

Eleven snap links.

Three 12-foot utility ropes (sling ropes).

One doughnut ring or similar anchoring device.

One rappel D-bag.

Eight 5-gallon containers and four duffel bags.

MAINTENANCE

Maintenance organizations function essentially the same as in other operations. The high humidity and temperature prevalent in jungle areas will increase maintenance requirements. Preventive maintenance on any item affected by moisture and heat is extremely important. Emphasis must be placed on on-site maintenance and the use of aircraft to transport maintenance contact teams and repair parts to unit level. To respond to the need for responsive maintenance support, the number of repair parts for immediate direct exchange must be increased.

TRANSPORTATION

Units fighting in the jungle should consider all types of transportation. Surface transportation facilities are poor in most jungle areas and cannot handle heavy military traffic without extensive improvements. An air line of communication can eliminate many of the problems associated with surface movement.

HUMAN PORTAGE

This is a basic means of moving supplies and equipment in jungle operations. At best, this method is slow, laborious, and inefficient. Yet, in certain extreme situations, it may be the only method available.

WHEELED AND TRACKED VEHICLES

Wheeled vehicles are normally restricted to roads and wider trails, and even these may prove impassable during heavy rains. Sometimes, goods must be transported by cross loading from wheeled to tracked vehicles. For example, large wheeled vehicles move the supplies as far forward as possible, where they are transloaded to tracked vehicles which move them cross-country. In rugged terrain, the supplies may have to be further transloaded to pack animals or native supply bearers.

INLAND WATERWAY

INLAND WATERWAYS

These channels are a very important part of a transport system, especially in remote areas. Large rivers often allow small ships and large boats to penetrate several hundred miles inland. Where smaller streams branch out, a unit may establish transfer points for the transloading of cargo into smaller watercraft. A river that is normally very shallow during the dry season will be deeper during the rainy season, permitting travel by larger craft.

A unit should establish supply points high enough above the level of the water to prevent damage in the event of flash flood.

FLXED-WING TRANSPORTAIRCRAFT

These aircraft can usually operate at greater distances without refueling than cargo helicopters. However, use of fixed-wing aircraft to airland supplies requires more landing strips than may be present. Construction and maintenance of airfields in jungles is a difficult engineer task, but open savanna may be large enough and firm enough to use as an airstrip.

Airdrop of supplies is an alternative to airlanding Airdrop by parachute is a rapid means of delivery and makes deliveries to isolated units possible without further transloading.

Disadvantages include the dispersion of supplies and the possibility of lost cargo in the jungle canopy, vulnerability to local enemy air defense, and requirement for at least local friendly air superiority.

A variation of airdrop by parachute is the low altitude parachute extraction system (LAPES). The parachute is used to pull the load from the rear ramp of an aircraft flying at a reduced speed just a few feet off the ground. LAPES solves dispersion and tree hangup problems, allowing use of the smaller drop zones more common in jungle areas. However, fragile equipment might not survive this type of drop. In areas where the aircraft can make a low-level approach, such as savannas, LAPES may be a valuable resupply technique. It is used with C-130 aircraft only.

LAPES

The container parachute delivery system provides single pass delivery of from 1 to 16 individually rigged A-21 containers into a small drop zone (DZ) at altitudes of 500 to 600 feet.

When using this system, selection of a DZ is critical to accurate delivery of the bundles. The nature of jungle terrain and the small DZs do not allow for accurate delivery of bundles. The DZ must be at least a 400-meter by 400-meter area. A straight stretch of road or a large field visible from the air may be used. If an open ground area cannot be located, the next step is to try to make the drop in a river, lake, or ocean. If dropped in the water, bundles must be rigged with adequate buoyancy (empty water cans will do). Recovery of the "bundles must be preplanned and executed immediately upon completion of the drop.

Another method which can be used for small loads of some sturdy types of supplies and equipment is the free drop. Here, the load is simply dropped, without parachute, from a low-flying aircraft.

Helicopters are very effective in this technique because they can hover low over a unit and drop the supplies exactly where desired. Because helicopters are vulnerable to small-arms fire, however, the unit on the ground must take adequate security measures. It should establish a perimeter defense around the area over which the helicopter will hover, and it should conduct patrols prior to the resupply to insure that the surrounding area is cleared of enemy.

FREE DROP

MEDICAL OPERATIONS

Medical support in jungle operations follows the basic principles of military medicine, but medical treatment and evacuation will often be complicated by the extended distances and inaccessibility due to terrain and vegetation. The manner in which medical units support tactical operations will depend on the employment of the supported unit and the type of jungle in which they are operating. The relatively high incidence of disability caused by heat, humidity, and insect-borne diseases will require increased emphasis on health and sanitation measures.

The evacuation of wounded in jungle warfare presents numerous problems. Because of the prevalence of rapidly debilitating tropical diseases and the likelihood of quick infection of wounds in the jungle, speedy evacuation is vital. Units must use all available methods to augment organic medical evacuation means. These may include tracked vehicles, pack animals, watercraft, helicopters, and litter bearers. Ground medical evacuation in the jungle will normally be along supply routes, as they are usually given all possible protection from the enemy.

The difficulties of overland medical evacuation in jungle operations emphasize the advantages of air evacuation. When weather, aircraft availability, and friendly air superiority permit their use, helicopters provide fast, comfortabe, and efficient transportation for casualties. The reduction of time between injury and treatment increases the chance of survival and raises morale. A helicopter equipped with a cable hoist system can evacuate casualties from thick jungle vegetation where no landing zones are available.

In some remote and densely foliaged jungles, the only means of evacuation maybe by litter. This is a slow and exhausting task requiring well-conditioned soldiers to carry the litters. At best, litter teams can carry patients only a few hundred meters over rough jungle terrain before needing rest or relief.

There are several types of litters. In very rugged terrain, the mountain-type metal basket litter offers most security to the patient. However, it is bulky to carry even during normal operations. The standard folding litter or field expedient variations made of cloth and poles are less bulky and get the job done just as well if patients are moved with great care.

Litter hauls should be as short as possible. Medical facilities should be as close as practicable to the troops.

PERSONNEL MA NAGEMENT CONSIDERATIONS

REPLACEMENTS

The jungle may appear as a fearsome place to the uninitiated. Individual replacements may doubt their ability to cope with tropical diseases; heat; oppressive humidity; entangling close vegetation; the constant threat of ambush and close combat; wildlife; and periodic isolation from support elements. Commanders must realize that these are natural fears and must provide a period for acclimation and psychological adjustment. All new arrivals must receive a complete orientation on the unit, its mission, and the enemy, and be trained in the routines of jungle living, working, and fighting.

GRAVE REGISTRATION

Direct support for the recovery, identification, and evacuation of deceased personnel will normally be available in the jungle. However, the hot and humid climate presents special problems, and it is important that the remains of the dead be removed rapidly to prevent their deterioration. *This may have profound effect on troop morale.* Hasty burials should be conducted as a last resort. Grave sites should be marked clearly and reported.

APPENDIX A

References

FIELD MANUALS (FM)

3-10		Employment of Chemical Agents
5-15		Field Fortifications
5-20		Camouflage
5-31		Boobytraps (U)
5-36		Route Reconnaissance and Classification
6-20	(HTF)	Fire Support in Combined Arms Operations
6-40		Field Artillery Cannon Gunnery
6-50		The Field Artillery Cannon Battery
7-7	(HTF)	The Mechanized Infantry Platoon and Squad
7-8	(HTF)	The Infantry Platoon and Squad (Infantry, Airborne, Air Assault, Ranger)
7-10	(HTF)	The Infantry Rifle Company
7-20	(HTF)	The Infantry Battalion (Infantry, Airborne, Air Assault, Ranger)
8-35		Evacuation of the Sick and Wounded
19-40		Enemy Prisoners of War, Civilian Internees, and Detained Persons
20-12		Amphibious Embarkation
20-32		Mine/Countermine Operations at the Company Level
20-33		Combat Flame Operations
21-10		Field Hygiene and Sanitation
21-11		First Aid for Soldiers
21-20		Physical Readiness Training
21-26		Map Reading
21-30		Military Symbols

21-31		Topographic Symbols
21-40		NBC (Nuclear, Biological, and Chemical) Defense
21-60		Visual Signals
21-75		Combat Training of the Individual Soldier and Patrolling
21-76		Survival, Evasion, and Escape Manual
23-9		M16A1 Rifle and Rifle Marksmanship
23-90		81-mm Mortar
24-18		Field Radio Techniques
24-20		Field Wire and Field Cable Techniques
24-21		Tactical Multichannel Radio Communications Techniques
29-51		Division Supply and Field Service Operations
30-5		Combat Intelligence
30-10		Military Geographical Intelligence (Terrain)
44-1	(HTF)	US Army Air Defense Artillery Employment
44-3	(HTF)	Air Defense Artillery Employment: Chaparral/Vulcan
57-38		Pathfinder Operations
71-1	(HTF)	Tank and Mechanized Infantry Company Team
71-2	(HTF)	The Tank and Mechanized Infantry Battalion Task Force
90-2	(HTF)	Tactical Deception
90-4	(HTF)	Airmobile Operations
90-6	(HTF)	Mountain Operations
90-10	(HTF)	Military Operations on Urbanized Terrain (MOUT)
90-13	(HTF)	River Crossing Operations
100-5	(HTF)	Operations
100-20		Low Intensity Conflict

TECHNICAL MANUALS (TM)

3-240	Field Behavior of Chemical, Biological, and Radiological Agents
5-277	Bailey Bridge
5-312	Military Fixed Bridges
9-1300-205	Ammunition and Explosive Standards
10-1670-262-12	Operator's and Organizational Maintenance Manual (Including Repair Parts and Special Tool Lists) for Personnel Stabo Extraction System and Cargo and Personnel Lowering Device
11-666	Antennas and Radio Propagation

APPENDIX B

Navigation and Tracking

Section I. GENERAL

This appendix assumes that the reader already knows the basics of map reading and land navigation. For a review of these subjects, the reader should refer to *FM 21-26.*

Section II. JUNGLE NAVIGATION

Navigating in the jungle can be difficult for those troops not accustomed to it. This appendix outlines techniques which have been used successfully in jungle navigation. With training and practice, troops should be able to use these techniques to navigate in even the thickest jungle.

NAVIGATION TOOLS

MAPS

Because of the isolation of many jungles, the rugged ground, and the presence of the canopy, topographic survey is difficult and is done mainly from the air. Therefore, although maps of jungle areas generally depict the larger features (hill, ridges, larger streams, etc.) fairly accurately, some smaller terrain features (gullies, small or intermittent streams, small swamps, etc.), which are actually on the ground, may not appear on the map. Also, many older maps are inaccurate. So, before going into the jungle, commanders and staff should bring their maps up to date.

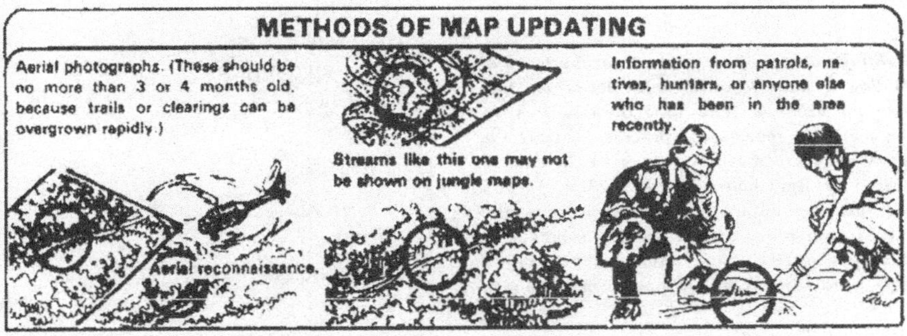

CONTENTS

COMPASS

No one should move in the jungle without a compass. It should be tied to the clothing by a string or bootlace. The three most common methods used to follow the readings of a compass are:

Sighting along the desired azimuth. The compass man notes an object to the front (usually a tree or bush) that is on line with the proper azimuth and moves to that object. *This is not a good method, in the jungle as trees and bushes tend to look very much alike.*

Holding the compass at waist level and walking in the direction of a set azimuth. *This is a good method for the jungle. The* compass man sets the compass for night use with the long luminous line placed over the luminous north arrow and the desired azimuth under the black index line. There is a natural tendency to drift either left or right using this method. Jungle navigators must learn their own tendencies and allow for this drift.

Sighting along the desired azimuth and guiding a man forward until he is on line with the azimuth. The unit then moves to the man and repeats the process. This is the most accurate method to use in the jungle during daylight hours, but it is slow. In this method, the compass man cannot mistake the aiming point and is free to release the compass on its string and use both hands during movement to the next aiming point.

The keys to navigation are maintaining the right direction and knowing the distance traveled. Skill with the compass (acquired through practice) takes care of the first requirement. Ways of knowing the distance traveled include checking natural features with the map, knowing the rate of movement, and pacing.

CHECKING FEATURES

Major recognizable features (hills, rivers, changes in the type of vegetation) should be noted as they are reached and then identified on the map. Jungle navigators must BE CAUTIOUS ABOUT TRAILS—the trail on the ground may not be the one on the map.

RATE OF MOVEMENT

Speed will vary with the physical condition of the troops, the load they carry, the danger of enemy contact, and the type of jungle growth. *The normal error is to overestimate the distance traveled.* The following can be used as a rough guide to the maximum distance covered in 1 hour during daylight.

DAYLIGHT MOVEMENT

TYPE TERRAIN	MAXIMUM DISTANCE (in meters per hour)
TROPICAL RAIN FOREST	1,000
DECIDUOUS FOREST, SECONDARY JUNGLE, TALL GRASS	500
SWAMPS	100 TO 300
RICE PADDIES (WET)	800
RICE PADDIES (DRY)	2,000
PLANTATIONS	2,000
TRAILS	1,500

PACING

In thick jungle, this is the best way of measuring distance. It is the only method which lets the soldier know how far he has traveled. With this information, he can estimate where he is at any given time-something that must be known to call for indirect fire support in a hurry. To be accurate, soldiers must practice pacing over different types of terrain. Each soldier should make a PERSONAL PACE TABLE like this

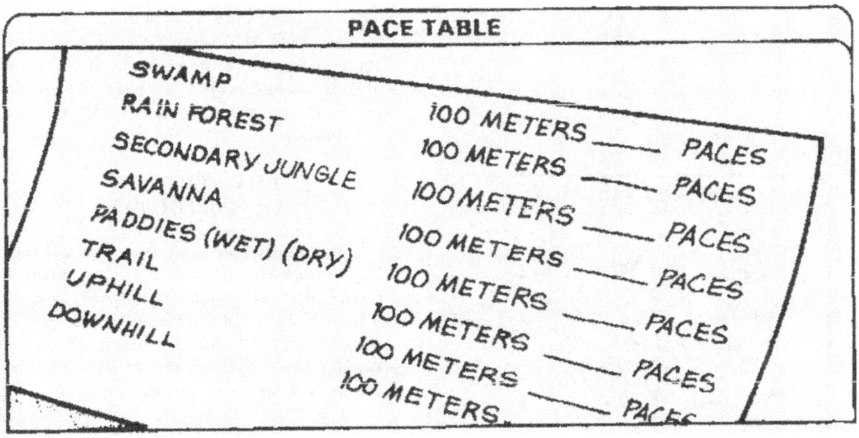

At least two men in each independent group should be compass men, and three or four should be keeping a pace count. The artillery fire support team (FIST) chief should keep an accurate fix on his location, as should the platoon forward observers (FO). He can be a great help to the rifle company commander in matters of navigation. When in doubt, the commander should **do a resection, using artillery marking rounds as outlined in** *FM 21-26, paragraph 10.*

NOTE: Jungle foliage will often require that artillery marking rounds be sensed by sound.

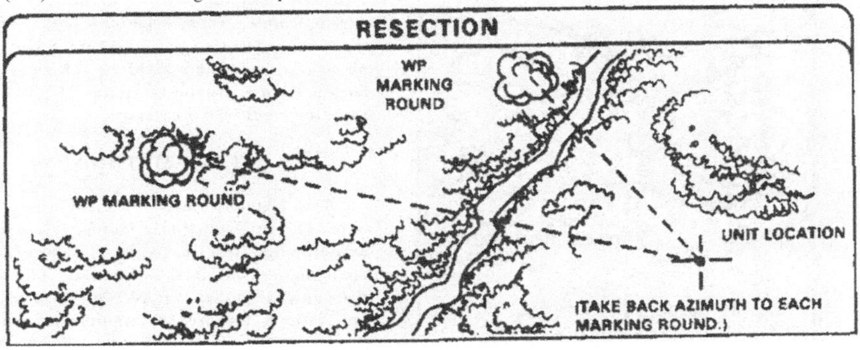

To locate a position by resection, the general location must be known. White phosphorus rounds (airbursts) are then called on two widely separated grids which are not on terrain features like the one the unit is occupying and which are a safe distance from the estimated location. A back azimuth to each of these rounds is taken and plotted on the map. The point where they intersect is the observer's approximate location.

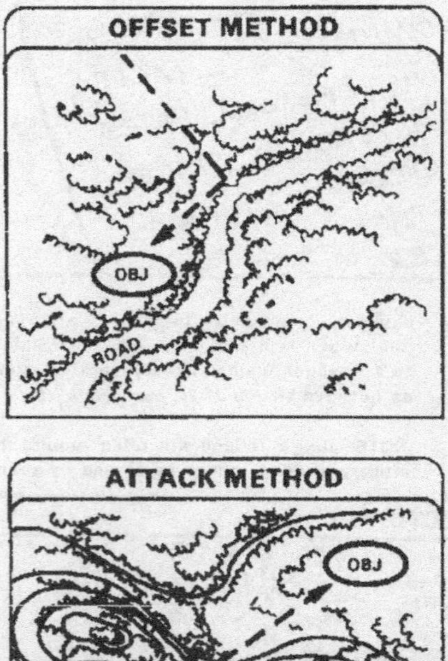

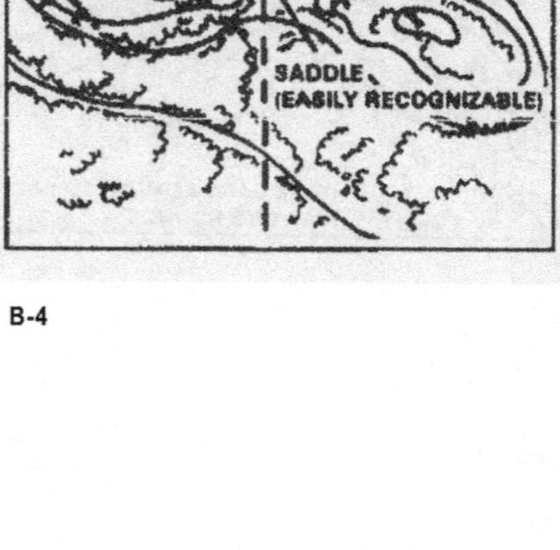

LOCATION OF AN OBJECTIVE

In open terrain, an error in navigation can be easily corrected by orienting on terrain features which are often visible from a long distance. In thick jungle, however, it is possible to be within 50 meters of a terrain feature and still not see it. Here are two methods which can aid in navigation.

OFFSET METHOD

This method is useful in reaching an objective that is not large or not on readily identifiable terrain but is on a linear feature, such as a road, stream, or ridge. The unit plans a route following an azimuth which is a few degrees to the left or right of the objective. The unit then follows the azimuth to that terrain feature. Thus, when the unit reaches the terrain feature, the members know the objective is to their right or left, and the terrain feature provides a point of reference for movement to the objective.

ATTACK METHOD

This method is used when moving to an objective not on a linear feature. An easily recognizable terrain feature is chosen as close as possible to the objective. The unit then moves to that feature. Once there, the unit follows the proper azimuth and moves the estimated distance to get to the objective.

WHAT TO DO IF LOST

Do not panic. Few soldiers have ever been permanently lost in the jungle, although many have taken longer to reach their destination than they should.

Disoriented navigators should try to answer these questions. (If there are other navigators in the group, they all should talk it over.)

What was the last known location?

Did the unit go too far and pass the objective? (They should compare estimates of time and distance traveled.)

Does the terrain look the way it should? (They should compare the surroundings with the map.)

What features in the area will help to fix the unit's location? (They should try to find these features.)

If the unit is unable to locate itself using these techniques, the leader can call for an air or artillery orienting round. However, this may cause a loss of security, particularly if the unit is moving by stealth.

An airspot can usually be obtained from Army or Air Force aircraft. The pilot can be contacted and guided to the general location by radio. A mirror, smoke, panels, or some other signal can be shown to the pilot. He will be able to determine the unit's location and report it to them.

This, also, is a loss of security and should only be a last resort.

NIGHT MOVEMENT

The principles for navigation at night are the same as those for day movement. The problem in night movement is one of control, not navigation. In clear weather, through sparse vegetation and under a bright moon, a unit can move almost as fast by night as by day. If the sky is overcast, vegetation is thick, or there is little or no moon, movement will be slow and hard to control. The following points can assist a unit during night movement.

Attach *luminous tape* to the back of each soldier's headgear. Two strips, side by side, each about the size of a lieutenant's bar, are recommended. *The two strips aid depth perception and reduce the hypnotic effect that one strip can cause.*

When there is no light at all, distance between soldiers should be reduced. When necessary to prevent breaks in contact, each soldier should hold on to the belt or the pack of the man in front of him.

The leading man should carry a long stick to probe for sudden dropoffs or obstacles.

In limited visibility conditions, listening may become more important to security than observing. A unit which hears a strange noise should halt and listen for at least 1 minute. If the noise is repeated or cannot be identified, patrols should be sent out to investigate. Smell, likewise, can be an indication of enemy presence in an area.

All available night vision devices should be used.

NAVIGATIONAL TIPS

- Trust the map and compass, but understand the map's possible shortcomings. Use the compass bezel ring, especially during night navigation.
- Break brush. Do not move on trails or roads.
- Plan the move, and use the plan.
- Do not get frustrated. If in doubt, stop and think back over the route.
- Practice leads to confidence.

Section III. TRACKING

Visual tracking is following the paths of men or animals by the signs they leave, primarily on the ground or vegetation. Scent tracking is following men or animals by the odors they leave.

Practice in tracking is required to achieve and maintain a high standard of skill. Because ofthe excellent natural concealment the jungle offers, all soldiers should be familiar with the general techniques of visual tracking to enable them to detect the presence of a concealed enemy, to follow the enemy, to locate and avoid mines or booby-traps, and to give early warning of ambush.

Tracking is important in counterguerrills operations where it is often difficult to locate the enemy. Guerrillas who conduct raids and ambushes will normally return to their bases as quickly as possible. Well-developed tracking skills will help units to maintain contact with the enemy.

SIGNS

Men or animals moving through jungle areas leave signs of their passage. Some examples of these signs are listed below.

TRACKING POINTS

SAVANNA	ROCKY GROUND
NOTE:If the grass is high, above 3 feet, trails are easy to follow because the grass is knocked down and normally stays down for several days. If the grass is short, it springs back in a shorter length of time. ■ Grass that is tramped down will point in the direction that the person or animal is traveling. ■ Grass will show a contrast in color with the surrounding undergrowth when pressed down. ■ If the grass is wet with dew, the missing dew will show a trail where a person or an animal has traveled. ■ Mud or soil from boots may appear on some of the grass. ■ If new vegetation is showing through a track, the track is old. ■ In very short grass (12 inches or less) a boot will damage the grass near the ground and a footprint can be found.	■ Small stones and rocks are moved aside or rolled over when walked on. The soil is also disturbed, leaving a distinct variation in color and an impression. If the soil is wet, the underside of the stones will be much darker in color than the top when moved. ■ If the stone is brittle, it will chip and crumble when walked on. A light patch will appear where the stone is broken and the chips normally remain near the broken stone. ■ Stones on a loose or soft surface are pressed into the ground when walked upon. This leaves either a ridge around the edge of the stone where it has forced the dirt out, ora hole where the stone has been pushed below the surface of the ground. ■ Where moss is growing on rocks or stones, a boot or hand will scrape off some of the moss.

TRACKING POINTS CONTINUED	
PRIMARY JUNGLES NOTE: Within rain forests and deciduous forests, there are many ways to track. This terrain includes undergrowth, live and dead leaves and trees, streams with muddy or sandy banks, and moss on the forest floor and on rocks, which makes tracking easier. ■ Disturbed leaves on the forest floor, when wet, show up a darker color when disturbed. ■ Dead leaves are brittle and will crack or break under pressure of a person walking on them. The same is true of dry twigs. ■ Where the undergrowth is thick, especially on the edges of the forest, green leaves of the bushes that have been pushed aside and twisted will show the underside of the leaf—this side is lighter in color than the upper surface. To find this sort of trail, the tracker must look through the jungle instead of directly at it. ■ Boot impressions may be left on fallen and rotting trees. ■ Marks may be left on the sides of logs lying across the path. ■ Roots running across a path may show signs that something has moved through the area. ■ Broken spiderwebs across a path indicate that something has moved through the area.	**SECONDARY JUNGLE** ■ Broken branches and twigs. ■ Leaves knocked off bushes and trees. ■ Branches bent in the direction of travel. ■ Footprints. ■ Tunnels made through vegetation. ■ Broken spiderwebs. ■ Pieces of clothing caught on the sharp edges of bushes. **RIVERS, STREAMS, MARSHES, AND SWAMPS** ■ Footprints on the banks and in shallow water. ■ Mud stirred up and discoloring the water. ■ Rocks splashed with water in a quietly running stream. ■ Water on the ground at a point of exit. ■ Mud on grass or other vegetation near the edge of the water.

DECEPTION

The enemy may use any of the following methods to deceive or discourage trackers. They may, at times, mislead an experienced tracker.

These deceptions include:

- Walking backwards. The heel mark tends to be deeper than that of the ball of the foot. The pace is shorter.
- More than one person stepping in the same tracks.
- Walking in streams.
- Splitting up into small groups.
- Walking along fallen trees or stepping from rock to rock.
- Covering tracks with leaves.

WARNING

A TRACKER SHOULD ALWAYS BE ALERT TO THE POSSIBILITY THAT THE ENEMY IS LEAVING FALSE SIGNS TO LEAD THE UNIT INTO AN AMBUSH.

APPENDIX C

Jungle Obstacles

Section I. GENERAL

Units operating in the jungle will have to cross many obstacles. The most difficult obstacles will be streams and cliffs. In addition, units operating in the jungle will frequently have to insert or extract soldiers and units in places where helicopters cannot land. This appendix covers the skills required to perform these tasks.

Before learning these skills, however, soldiers must be familiar with ropes and knots.

Section II. CHARACTERISTICS AND USE OF EQUIPMENT.

ROPES AND KNOTS

TYPES OF ROPE

Nylon. Nylon rope is most commonly used in climbing and rappelling. The rope is seven-sixteenths of an inch in diameter and is issued in 120-foot lengths. Its dry breaking strength averages 3,840 pounds (plus or minus 5 percent). Strength is reduced by about 20 percent when the rope is wet. It will also stretch about one-third of its length when wet. Nylon sling (utility) ropes are commonly prepared by a unit in 12-foot lengths from older ropes that are no longer used for climbing or rappelling.

Vegetable Fiber. This is readily available in jungle areas as it is made primarily from the fibers of tropical plants.

Manila rope is made from the fibers of the leaves of a banana tree. The lighter the color of the rope, the better the quality. This rope is superior to nylon rope for suspension traverses and rope bridges because it does not stretch as much as nylon, and it is not weakened when wet.

CONTENTS

The breaking strength and safe load capacity (respectively) for the sizes of manila rope most often used by jungle troops are:

1 -inch-diameter rope-9,000 pounds/ 2,250 pounds

1 /2-inch-diameter rope-3,650 pounds/ 660 pounds

Hemp rope is made from the fibers of the hemp plant. This is the strongest of the fiber ropes. It is usually soaked in tar to preserve the rope from damage caused by dampness, but this tar tends to reduce the rope's strength. Also, because of its greater weight, tarred hemp is not practical for use by infantry troops.

NOTE: The breaking strength of a rope is always greater than its safe working capacity. The difference is a "safety factor. " Individual ropes can vary greatly in minimum breaking strength. Even though a rope may not break under this load, the fibers are stretched beyond their elastic limit. Thereafter the strength of the rope is permanently reduced. Exposure, wear, use, and bending decrease a rope's strength over a period of time. This should be allowed for in estimating the strength of a used rope. The strength of a rope that is slung over a hook or contains a knot is reduced by about 30 percent; sharp bends over corners will cut strength by 50 percent; sand or grit between the fibers will quickly cut the fibers, and sharply drop the overall strength of the rope.

CARE OF A ROPE

Clean a muddy rope by washing it in water, but not in salt water.

Do not pull a rope oversharp edges. Place layers of heavy cloth or grass between the rope and any sharp edge to prevent the cutting of fibers.

Do not drag a rope through sand and dirt, or step on it, or drive over it.

Keep a rope dry. If it gets wet, dry it as soon as possible to prevent rotting. (A mildewed rope will have a musty odor and inner fibers will have a dark, stained look.)

Do not leave a rope knotted or tightly stretched any longer than needed.

Never splice a climbing or rappelling rope.

Inspect a rope often, both the outside and the inside. Untwist a few strands at different points to open the rope to check the inside.

Melted nylon and dark streaks indicate burns. Nylon rope burns when it rubs against other nylon ropes. Nylon ropes should never be tied in such away that there is rope-to-rope friction.

Dirt and sawdust-like material inside the rope indicates damage.

A rope should be checked at a number of different places—any weak point in it weakens the entire rope.

Whenever any unsafe conditions are found in a rope, it should be destroyed or cut up in short pieces. This w-ill prevent use of the rope for hoisting. The short pieces can be used for toggle ropes and for other purposes which do not involve load bearing operations.

KNOT-TYING TERMS

Knots to tie the end of a rope to an object (anchor knots).

The bowline will not slip under strain, yet it is easily untied.

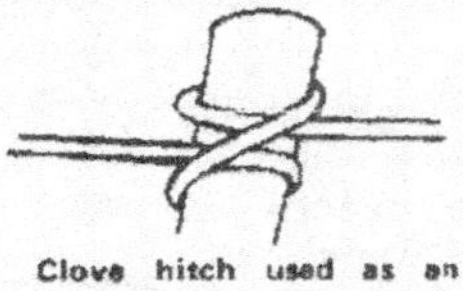

Clove hitch used as an intermediate anchor with tension applied at all times to prevent slipping.

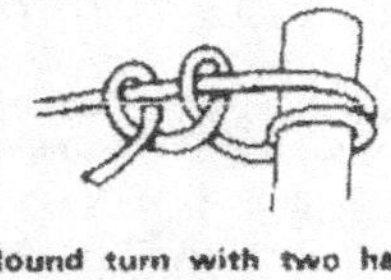

Round turn with two half-hitches.

Knots to tie the ends of two ropes together.

A square knot is used to tie ropes of equal diameter together.

The double sheet bend knot is used to tie ropes of unequal diameter together.

Middle-of-rope knots.

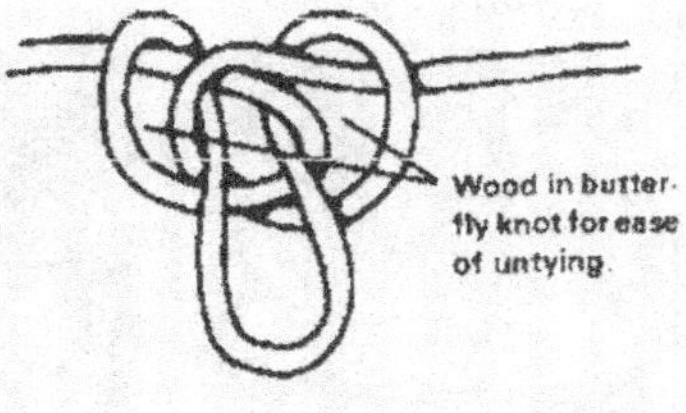

The butterfly knot is used to pull a line tight.

NOTE: After a butterfly knot is tied for a rope bridge, or for any other purpose where a great amount of strain or tension is applied to the knot, it becomes almost impossible to untie. Pieces of wood or pipe inserted through the two loops or wings of the knot will make it easier to untie.

Special knots.

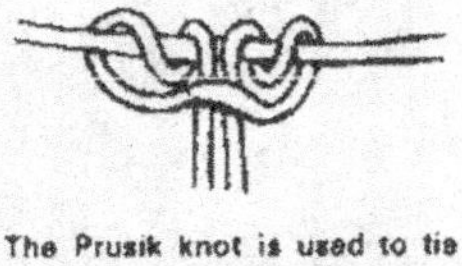

The Prusik knot is used to tie one rope around another. One rope will slide along the other if there is no tension, and it will hold if tension is applied. This knot should be tied off with a bowline.

Bowline on a bight forms a double loop.

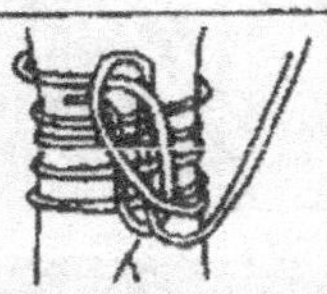

Bowline on a coil can be used by the first and last man on a climbing rope to take up unnecessary slack. A half-hitch must be employed behind the knot.

Section III. MOVING DOWN HILLS AND CLIFFS

RAPPELLING

Rappelling is a means to move quickly down very steep hills and cliffs. Rappelling involves sliding down a rope which has been anchored around a firm object (anchor point) such as a tree, projecting rock, or piton.

ROPE AT ANCHOR POINT

When setting up a rappel site:

Be sure the rope reaches the bottom of the site.

Test the anchor point carefully. The rope should run around the anchor point when pulled from below to recover the rope. Insure that the area is clear of loose rocks.

CAUTION:

GLOVES MUST BE WORN DURING ALL RAPPELLING TO PROTECT THE HANDS FROM SEVERE ROPE BURNS.

HASTY RAPPEL

The hasty rappel is a fast, easy way to get down a moderately steep slope or cliff. A soldier must wear a shirt to do a hasty rappel.

HASTY RAPPEL

To conduct a hasty rappel:

Face slightly sideways to the anchor point and place the ropes across the back. The hand nearest the anchor is the guide hand, and the lower hand is the brake hand.

Walk sideways down the hill or cliff, letting the rope move through the hands and across the back.

To stop, bring the brake hand across in front of the body, locking the rope. At the same time, turn and face up toward the anchor point.

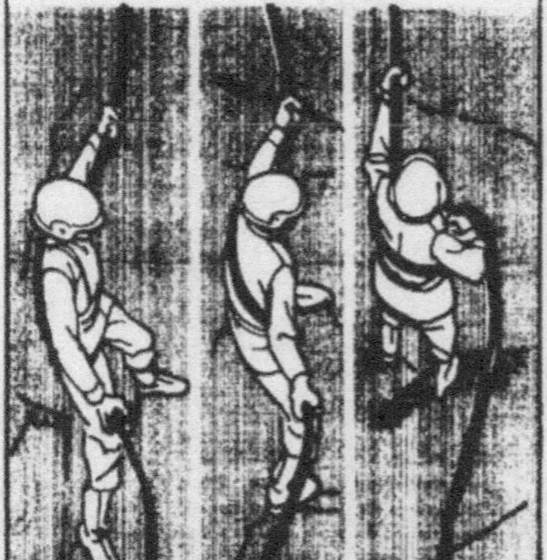

SEAT-HIP RAPPEL

The seat-hip rappel is a fast method of getting down a steep hill or cliff, and it is also used to rappel from helicopters. In this rappel, friction is taken up by a snaplink inserted in a rappel seat fastened to the body.

SEAT-HIP RAPPEL SEQUENCE

Tuck the fatigue jacket into the trousers. Any loose clothing or equipment around the waist can be pulled into the snaplink during the rappel, and present a safety hazard.

Place the center of the sling rope on the hip opposite to the brake hand. (For right-handed soldiers, the right hand is the brake hand. For left-handed soldiers, the left hand is the brake hand.)

Wrap the rope around the waist (keeping the center of the rope on the hip) and tie an overhand knot in front of the body.

Bring the ends of rope between the legs (front to rear), up and around the legs (rear to front), and under the waist loop.

Tie the ends of the rope with a square knot and two half-hitches on the side opposite the brake hand (secure any loose ends).

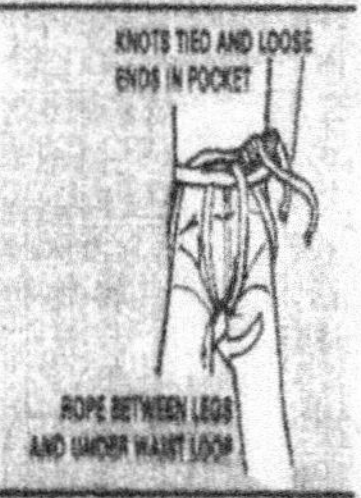

Place the snaplink through the single rope around the waist and through the two ropes forming the overhand knot. (Insert the snaplink with the gate down and the opening toward the body.)

Then rotate the snaplink one-half turn so that the gate is up and opens away from the body.

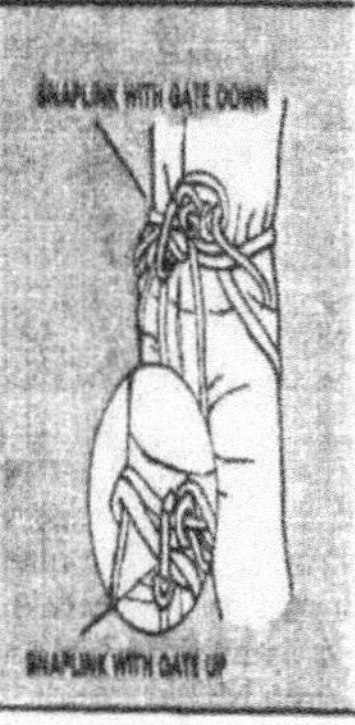

Stand to one side of the rappelling ropes (when braking with the right hand, stand on the left side, and when braking with the left hand, stand on the right). Place the ropes into the snaplink.

Take up some slack in the ropes between the snaplink and the anchor point and bring it underneath, around, over, and back through the snaplink. (This results in a turn of the ropes around the solid shaft of the snaplink which does not cross itself when under tension. When using a single rope, make two turns.)

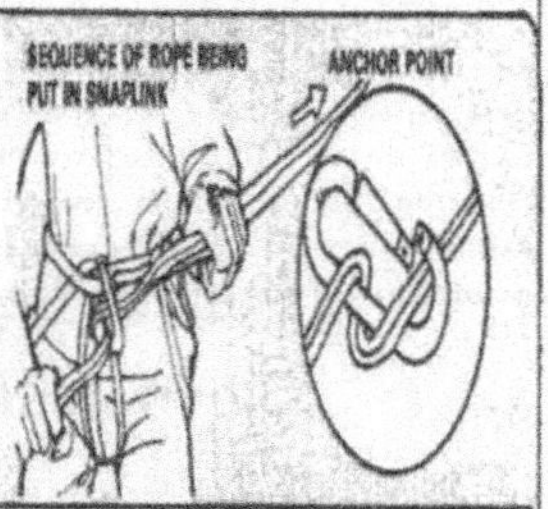

Facing the anchor point, back carefully over the edge of the obstacle and lean well out, almost at a right angle to the surface (the "L" position).

Shout "On rappel."

Go down using the upper hand as a guide and the lower hand as the brake hand. Hold the brake hand behind and slightly above the hip. Brake by closing the brake hand and pressing the rope against the body.

Continue to "walk" down, looking at the ground over the brake hand. To go faster, push off the cliff surface and go down in "bounds" with the brake hand extended toward the ground.

NOTE: For clarity, the load-carrying equipment and weapon are not shown.

Upon reaching the bottom, get off the rope, straighten it, and give the signal, "Off rappel." (When silence is required, use a prearranged signal such as pulling on the rope a set number of times.)

When the last man is down, recover the rope by pulling one end of it until the entire rope has pulled free of the anchor point. This can only be done if the rope is not tied to the anchor point. Pull the rope smoothly to prevent it from getting tangled.

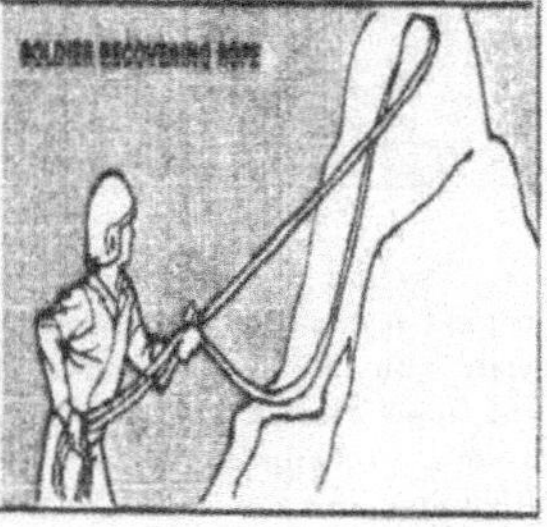

Section IV. MOVING BY HELICOPTER

RAPPELLING FROM HELICOPTERS

In the jungle, there are many places where the vegetation or the ruggedness of the terrain will not permit a helicopter to land. Therefore, it maybe necessary to rappel from a helicopter to get on the ground. Special equipment is required to rappel from helicopters. There are no safe field expedients. The following paragraphs describe the equipment and procedure used in rappelling from helicopters and furnish information on other extraction means.

EQUIPMENT

The *donut ring* is the primary anchoring device (anchor point) inside a helicopter. The floating safety ring is the secondary anchor point. The donut ring consists of a 12-inch solid ring of VS-inch cold-rolled steel cable; seven parachute static line snap hooks; four Ma-inch U-bolts; and 12 inches of chain or '/2-inch cable. Instructions for making a donut ring and a floating safety ring are found in FM 57-38.

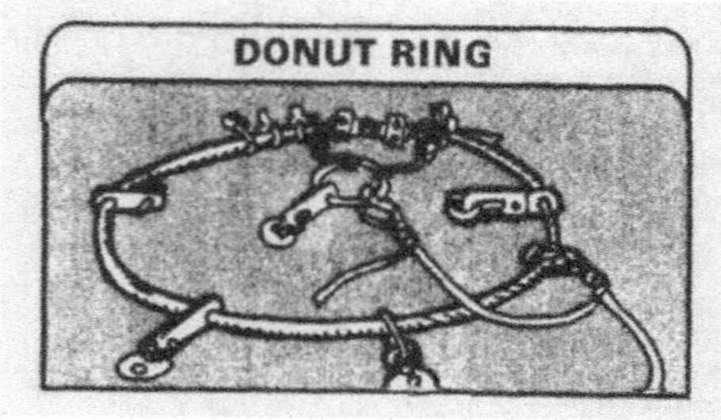

The *log coil* helps the double rappelling rope fall clear of the aircraft. To prepare it, start with the running end of the rope and coil the rope evenly and tightly around the log. Use a log approximately 2 to 3 inches in diameter and 16 to 24 inches long.

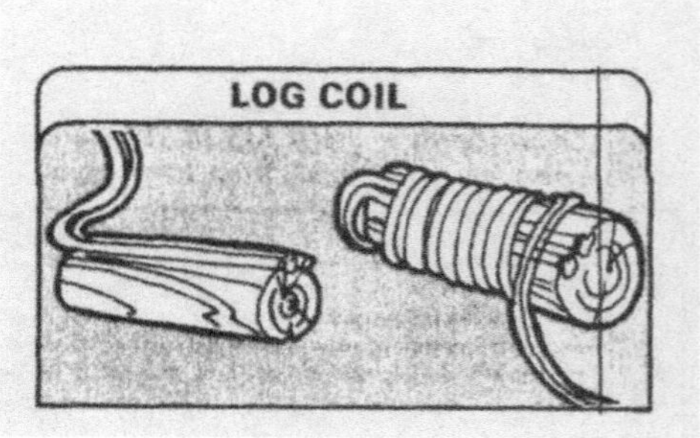

A system that has proven to be almost ideal for rope deployment in helicopter rappel operations is that of the rope deployment bag. The rope deployment bag is issue as a component of the Stabo extraction system The ropes are prepared in the normal rappel configuration and stretched to full length The D-bag is placed at the loose ends of rope opposite the snaplinks. After insur that all rubber bands are present all stowing lines of the D-bag, a bight is formed in the two runnings ends of the rappel ropes and then placed in the center retainer band just above the stow pocket. The rope is then folded and stowed in the retainer bands working from side to side of the D-bag, while making sure that folds do not extend past either side of the D-bag. Six to eight folds of rope are placed in each retainer band, working towards the top of the D-bag. Once the top of the bag is reached, a bight is formed in the climbing ropes 24 inches below the first snaplink and stowed in the top center retainer band. After the D-bag is inspected, the bag is rolled, going from bottom to top, leaving the snaplinks exposed. Tape is used to secure the top flap of the bag.

PROCEDURES

The rappelling rope is connected to the donut ring and the floating safety ring, in the following order, and in the manner described.

CONNECTION OF A RAPPELLING ROPE

The No. 1 snaplink is attached to the donut ring in the following manner:

Take a bight (loop of rope) approximately 5 feet from the end of the working end of the rope.

Insert the rope into the snaplink.

Make one turn through the snaplink, forming a round turn.

Secure the round turn to the snaplink with two half-hitches.

NOTE: Again, for clarity, a single rope is shown. But a double rope would be actually used.

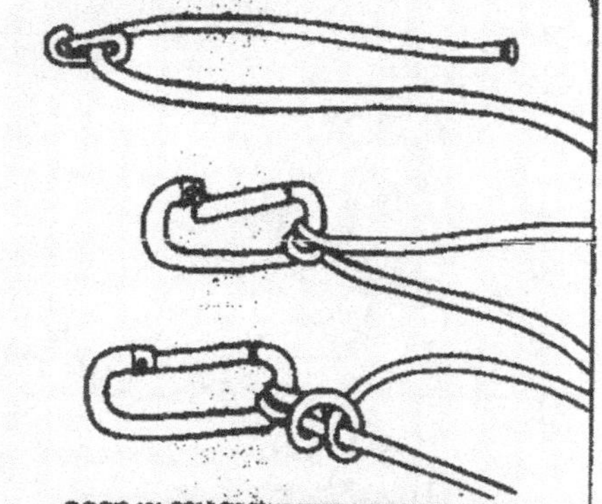

ROPE IN SNAPLINK WITH ROUND TURN AND TWO HALF-HITCHES

Snap the snaplink (gate upward and facing away from the knot) to the donut ring cable.

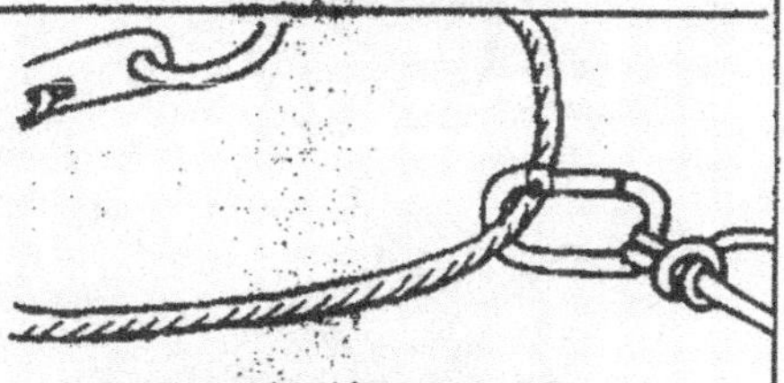

SNAPLINK SNAPPED TO CABLE

The No. 2. snaplink is attached to the floating safety ring in the same manner as the No. 1 snaplink, with the following exceptions:

Take a bight approximately 2 feet from the end of the working end of the rope.

Connect the snaplink to the rope in the same manner as the first connection. Tape the end of the working end of the rope and the knots with masking tape to secure them in place.

Snap the snaplink (gate upward and facing away from the knot) to the floating safety ring.

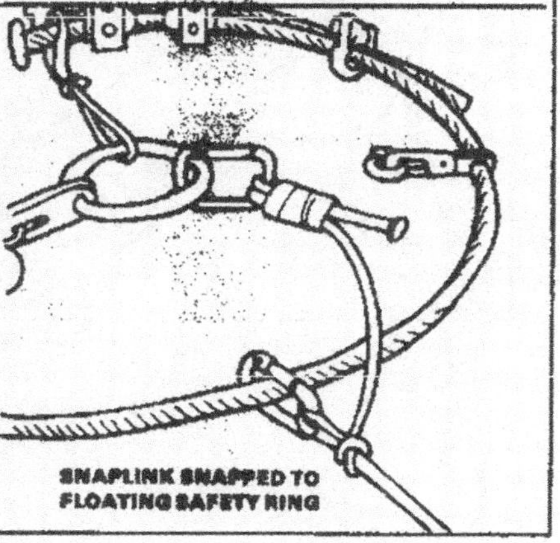

SNAPLINK SNAPPED TO FLOATING SAFETY RING

To rappel from a helicopter,

. Hook into the rope, as in the seat-hip rappel, upon entering the helicopter. Sit on the floor, keeping the brake hand firmly in the small of the back.

• Place the rope in lap.

• Upon the command, "GET READY," look toward the donut ring and pull on the rope to check the anchor point connection. Check the rappel seat and snaplink to insure that the rope is properly inserted. Conduct a final visual inspection of the hookup.

• Upon the command, "SIT IN THE DOOR," swing the feet out to the helicopter skid, keeping the brake on.

• Upon the command, "DROP ROPE," drop the rope with the guide hand, insuring that the rope does not fall between the cargo compartment and the skid and that the rope is not tangled or fouled.

• Upon the command, "POSITION," using the guide hand to assist, pivot 180 degrees on the helicopter and skid bar. Face the inside of the helicopter. Spread the feet shoulder-width apart; lock the knees; and bend forward at the waist, forming an "L" body position.

• Upon the command, *"Go,"* flex the knees and jump backward, letting the rope run through both the brake hand and the guide hand. Descend 5 to 10 meters at a time, looking at the ground over the brake hand. Keep the feet together and legs straight, while maintaining the "L" body position.

• Upon reaching the ground, back all the way out of the rope and move quickly away from beneath the helicopter.

OTHER MEANS OF ENTERING AND LEAVING HELICOPTERS

The *troop ladder* is a good method for larger groups to enter or leave an area where the helicopter cannot land. Installation of the ladder is an aviation responsibility. The crew chief will control the number of troops on the ladder. Only five or six troops will be allowed on the ladder at a time.

Use the legs for climbing and descending; the arms for stability and holding the ladder close to the body.

If possible, each soldier, and especially those carrying heavy loads, should tie a rappel seat with a snaplink attached before ascending. Then, if he becomes tired, he can "snap in," avoiding the chance of a fall.

When going down a ladder, the first soldier on the ground steadies the ladder for the remaining troops. When climbing up a ladder, the soldier designated to hold the ladder steady is the last one up the ladder. If the helicopter starts going up before everyone is loaded, the soldier holding the ladder on the ground should release it at once.

When on the ladder, remain calm at all times. If the helicopter should start settling to the ground, stay calm, watch the ground, and stay on the ladder until reaching the ground.

Once on the ground, move from underneath the helicopter.

The *jungle penetrator* is a metal seat rescue assembly attached to a winch on a helicopter. The seats remain folded when the device is being lowered. The seats are unfolded when the device is on the ground. A conical nose allows it to penetrate the jungle foliage. One to three soldiers (two is a standard load) can be placed on it. Nylon straps are placed under the arms of the soldier(s) to be hoisted. The device lifts the soldier(s) into the helicopter.

TROOP LADDER

JUNGLE PENETRATOR

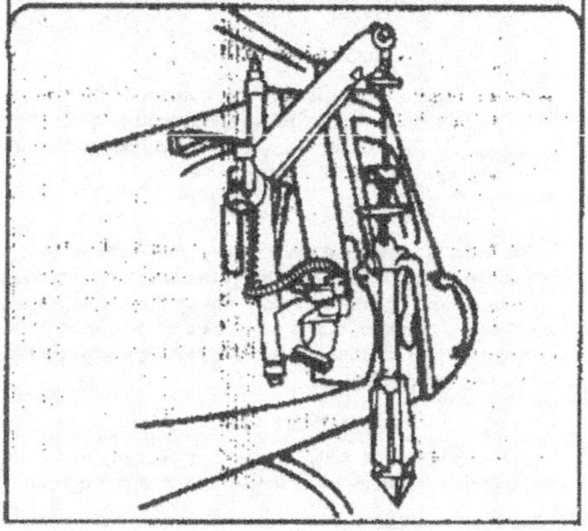

WARNING:

WHILE THE PENETRATOR IS BEING LOWERED. STATIC ELECTRICITY CAN BUILD UP ON THE METAL SURFACE. THE PENETRATOR SHOULD TOUCH THE GROUND BEFORE ANYONE TOUCHES IT, TO ALLOW THE STATIC ELECTRICITY TO GROUND ITSELF

OTHER EXTRACTION MEANS

Other means which can be used for extraction of troops where landing zones (LZ) are not available include such devices as the personnel Stabo extraction system, the Maguire rig, the Palmer rig, or the jungle operations extraction system (JOES).

STABO SYSTEM

This system provides a means for rapid pickup of soldiers by helicopter from areas where the helicopter cannot land. The system consists of the *personnel harness,* the *bridle, the suspension rope,* the *safety rope,* and the *deployment bag.*

As the pickup process is initiated, the helicopter hovers over the pickup zone at

STABO SYSTEM

PERSONNEL HARNESS

The personnel harness is designed to be worn by the user in the field and to partially replace the soldier's load-carrying equipment. It is made of nylon webbing and stitched to a standard, medium, or large web pistol belt. The two V-rings at the top of each of the harness shoulder straps connect to the bridle snaphooks and provide a lift point for the harness. The leg straps are adjustable for comfort. When the user is in the field, leg straps may be disconnected, folded, and secured to a suitable point on the harness.

BRIDLE

The bridle is also made of nylon webbing. During use, the D-ring is connected to a suspension rope snaphook while the two bridle snaphooks are attached to the two personnel harness lift V-rings.

SUSPENSION ROPE

The suspension rope is the same nylon rope that is used for rappelling. It has snaphooks attached to both ends: one to connect to the bridle and one to connect to the anchor point in the helicopter.

SAFETY ROPE

The safety rope is 3 meters long and made of nylon. Each end is looped and spliced, with a snaphook attached to each loop. It is used when two or three soldiers are extracted together, and it provides a means for them to hold together to minimize wind buffeting and oscilation during pickup and flight.

DEPLOYMENT BAG

The deployment bag is made of cotton duck and contains the bridle, suspension rope, and safety rope.

altitudes up to 150 feet. A member of the helicopter crew drops the extraction system deployment bag from the left door of the helicopter. (A maximum of three units may be connected and dropped simultaneously.) As the deployment bag descends, the suspension rope deploys until the bag reaches the ground. The soldier to be extracted then attaches the bridle snaphooks to the lift V-rings on his harness. After insuring that the leg straps are connected and tight the soldier notifies the helicopter by radio or hand signals that liftoff may begin. The helicopter then lifts the soldier from the area and, carrying him suspended beneath the helicopter, moves to an area where a safe landing can be made. The helicopter then lowers the suspended soldier to the ground, lands nearby, and allows him to board the helicopter. For details on employing this system, consult *TM 10-1670-262-12.*

SOLDIER IN HARNESS BEING LIFTED OFF

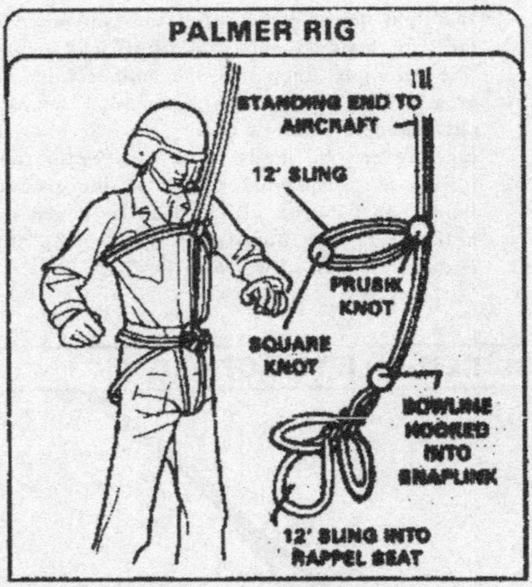

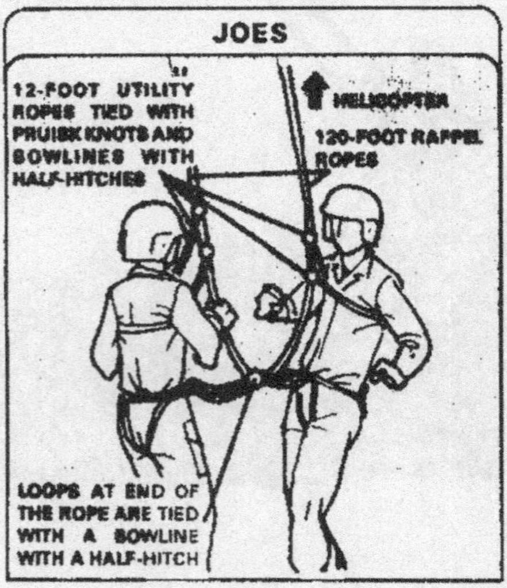

PALMER RIG

This rig is constructed with the 120-foot nylon rope and two 12-foot nylon sling ropes. One end of the 120-foot rope is secured to the donut ring. The running end of the 120-foot rope is tied with an end of the rope bowline knot at the end of the rope. One of the 12-foot sling ropes is tied 3 feet above the bowline with a Prusik knot, and the loose ends are tied off with a square knot to form a loop. The rider uses the remaining sling rope to form a rappel seat. He then fastens a snaplink to the rappel seat and the bowline knot; places his arms up and through the upper loop; and is extracted. The Palmer rig is relatively safe and may be used when evacuating wounded personnel.

MAGUIRE RIG

This rig is also simple and easy to construct. It is made with an 8-foot by 2-inch piece of nylon webbing sewn together at the ends to form a loop containing a D-ring. A smaller slip loop (wrist loop) is sewn 12 inches down from the top of the larger loop. During extraction, the rider simply sits in the seat of the large loop. He then places his wrist in the slip loop and tightens the loop, insuring that he does not fall from the larger loop during extraction.

JUNGLE OPERATIONS EXTRACTION SYSTEM

This system was developed after careful evaluation of the three previously mentioned systems, while considering those items of equipment available to the individual soldier. The JOES can be quickly constructed from components readily available from the supply system. The individual soldier requires very little (10 to 15 minutes) training in what he will be required to do if he needs to be extracted, and only a maximum of 5 minutes is required for him to actually prepare for extraction.

CONSTRUCTION OF A JOES FOR TWO INDIVIDUALS

Material needed:

- Anchor system for helicopter (donut ring).
- Two 120-foot rappel ropes.
- Three military snaplinks,
- Two 12-foot utility ropes.
- One deployment system.

Once all equipment has been thoroughly inspected, the JOES is prepared in the following manner.

- Lay out both 120-foot rappel ropes side by side.
- Move to one end and, taking both ropes as one, tie a round turn and two half-hitches around a snaplink approximately 5 feet from the end.
- From the first knot, move 18 inches toward the short end of the rope and tie another round turn and two half-hitches around another snaplink. Tape any excess rope. This now becomes the fixed end of the rope and will be attached to the anchor system in the helicopter.
- Move to the working end of the rope and, taking both ropes as one, tie a bowline with a half-hitch. The loops formed by this bowline should be approximately 12 inches in diameter, Any excess over 6 inches should be taped to help eliminate confusion during hook up.
- Approximately 18 inches from the top of the bowline knot toward the fixed end of the rope, take one end of a 12-foot utility rope and tie a Prusik knot and a bowline with a half-hitch around one 1 20-foot rappel rope.
- On the same rope, using the other end of the 12-foot utility rope, tie the same knot directly below the first.
- Repeat steps 5 and 6 with another sling rope on the other 120-foot rappel rope.
- Pack JOES in D-bag.

Each individual who is to be extracted must have a 12-foot utility rope and a snaplink. If the individual does not have these items, they can be dropped to him with the JOES. The individual to be extracted makes a rappel seat out of the 12-foot utility rope and installs the snaplink as if he were going to make a rappel.

When JOES is dropped, he and his buddy move to the bag. Each individual hooks his snaplink into one loop at the end of the rope bowline, and places the loop formed by one of the 12-foot utility ropes over his shoulders and under his armpits. He and his buddy stay as far back from the helicopter as practical until it starts to lift up, and as tension is put on the rope, they move forward until they are directly underneath it and linked up with other personnel. All soldiers should link up by holding onto the adjacent person's equipment.

Section V. MOVING ACROSS WATER OBSTACLES

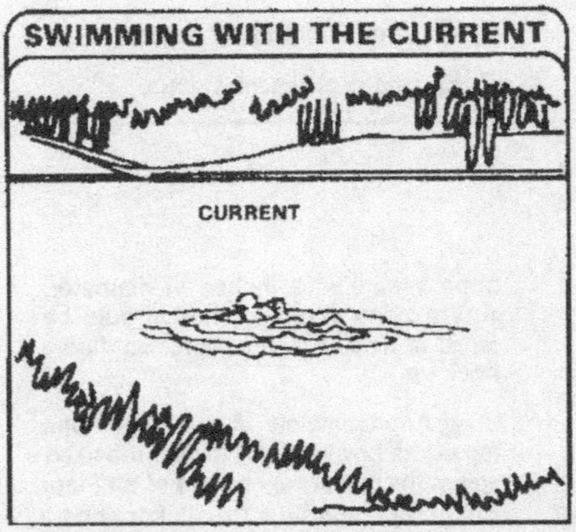

CROSSING RIVERS AND STREAMS

There are several expedient ways to cross rivers and streams. The ways used in any situation depends on the width and depth of the water, the speed of the current, the time and equipment available, and the friendly and enemy situation.

There is always a possibility of equipment failure. For this reason, every soldier should be able to swim. In all water crossings several strong swimmers should be stationed either at the water's edge or, if possible, in midstream to help anyone who gets into trouble.

If a soldier accidentally falls into the water, he should swim with the current to the nearer bank. Swimming against the current is dangerous because the swimmer is quickly exhausted by the force of the current.

FORDING

A good site to ford a stream has these characteristics:

- **Good concealment on both banks.**
- **Few large rocks in the river bed. (Submerged large rocks are usually slippery and make it difficult to maintain footing.)**
- **Shallow water or a sandbar in the middle of the stream. Troops may rest or regain their footing on these sandbars.**
- **Low banks to make entry and exit easier. High banks normally mean deep water. Deep water near the far shore is especially dangerous as the soldiers may be tired and less able to get out.**

A unit should cross at an angle against the current. Each soldier should keep his feet wide apart and drag his legs through the water, not lift them, so that the current will not throw him off balance. Poles can be used to probe in front of the troops to help find deep holes and maintain footing.

FLOATING AIDS

For deeper streams which have little current, soldiers can use a number of floating aids such as the following:

- The standard air mattress
- Trousers

NOTE: Trousers must be soaked in water before using.

- Canteen safety belt
- Poncho life belt
- Water wings
- Poncho brush raft
- Australian poncho raft
- Log rafts

TROUSERS AS A FLOATING AID

Prepare the trousers by tying the bottoms of the legs tightly.

Enter the water to waist depth and hold the trousers behind the shoulders, with the waist open.

Bring the trousers quickly over the head and bang them onto the water in front. This action fills them with air.

Squeeze and hold the waist together, lie over the trousers, and float as if on water wings.

CANTEEN SAFETY BELT

Attach at least eight empty plastic canteens to a pistol belt (or tie them to a rope which can then be used as a belt). Insure that the caps are screwed on tightly.

PONCHO LIFE BELT

Roll green vegetation tightly inside a poncho and fold the ends over to make a watertight life belt. Roll up the life belt like a big sausage at least 8 inches in diameter and tie it. Wear it around the waist or across one shoulder and under the opposite arm like a bandoleer.

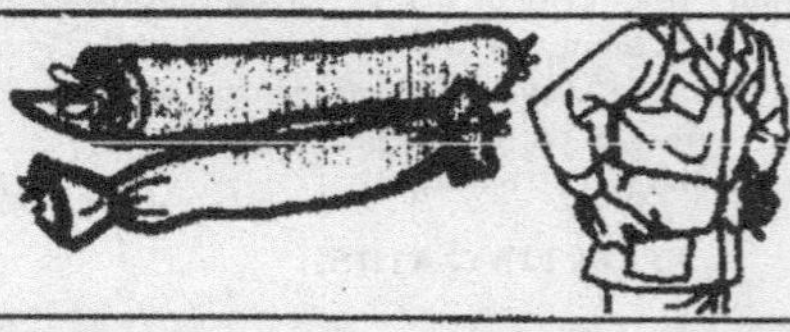

WATER WINGS

Two or more air-filled plastic bags, securely tied at the mouth, can be used as expedient water wings. Other expedients include empty water or fuel cans and ammunition canisters.

CONSTRUCTION OF PONCHO BRUSH RAFT

Use two ponchos, and tie the neck of each tightly by using the drawstring.

Spread one poncho on the ground with the hood up so that it will end up inside the raft.

Cut fresh, green brush (avoid thick branches or wood stakes) and pile it on the poncho to a height of 18 inches.

Place an X-frame made of small saplings (1 to 1 1/2 inches in diameter and 3 to 4 feet long) on the brush. Anchor this frame by tying the drawstring of the poncho to the center of the X-frame.

Pile another 18 inches of brush on top of the X-frame.

Compress the brush slightly and fold up the poncho, tying ropes or vines diagonally across the corner grommets and straight across from side grommets. The sides of the poncho should not touch.

Spread the second poncho on the ground, with hood up, next to the bundle made of the first poncho and brush. Roll the bundle over onto the center of the second poncho and tie the second poncho across the sides and diagonally across the corners. This raft will safely float 250 pounds and is very stable.

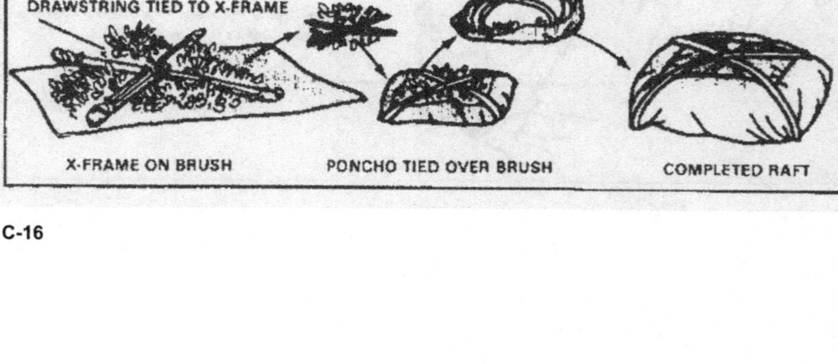

CONSTRUCTION OF AUSTRALIAN PONCHO RAFT

When there is not enough time to gather a lot of brush, this raft is made by using a soldier's combat equipment for bulk. Normally, two soldiers make this poncho together. It is more waterproof than the poncho brush raft but will float only about 80 pounds of weight. Two soldiers make this raft as follows.

Place one poncho on the ground with the hood facing up. Close the neck opening by tying it off with the drawstring.

Place two poles (or branches), about 1 to 1 1/2 inches in diameter and 4 feet long, in the center of the poncho about 18 inches apart.

Next, place the rucksack, and any other equipment desired, between the poles.

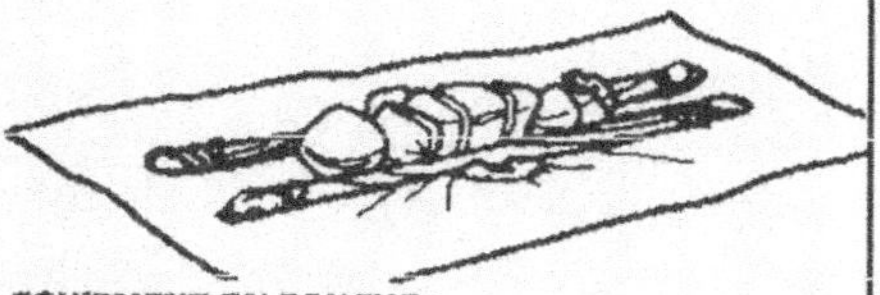

EQUIPMENT ON PONCHO

Snap the poncho together. Hold the snapped portion of the poncho in the air and roll it tightly down toward the equipment. Roll from the center out to both ends. At the ends, twist the poncho to form "pigtails." Fold the pigtails inward toward each other and tie them tightly together with boot laces, vines, communication wire, or other available tying material.

PONCHO ROLLED WITH PIGTAILS

Spread the second poncho on the ground, neck closed and facing up.

Place the equipment bundle formed with the first poncho, with the seam (tied pigtails) facing down, on the second poncho.

Roll and tie the second poncho in the same way as the first.

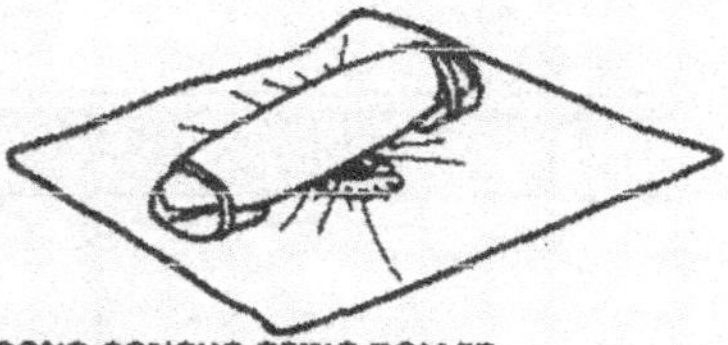

SECOND PONCHO BEING ROLLED

An empty canteen tied to one end of a rope with the other end tied to the raft helps in towing. One soldier pulls on the rope while the other pushes the raft. Place weapons on top of the raft and secure them with ropes. The weapons should be secured to the raft by the use of quick releases. The raft is now ready for the water.

SOLDIERS MOVING RAFT IN WATER

When launching any poncho raft or leaving the water with it, take care not to drag it on the ground as this will cause punctures or tears.

CONSTRUCTION OF LOG RAFT

Logs, either singly or lashed together, can be used to float soldiers and equipment. Be careful when selecting logs for rafts. Some jungle trees will not float. To see whether certain wood is suitable, put a wood chip from a tree in the water. If the chip sinks, so will a raft made of that wood.

ROPE BRIDGES

For crossing streams and small rivers quickly, rope bridges offer a suitable temporary system, especially when there is a strong current. Because of the stretch factor of nylon ropes, they should not be used to cross gaps of more than 20 meters. For larger gaps, manila rope should be used.

DETERMINATION OF RIVER WIDTH

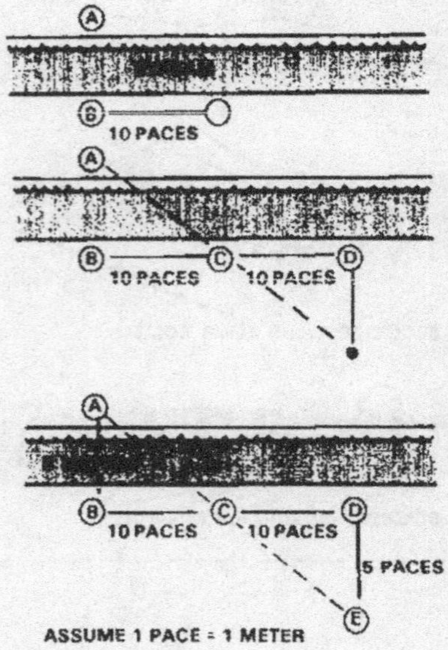

A method used to measure the width of a river or stream is described below.

Select a straight section of the stream.

Pick two points opposite each other (A and B).

Stand at B; turn in a direction parallel to the stream; walk off 10 paces. Mark that point as point C (B to C = 10 paces).

Continue walking in the same direction 10 more paces. Mark that point as point D (C to D = 10 paces).

Turn at a right angle away from the stream and walk until you are on line with points C and A. Mark this point as point E. Determine the distance between D and E by converting the pace count into meters. In this example, 1 pace is equal to 1 meter and the pace count is 5 paces. Therefore, the distance between D and E is 5 meters.

The distance from D to E is equal to the distance from A to B. Therefore, the width of the stream is also about 5 meters.

In order to erect a rope bridge, the first thing to be done is to get one end of the rope across the stream. This task can be frustrating when there is a strong current. To get the rope across, anchor one end of a rope that is at least double the width of the stream at point A. Take the other end of the line upstream as far as it will go. Then, tie a sling rope around the waist of a strong swimmer and, using a snaplink, attach the line to him. He should swim diagonally downstream to the far bank, pulling the rope across.

SWIMMER PULLING ROPE ACROSS

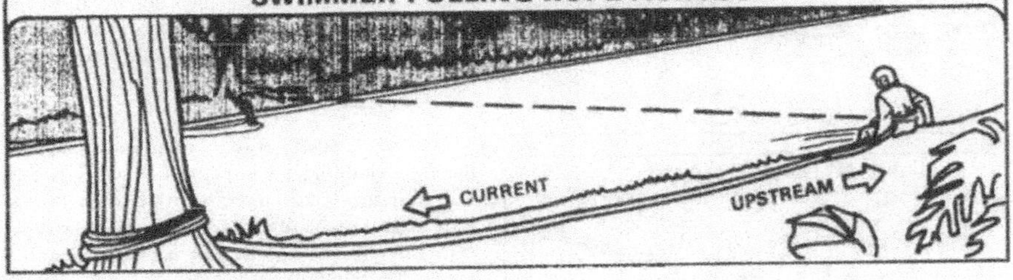

One-Rope Bridge. A one-rope bridge can be constructed either above water level or at water level. The leader must decide which to construct. The bridge is constructed the same regardless of the level.

CONSTRUCTION OF ONE-ROPE BRIDGE

To construct this bridge:

The bridge rope is pulled around the upstream side of the far side anchor point and temporarily secured without tying a knot. On the near side, a transport-tightening system is placed in the bridge rope by tying a double butterfly knot and placing two snaplinks in the butterfly. The running end of the bridge rope is then passed around the downstream side of the near side anchor point and through the two snaplinks.

When this transport-tightening system is prepared, the soldier on the far side pulls the butterfly knot approximately a third of the distance across the river. He then secures the bridge rope to the far side anchor point using a round turn and two half-hitches. Soldiers on the near side then pull the slack out of the bridge rope until the butterfly knot is back on the near side. (The bridge rope must be as tight as possible so that it will not sag when used.) The bridge rope is then tied off against itself using two half-hitches with a quick release in the last half-hitch.

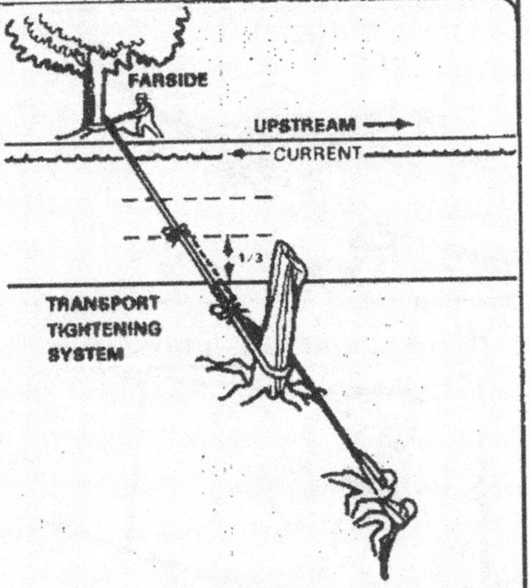

Crossing Method above Water Level. Use one of the following methods.

Commando crawl. *Lie* on the top of the rope with the instep of the right foot hooked on the rope. Let the left leg hang to maintain balance. Pull across with the hands and arms, at the same time pushing on the rope with the right foot. (For safety, each soldier ties a rappel seat and hooks the snaplink to the rope bridge.)

Monkey crawl. Hang suspended below the rope, holding the rope with the hands and crossing the knees over the top of the rope. Pull with the hands and push with the legs. (For safety, each soldier ties a rappel seat and hooks the snaplink to the rope bridge.) This is the safest and the best way to cross the one-rope bridge.

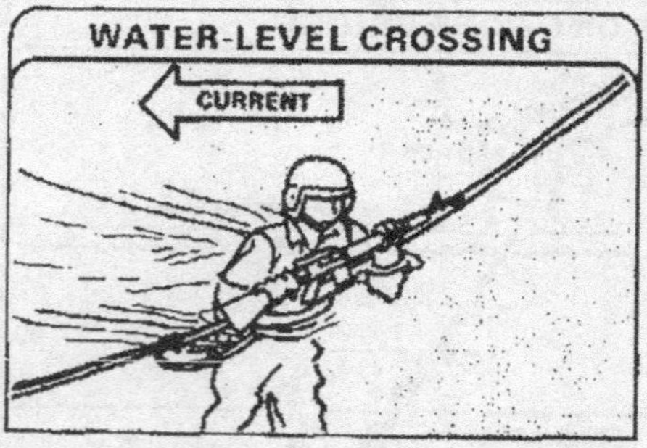

Crossing Method at Water Level. Hold onto the rope with both hands, face upstream, and walk into the water. Cross the bridge by sliding and pulling the hands along the rope. (For safety, each soldier ties a sling rope around his waist, leaving a working end of about 3 to 4 feet. He ties a bowline in the working end and attaches a snaplink to the loop. He then hooks the snaplink to the rope bridge.)

To recover the rope, the last soldier unties the rope, ties it around his waist and, after all slack is taken up, is pulled across.

Two-rope bridge. Construction of this bridge is similar to that of the one-rope bridge, except two ropes, a hand rope and a foot rope, are used. These ropes are spaced about 1.5 meters apart vertically at the anchor points. (For added safety, make snaplink attachments to the hand and foot ropes from a rope tied around the waist. Move across the bridge using the snaplink to allow the safety rope to slide.) To keep the ropes a uniform distance apart as men cross, spreader ropes should be tied between the two ropes every 15 feet. A sling rope is used and tied to each bridge rope with a round turn and two half-hitches.

OTHER CROSSING MEANS

Suspension traverses, bridges, and cableways can be used to move large numbers of soldiers or heavy equipment over wide rivers and ravines, or up and down cliffs in a short period of time. Because heavy or bulky material or equipment is needed to construct these expedients, their use is practical only if the needed items can be transported to the site by air or surface means (watercraft, pack animals, etc.).

APPENDIX D

Waterborne Operations

Section I. GENERAL

The inland waterways and jungle coastal or delta regions are land environments dominated by water routes. There may be one or more major waterways and an extensive network of smaller waterways. Usable roads are scarce, and cross-country movement is extremely difficult. The following describes jungle waterways.

UPPER SECTOR (HEADWATERS)

The headwaters of a waterway are usually formed in a mountainous region. The headwaters consist of numerous tributaries which merge to form a river system as the water flows down to the valley. Headwaters are characterized by waterfalls, rapids, and variations in water depth, all of which restrict the use of watercraft.

MIDDLE SECTOR (CENTRAL VALLEY)

When the waterway reaches the central valley, it has formed a broad river which is usually navigable for great distances inland. This river is usually fed by numerous tributaries. In those jungles where there are definite dry and rainy seasons, many of the tributaries found during the rainy season may not exist during the dry season. The river in the valley is wide, slow, and often meanders. During periods of heavy rainfall, the course of the river may change. The jungle vegetation grows up along the riverbanks to form an almost solid wall. The banks of the river are often steep and slippery. Many of the navigable tributaries feeding the major river will often be completely overgrown with vegetation and contain obstacles such as fallen trees.

CONTENTS

LOW SECTOR (DELTA)

When the river reaches the low coastal area, it spreads over a flat, alluvial plain and becomes a number of river tributaries (small streams or channels spreading fanlike from the main channel) disbursing a great amount of sediment into a gulf, bay, or ocean. Usually, there are many large and small tidal streams and channels, whose current may change speed or reverse with the tide in a predictable manner. Bottoms of the tributaries normally slope up to a crest or bar at the river's mouth. In some instances, only watercraft with a draft (that part of the craft under water) of 1 to 2 meters will be able to cross the crest or bar at high tide.

section II. PREPARING FOR OPERATIONS

WATERBORNE COMBAT OPERATIONS

The fundamentals and tactics applicable in conventional ground operations apply in waterborne operations. However, special organization, equipment, and techniques are required when ground forces are supported by Navy ships and craft. The waterborne force should be employed with all available modes of transportation to seek out and destroy the enemy and his installations. One portion of the force may enter the area by watercraft; another may enter by helicopters; still another may enter the area by moving overland. All units then maneuver to attack the enemy. All available fire support should be used in the operation: close air support, attack helicopters, waterborne and land-based artillery, and naval gunfire. Special considerations in the conduct of jungle waterborne operations include the following:

- The heavy vegetation along the banks of inland waterways offers excellent concealment and enhances the effectiveness of ambushes against watercraft. Therefore, counterambush measures must be planned in conjunction with all water movements. Steep, slippery river banks coupled with dense vegetation often make committing the waterborne force in a coordinated assault landing extremely difficult.

- Security measures during the movement phase along a jungle waterway include proper watercraft formations, constant water patrolling, and air observation, when possible. Fire support to include mortar, artillery, close air support, and available naval gunfire must be preplanned for all water movements.

- Intelligence is critical along jungle waterways. While aerial reconnaissance yields a considerable amount of information, it will have to be supplemented by reconnaissance by boat, especially in areas where tributaries are overgrown by vegetation.

Section III. USING JUNGLE WATERWAYS

SMALL BOAT HANDLING

The use of inland and coastal waterways can add flexibility, surprise, and speed to tactical operations in jungle areas. Use of these waterways will also increase the load-carrying capacity of units which normally operate dismounted. Thus, every combat leader should be familiar with the tactical and technical aspects of small boat handling.

EQUIPMENT

There are several types of small craft available.

RB-3

Reconnaissance Boat, Inflatable: RB-3

Length: 3 meters

Width: 1.3 meters

Weight: 24 pounds, 33 pounds w/paddles and pump

Load: A two-man crew and 500 pounds or a three-man crew and 300 pounds.

Powered by: Paddles.

RB-15

Assault Boat, Pneumatic: RB-15

Length: 5.2 meters

Width: 1.8 meters

Weight: 260 pounds

Crew: 1 coxswain, 10 paddlers

Load: A maximum of 15 men (11-man crew and 4 passengers) with equipment or 3,300 pounds.

Powered by: Paddles or a 25-horsepower outboard motor.

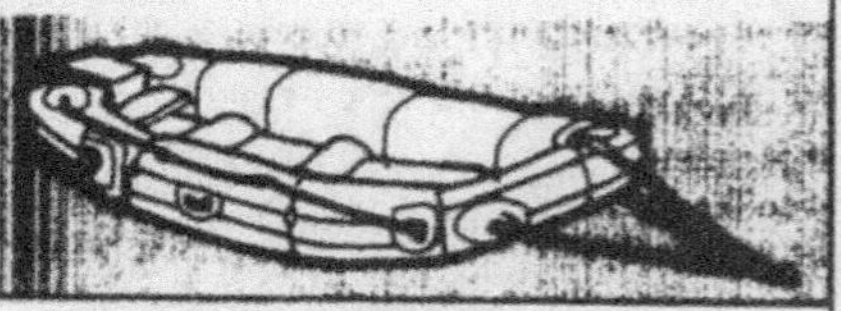

RB-7

Landing Boat Pneumatic: RB-7

Length: 4 meters

Width: 2 meters

Weight: 150 pounds

Crew: 1 coxswain, 6 paddlers

Load: 10 men (7-man crew and 3 passengers) with equipment.

Powered by: Paddles.

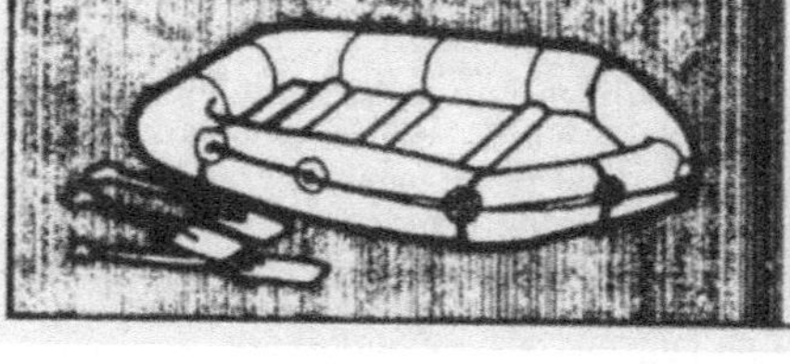

PAB

Plastic Assault Boat, M3: PAB

Length: 4.9 meters

Width: 1.7 meters

Weight: 300 pounds

Crew: 1 coxswain, 10 paddlers

Load: A maximum of 15 men (11-man crew and 4 passengers) with equipment or 3,200 pounds.

Powered by: Paddles or a 25-horsepower outboard motor.

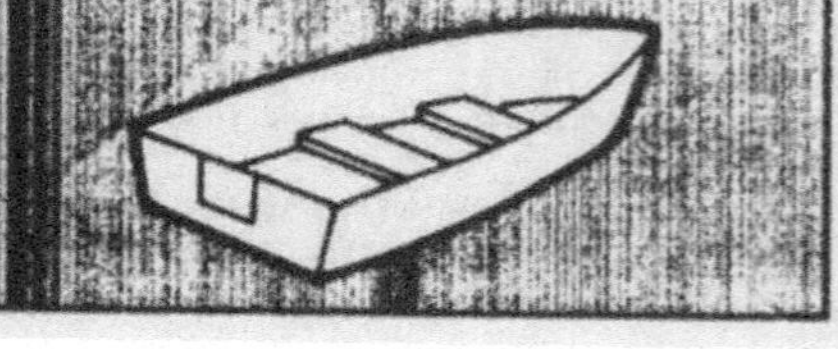

Boats may be powered by outboard motors, paddles, or oars. The mission, availability, and the river itself dictate the method of propulsion to be used.

Motors are noisy. On the other hand, they provide speed, reduce fatigue, and free personnel for security missions. The noise form motors can be heard for distances of 500 to 1,000 meters by day and up to 5,000 meters at night (sound carries better in the quieter, cooler night air). Provided the craft is not seen and the motor does not change pitch, however, it is difficult to estimate the direction of the sound and exactly how far away it is. Bearing these factors in mind, troops may be able to disguise a movement by deceptive tactics such as having other craft work the area. Stopping the motor when traveling downstream will also aid the security of movement.

Paddling is a slow and tiring process, but it is quieter than using motors. (With plastic, wooden, or metal craft, paddles may need to be wrapped with cloth to reduce noise when the paddles strike the craft.)

ORGANIZATION

Before a waterborne operation, each person in a boat is assigned a specific boat position and a corresponding number. (This is the long count method of organization.)

NOTE: The unit, normally a squad, that uses the RB-15 for transportation actually comprises the crew that operates it. All others that do not operate the boat are passengers.

First, the crewmembers are assigned their positions. Next, the passengers are assigned their positions. When using an RB-15, for example, the crewmembers are assigned positions 1 through 11, and the passengers are assigned positions 12 through 15. One person is designated as the boat commander (normally the coxswain). Two persons are designated as a navigator-observer team.

For operational purposes, the crew is organized into pairs. Passengers are not numbered in this method. (This is the short count method of organization.) When using the RB-15, for example, the crewmembers (in pairs) are assigned to operational positions 1 through 5.

PADDLES WRAPPED TO REDUCE NOISE

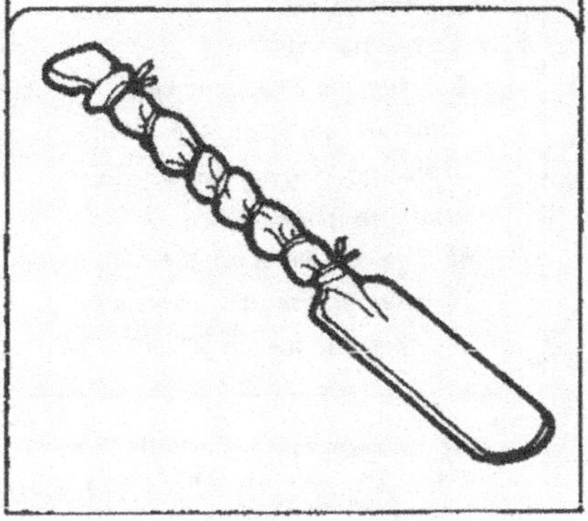

RB-15 WITH POSITIONS NUMBERED—LONG COUNT

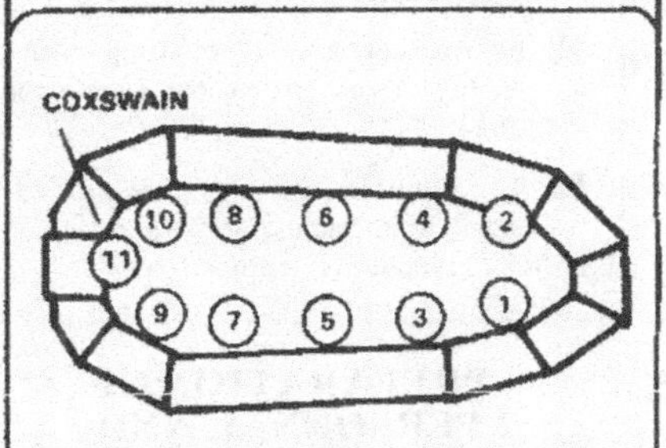

NOTE: PASSENGERS ARE NUMBERED FROM FRONT TO REAR AFTER CREW IS NUMBERED.

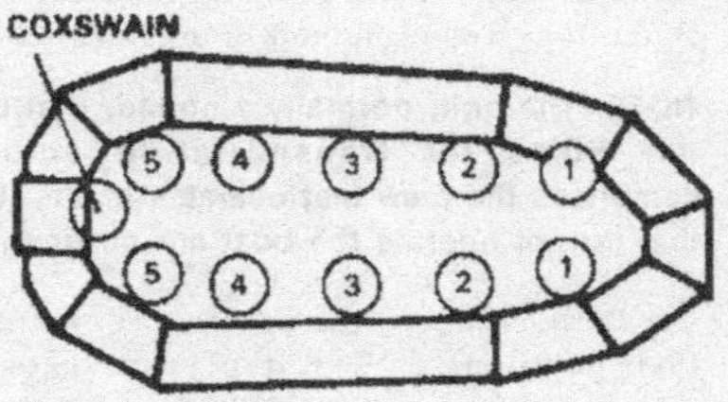

Crew duties:

- The coxswain is responsible for the control of the boat and action of the crew. He supervises the loading, lashing, and distribution of equipment. He also maintains the course and speed of the boat.
- The number 1 paddler (long count method) is the observer and is responsible for the storage and use of the bowline.
- The number 2 paddler (long count method) is responsible for setting the stroke.
- All paddlers are responsible for loading and lashing the equipment in their respective compartment.

PREPARATION OF PERSONNEL AND EQUIPMENT

Each crewmember and passenger must wear a life preserver.

The load-carrying equipment harness is worn unbuckled at the waist.

The rifle is slung outside of thelife preserver, opposite the outboard side, with the muzzle down.

Crew-served weapons, radios, ammunition, and other bulk equipment are lashed securely to the boat to prevent loss if the boat should overturn.

Radios, batteries, and unboxed ammunition are waterproofed.

Hot weapons are cooled prior to being placed in the boat to prevent damage to the boat or injury to personnel.

Pointed objects are padded to prevent puncture of the boat.

The most effective equipment-lashing system that has been developed is the RB-15 lashing system. This system is quick and easy to install, requires no special equipment, and prevents loss of equipment in the event the craft is capsized. It also allows the craft to be easily righted.

The equipment needed in this lashing system is:

Ten sling ropes-

- Three for capsize lines.

One for securing the M60.

- Six for rigging the RB-15.

Nine snaplinks-

- Two for securing the M60.
- Four for securing rucksacks.
- Three for the center line.

The average squad can fully rig and lash an RB- 15, using this system, in approximately 15 minutes.

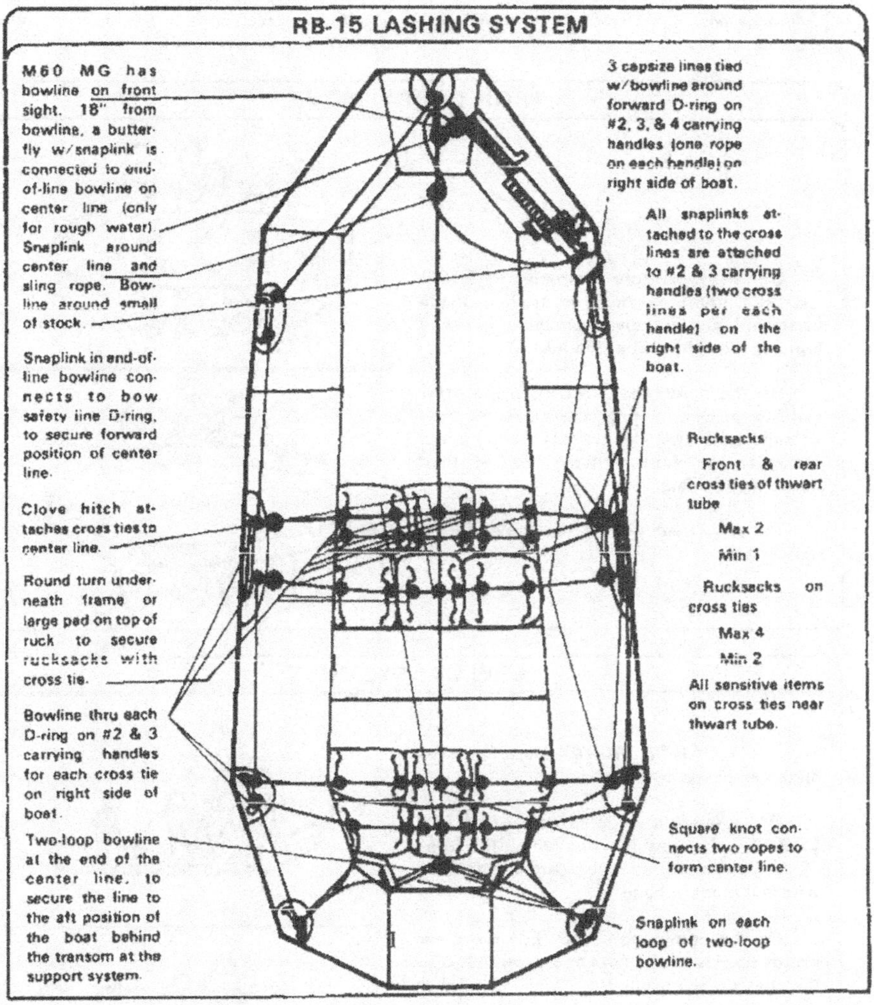

COMMANDS

"short Count, *count off.* " Crew counts *off* their positions by pairs, for example, 1,2,3,4, 5, coxswain (RB-15).

"L"*S count, count off.* " Crew counts off their positions by individuals, for example, 1, [illegible], 7, 8, 9, 10, coxswain (RB-15).

"Boat stations." Crew takes position along side of boat.

"High carry, mole" (used for long distance moves overland).

HIGH CARRY, MOVE

On the preparatory command of "**HIGH CARRY,**" the crew faces to the rear of the boat and squats down grasping carrying handles with the inboard hand.

On the command, "**MOVE,**" the crew swivels around, lifting the boat to their shoulders, so that the crew is standing and facing to the front with the boat on their inboard shoulders.

The coxswain guides the crew during movement.

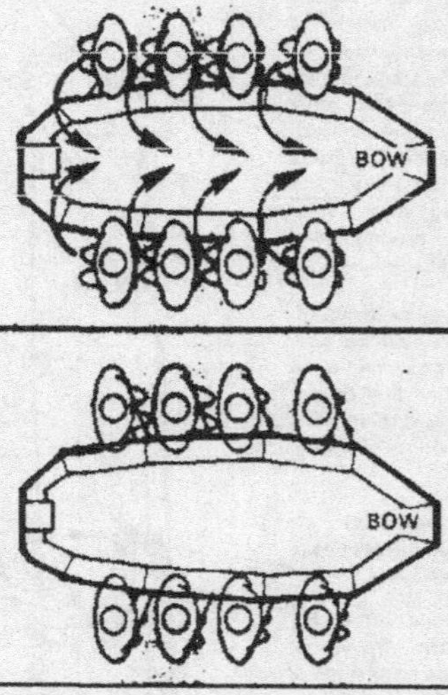

LOW CARRY, MOVE

"**LOW CARRY, MOVE**" (used for short distance moves overland).

On preparatory command of "**LOW CARRY,**" the crew remains facing the front of the boat and grasps the carrying handles with the inboard hand.

On the command, "**MOVE,**" the crew stands up raising the boat approximately 6 to 8 inches off the ground.

The coxswain guides the crew during movement.

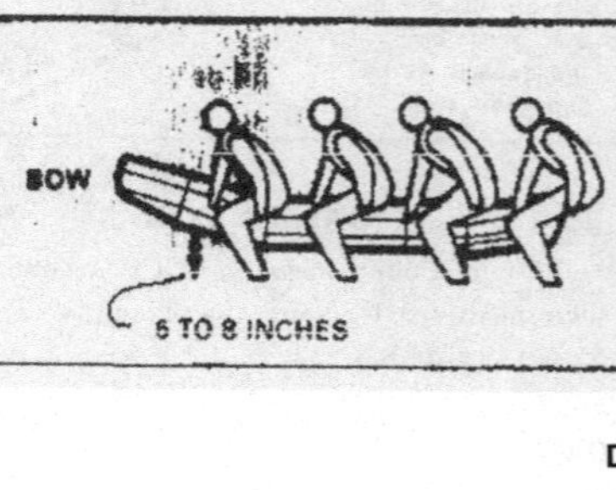

"Lower the boat, move." Crew lowers the boat gently to the ground using carrying handles.

"Give way together." "Crew paddles to the front, with Number 2 setting the stroke for the rest of the crew.

"Hold." Entire crew keeps paddles motionless in the water, thereby stopping the boat.

"Hold left (right)." Left crewmembers hold, right crewmembers continue with previous command.

"Back paddle." Entire crew paddles backward. This action propels the boat to the rear.

"Back paddle left (right)." Left crewmembers back paddle causing the boat to turn left, right crewmembers continue with previous command.

"Rest paddles." Crewmembers place paddles on their laps with blades outboard. This command may be given to pairs, i.e., *"Number l's rest paddles."*

LAUNCHING AND LANDING

When launching, the crew maintains a firm grip on the boat until they are inside the boat; similarly, when landing, they hold onto the boat until it is completely out of the water.

The crew stays as low as possible when entering and leaving the boat to avoid capsizing it.

Crewmembers can load or unload a boat either by individuals or in pairs. They load and unload a boat by individuals at steep river banks and along shoreline where the water is deep near the shore. They also load or unload from or into a larger vessel such as a landing craft, mechanized (LCM), by individuals.

The crewmembers load or unload individually according to their number in the long count method. The coxswain directs them by saying, *"One in (out), two in (out),"* etc.

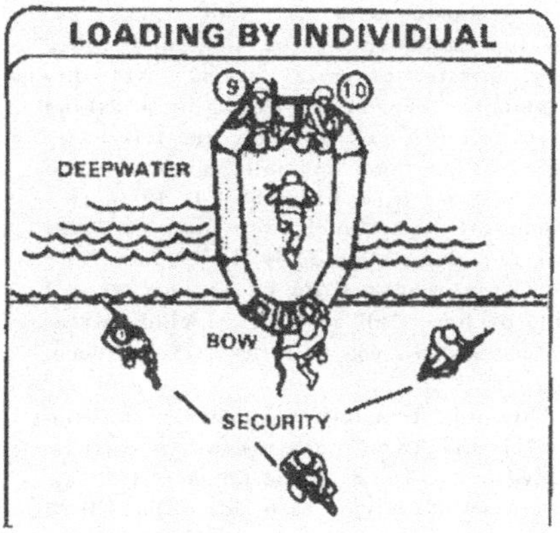

The crewmembers load and unload a boat in pairs when at shallow water riverbanks. They load or unload in pairs according to their number in the short count method. The coxswain directs them by saying, *"Ones in (out), twos in (out),"* etc.

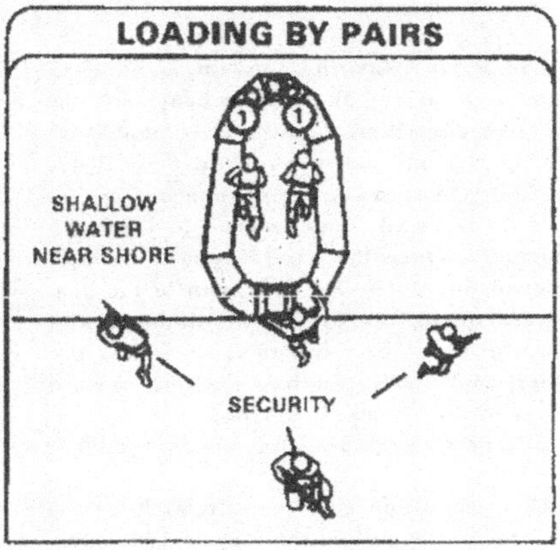

HELOCASTING

Helocasting is an excellent method for deploying troops and equipment in any terrain in which water courses exist. This technique involves a CH-47 helicopter, a 15-man rubber boat (RB-15), and a squad. The RB-15 is loaded with the squad's rucksacks, crew-served weapons, radios, and other heavy mission-essential items. All this equipment is kept in the RB-15 by a lashing system. At the desired time, the RB-15 is pushed off the ramp and into the water. The squad follows it, exiting the CH-47 in two columns off the tailgate. Drop speed is 20-25 knots and drop altitude is 10-20 feet. The water should have little or no current and should be free of all obstacles, including seaweed and stumps, and be at least 15 feet deep.

The only preparation necessary for the CH-47 is that two lengths of rollers must be installed in the center of the tailgate. The two rearward set of seats on each side of the CH-47 must be raised to fit the RB-15 inside. Personnel wear fatigues (boots unbloused, shirts out, sleeves rolled down, top button fastened), load-carrying equipment, and an individual life preserver. The individual weapon is tied to the individual in such a manner that it can be raised overhead when the individual exits the CH-47.

The RB-15 lashing system is used in rigging and lashing the rubber boat with the respective equipment. The boat is placed on the roller system and moved into the CH-47, bow first. On signal, it is pushed out by the castmaster and coxswain or RB-15 commander. Once the RB-15 is in the water, the squad follows it. The first man to the boat makes a quick inspection for damage and accountability of equipment, frees the paddles, and starts paddling the boat toward the rest of the squad. If the boat capsizes, it can be easily righted using standard drills.

This technique can be used without the RB-15 to cast scouts along a riverbank. They would swim to a designated shore or to a designated point to conduct their mission.

RB-15 helocasting is also an effective means of resupplying a company operating along a water obstacle. Over 1,000 pounds of rations, ammunition, and supplies can be placed in each boat.

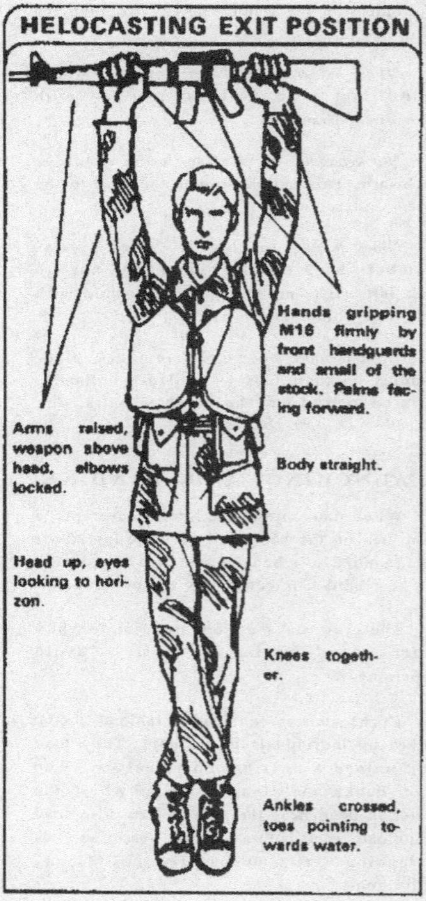

RIVER MOVEMENT

RIVER CHARACTERISTICS

BEND		A bend is a turn in the river course.
REACH		A reach is a straight portion of river between two bends.
SLOUGH		A slough (pronounced slue) is a dead end branch from a river, it is normally quite deep and can be distinguished from the true river by its lack of current.
DEAD WATER		Dead water is a part of the river, due to erosion and changes in the river course, that has no currents. Dead water is characterized by excessive snags and debris.
ISLAND		An island is usually a tear-shaped mass of land in the main current of the river. Upstream portions of islands usually catch debris and should be avoided.
CURRENT		The current in a narrow part of a reach is normally faster than in a wider portion. The current is fastest on the outside of a curve; the sandbars and shallow water are found on the inside of the curve.
SANDBARS		Sandbars are located at those points where a tributary feeds into the main body of a river or stream.

NAVIGATION

There are two acceptable methods of river navigation.

Checkpoint and General Route Method This method is used when the landing site is marked by a well-defined terrain feature and the waterway does not have many branches and tributaries. The navigator uses a strip map, with the route drawn on it, and looks for prominent checkpoints along the way. It is best used during daylight hours and for short distances. Except for those periods when the navigator is right at a checkpoint, this method is not completely accurate. It is, however, the easiest means of river navigation.

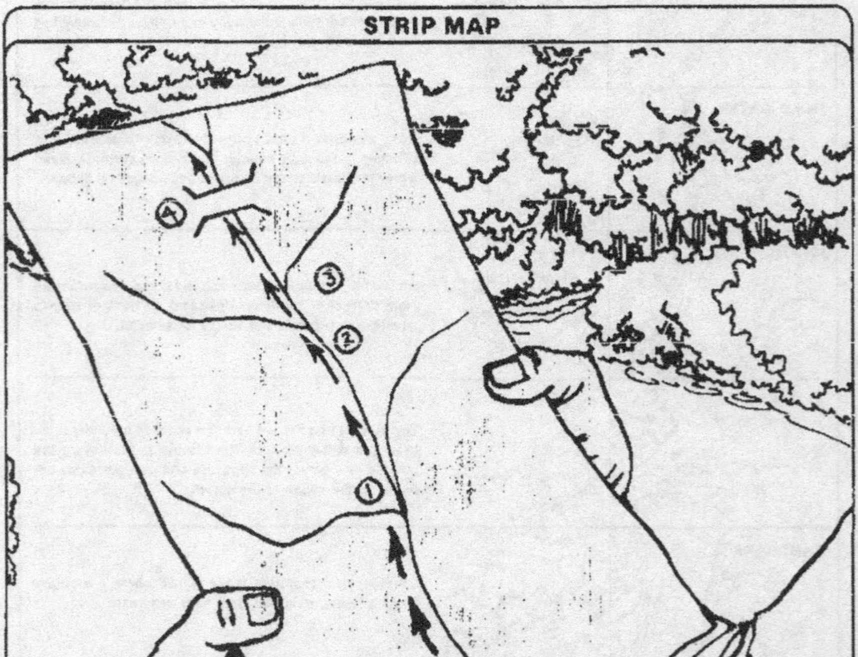

Navigator-Observer Method, This is the most accurate means of river navigation and can be used effectively in all light conditions.

The navigator is positioned in the center of the boat and does not paddle. During hours of darkness, he uses his flashlight under a poncho to check his map.

The navigator keeps his map and compass oriented at all times.

The navigator keeps the observer informed of the configuration of the river by announcing bends, sloughs, reaches, and stream junctions as shown on his map.

The observer compares this information with the bends, sloughs, reaches, and stream junctions he actually sees. When these are confirmed the navigator notes the boat's location on his map.

The navigator also keeps the observer informed of the general azimuths of reaches as shown on his map. The observer confirms these with actual compass readings.

The navigator announces only one configuration at a time to the observer and does not announce another until the first is confirmed and noted.

At night, a strip map drawn on clear acetate backed with luminous tape may be used instead of a map. It should be to scale or a schematic. It should show all curves and the azimuth and distance of all reaches. It should also show terrain features, streams, junctions, and sloughs.

TACTICAL MOVEMENT

The techniques of tactical river movement are very similar to those employed on land. As on land, movement techniques depend primarily on the likelihood of enemy contact and must be based on the concepts of traveling, traveling overwatch, and bounding overwatch.

OVERWATCH

Distances between elements will vary depending upon observation, range of weapons, and means of communication used. Each boat must maintain visual contact with the boat to its front.

Boats move close to the shoreline, taking advantage of the natural concealment.

When bends in the river deny observation, a unit sends a reconnaissance team ashore to reconnoiter the river beyond the bend. When the reconnaissance team determines that the area is clear, it signals the boats to move forward.

Troops in the boats are assigned specific sectors in which to observe and fire.

One person is appointed in each boat as an air guard. If an enemy aircraft is sighted, the boats immediately move close to shore for concealment. Troops sit quietly in the boats until all is clear. If the aircraft makes a firing pass, the unit beaches the boats and takes the appropriate defensive actions for an air attack.

Actions taken on enemy contact resemble those taken ashore. The elements caught in the enemy's fire return fire, beach the boats, seek cover, and continue to fire. Other elements beach their boats and maneuver ashore to destroy the enemy. When a patrol is inserted by boat, the landing site must be secured before all elements of the patrol disembark. A suggested technique is to have the lead boat unload its personnel at the landing site while the other boats cover them from a distance. After the site is secured, the other boats are landed on signal. After the boats have landed, the crews either hide the boats or have them removed from the area. In either case, the crews remove any signs of activity on the landing site.

CAPSIZE PROCEDURE

RIGHTING BOAT

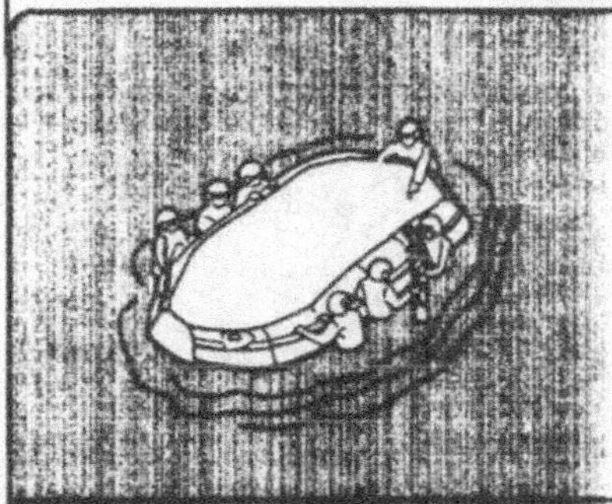

During small boat operations, there is always a chance of a boat being capsized (overturned) unexpectedly. In order for a unit to continue its mission, it must right the boat. To do this, the unit must follow this procedure:

The coxswain must first account for his men. He does this by having them count-off using the long count method. This will tell him who, if anyone, is injured or missing.

He next has the men hold on to the boat. They then pass their paddles to one man (designated by the coxswain before the operation). This man secures the paddles while the others right the boat.

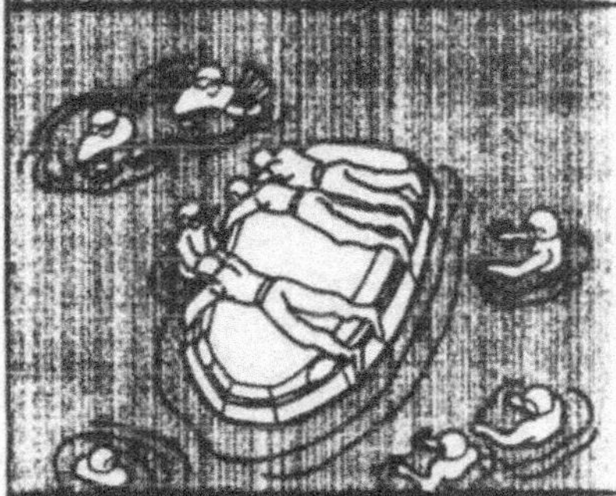

Next, three men are helped onto the boat. These men (all on the same side of the boat) grasp the bowline (which has been secured to one side of the boat), stand up, and prepare to pull the boat over. All but one of the men remaining in the water release the boat and move out to about 3 meters from it. The one man (in the water) holding on prepares to be pulled into the boat once the boat is pulled over.

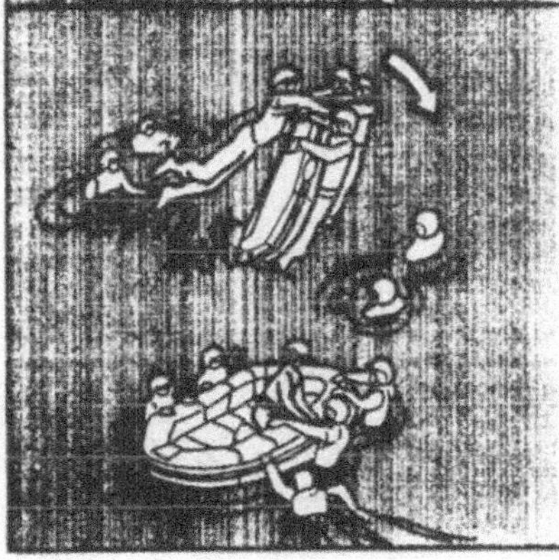

The three men on the boat lean backwards and pull the boat over (upright). This also pulls the one man (in the water) into the boat.

Once the boat is upright, all of the men move to it. The one man in the boat helps the others into the boat. The man with the paddles passes them up to the other crewmembers and then climbs into the boat.

Once the crewmembers are in their positions, the coxswain again has the men count-off using the long count method. The men also check to see that their equipment is accounted for. The coxswain then gives the crew the appropriate orders for continuing the mission.

INTENTIONAL CAPSIZE

There may be times when a unit will need to capsize the boat intentionally. This may be necessary when the boat is full of water due to rough seas or heavy rainfall. To capsize the boat, the unit follows this procedure:

The coxswain must first have the men insure that all equipment is secured to the boat. He then gives the command, "**Pass paddles.**" The men then pass their paddles to one man—normally the man to either the left or right side of the coxswain. The coxswain also gives the command, "**Pass the bowline.**" The men pass the bowline and secure it along one side of the boat and in the rear.

The coxswain designates three men to remain in the boat. (They will capsize the boat after the others are in the water.) He then orders the other members out of the boat. He does this by giving the commands, "**Ones out, twos out,**" etc., until only three men remain in the boat. Once out of the boat, the men move out to about 3 meters from it.

The coxswain then designates one of the men in the water to hold onto the boat in order to be pulled over onto the boat once it is capsized. The three men in the boat grasp the bowline, stand up, and lean backwards until the boat is capsized. This will pull the one man onto the boat when it is capsized.

The man now on top of the boat helps two other men onto the boat to help in righting it. They then follow the procedure outlined previously for righting a boat and continue the mission.

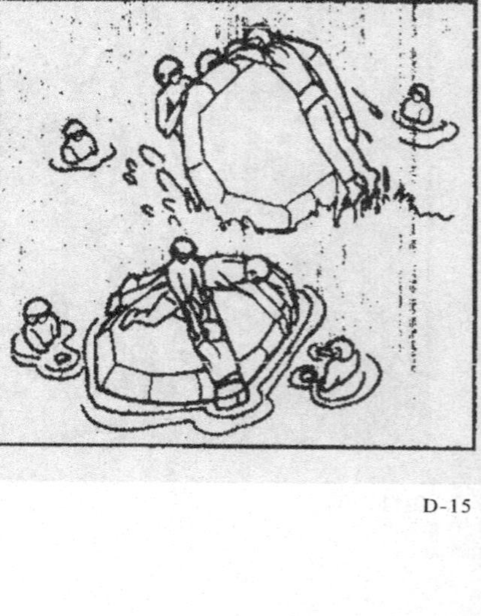

APPENDIX E

Fighting Positions and Shelters

Section I. GENERAL

Although jungles have excellent concealment, the jungle fighter must still prepare fighting positions to get as much cover as possible. Although not as important as fighting positions, jungle shelters provide shelter from the elements and make life in the jungle more comfortable and healthful.

Section II. PREPARING POSITIONS AND SHELTERS

FIGHTING POSITIONS

A jungle fighting position should provide cover, permit effective engagement of the enemy, and blend in with its surroundings. Swamps and low areas should be avoided because they normally do not afford good fields of fire and are too wet to permit the digging of fighting positions.

PROTECTION FROM ENEMY FIRE

A good fighting position provides protection from direct small-arms fire and indirect fire fragmentation. Protection from small-arms fire is provided by frontal cover. Natural frontal cover (trees, rocks, logs, rubble, etc.) is best as long as it is strong enough to stop bullets. Natural cover blends with the surroundings; therefore, it is hard for the enemy to identify positions. It maybe necessary, however, to use the dirt from the hole to build frontal cover when natural frontal cover is not available. Frontal cover must be wide enough to provide room for two soldiers, thick enough (at least 46 centimeters [18 inches] of dirt) to stop enemy small-arms

CONTENTS

fire, high enough to cover the heads of the soldiers firing from behind it, and far enough in front of the hole to provide room for elbow rests and sector stakes.

Protection from indirect fire fragmentation is gained by getting the soldiers below ground level and by building overhead cover. The logs cut to build overhead cover must be strong. Soft wood will collapse when struck by shell fragments.

To insure complete protection for fighting positions, it is necessary to build flank and rear cover. This cover protects against the effects of indirect fire that bursts to the flanks or rear of the position and against the effects of friendly supporting weapons located in the rear.

To conceal the fighting position, troops should take advantage of the jungle's natural cover. They should use large fallen or standing trees and depressions in the ground, and work to improve upon the natural cover provided. Some logs, such as palm tree logs, are too soft to stop bullets. If forced to use soft wood in building cover, a 10-inch layer of dirt or sandbags will have to be added to the protection provided by the wood.

FIGHTING POSITION

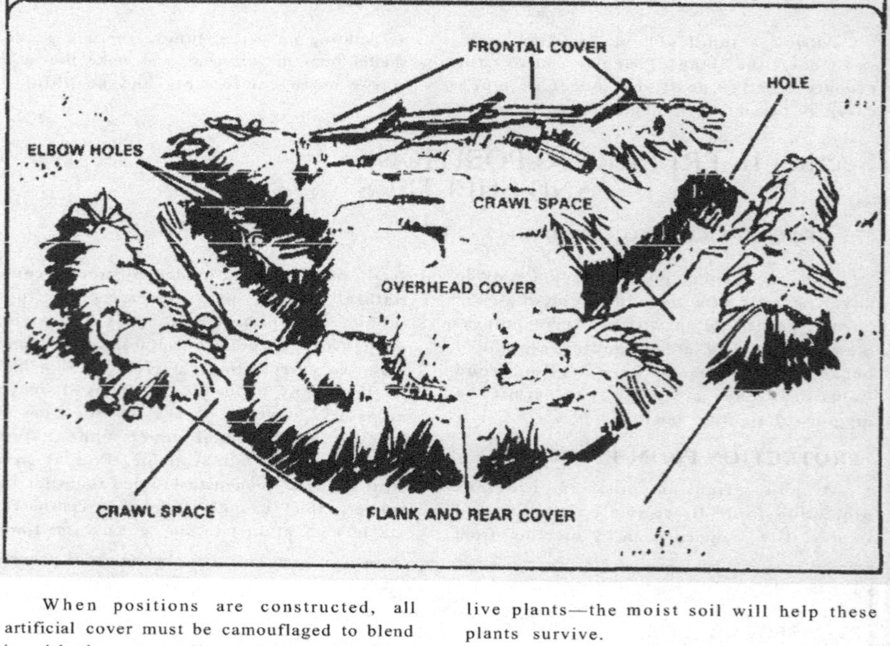

When positions are constructed, all artificial cover must be camouflaged to blend in with the surrounding terrain so it cannot be detected by the enemy. Camouflage should be kept fresh-dead or rotting vegetation will give away positions. Troops can transplant live plants—the moist soil will help these plants survive.

All positions should provide for drainage during tropical rainstorms. The floor of the position should slope from the center of the position toward the grenade sumps. This is important during the rainy season to avoid immersion foot.

If time permits, an elevated floor stand of saplings can be built so that soldiers do not have to stand in mud or water. Grenade sumps must remain clear.

Jungle soils are soft and erode quickly during the rainy season. Revetments will prevent the walls of a position from caving in. The stakes should be driven all the way into the ground, once the anchor lines are attached. They must then be concealed so that they cannot be confused with the firing stakes.

OBSERVATION AND FIELD OF FIRE

Soldiers should clear away only what is absolutely necessary, in order to retain as much natural concealment as possible. In properly constructed positions, the soldier's eyes and weapon will be at ground level. In this way, the soldier is concealed from the enemy, is protected by cover, but still can detect any approaching enemy. If a small tree is in a soldier's field of fire, it should not be chopped down completely. Only the branches that deny observation should be removed. The cuttings on the tree should be darkened with mud. Overly cleared areas or fresh, improperly concealed cuttings are easy to spot.

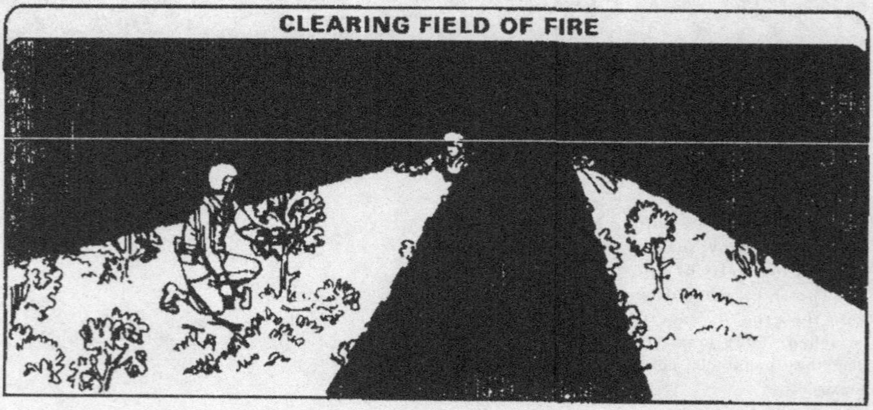

CONCEALMENT FROM ENEMY AIR OBSERVATION

Positions should be placed under the jungle canopy for its natural concealment from air observation. If this is not possible, positions should be covered with camouflage nets or with expedient covers of loosely woven vines and branches.

For more detailed information on how to prepare jungle fighting positions, *see FM 7-8.*

SHELTERS

Shelters are made when the tactical situation permits. The poncho can be used as a roof for the shelter. It reduces the need for extensive concealing foliage and provides protection from the elements. However, natural camouflage must still be applied to break up the outline of the poncho and reduce the glare of a wet poncho.

EXAMPLES OF PONCHO SHELTERS

HASTY SHELTER-CANOPY FASHION

A hasty shelter is made by suspending the poncho from low underbrush. Due to its simplicity, it can be easily erected at night, especially if heavy strings have already been tied to the corners of the poncho.

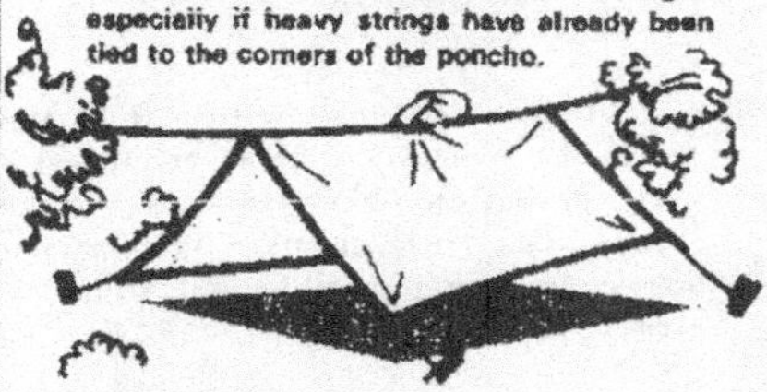

HASTY SHELTER-CANOPY FASHION

This is another hasty shelter pitched canopy fashion.

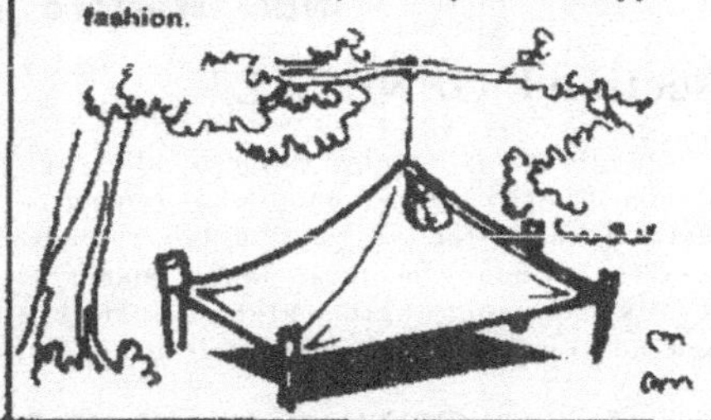

PONCHO AND SPREADER BARS

This is a hasty shelter using a poncho and two branches for spreader bars.

LOW SILHOUETTE SHELTER

This low silhouette shelter can be used while improving fighting positions. It can be lowered by removing the front upright supports.

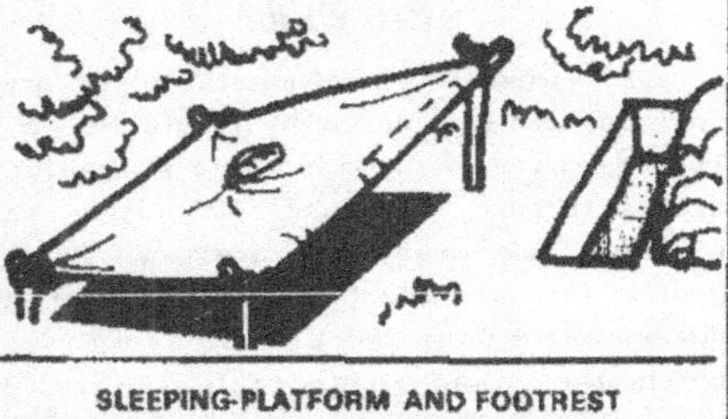

PONCHO SHELTER

Two ponchos fastened together will shelter four soldiers from the rain. Extra ponchos can be used as ground sheets.

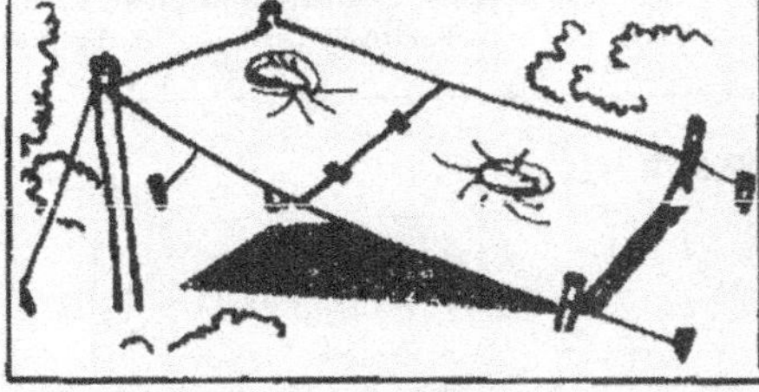

SLEEPING-PLATFORM AND FOOTREST

The following type of shelter may be used for a longer stay in more secure areas. A sleeping-platform and footrest protect from dampness and insects.

APPENDIX F

NBC Warfare in Jungle Areas

Section I. GENERAL

Jungle areas require exceptionally high standards of discipline and conditioning to maintain an effective NBC defense readiness.The demand for these higher standards results from the jungle's high temperature and humidity and heavy rainfall. Sunlight and wind are reduced within thick jungles due to the vegetation. These conditions affect NBC operations because they reduce the chances that NBC agents will disperse, thus increasing their impact on operations in that area.

Section II. CONDITIONS CREATED BY NBC AGENTS

SPECIAL CONDITIONS

NUCLEAR

The initial effects of nuclear blasts are not significantly reduced by jungle foliage. The canopy may provide some protection against thermal radiation. The blast wave will blow down many trees and cause flying debris. These fallen trees may restrict movement, but may also improve observation and fields of fire of some weapons. Fallout may be temporarily retained in the jungle canopy, reducing the immediate hazard. Later rains, however, will wash these particles to the ground and concentrate them in low areas. These areas are likely to become radiation "hotspots."

BIOLOGICAL

Jungle climates favor the use of biological agents. Biological agents thrive in heat, humidity, and shade. As with chemical agents, downwind and spray hazards will be reduced by the lack of wind in the jungle. Strict supervision of field sanitation and adequate purified water supplies will provide the best defense against biological agents as well as control natural diseases in the jungle.

CONTENTS

CHEMICAL

If jungle enemies use chemical weapons, they can cause serious problems for US troops. Persistent agents delivered by means of artillery shells will remain effective for long periods under the canopy. The jungle heat will vaporize chemical agents that are delivered in liquid form. Because of low wind speeds, these vapors will hang in the air and remain hazardous for long periods. The downwind hazard, however, will be reduced. Likewise, the danger from sprayed agents will be reduced. Protective masks and clothing will be uncomfortable in high heat and humidity. Commanders must plan for a decrease in performance by soldiers in this environment. They should also plan for heat casualties. Finally, NBC equipment must be maintained so that it does not rot, mildew, or rust.

Section III. ACTIONS TAKEN TO PROTECT TROOPS

JUNGLE MISSION ORIENTED PROTECTION POSTURE

When temperatures rise to the 85 to 100 degree Fahrenheit range, troops can be expected to continue moderate or heavy workloads only if they are permitted to reduce their mission oriented protection posture (MOPP) level. This increases the risk of chemical casualties if the unit comes under chemical attack. Vision, especially downward vision, is limited while wearing the mask. While these factors tend to discourage the wearing of protective equipment, leaders must balance the need to accomplish their mission as opposed to the NBC threat.

As in any other environment, the commander must plan his MOPP level based upon the mission and the estimate of the chemical threat. The level and the length of exertion are important factors. Long-term efforts, such as a road march, are more taxing than short-term, high-energy tasks, such as an assault.

A commander considering protective measures should keep two principles in mind. The first is that although troop safety is an important consideration, the accomplishment of the mission is of greater importance. The second is that even though a higher level of protection increases the risk of heat casualties, these casualties are not as hazardous or as long lasting as are chemical casualties. Heatstroke, for example, is fatal about 50 percent of the time, but makes up a very small percentage of all heat casualties. Physical conditioning and acclimatization of soldiers have an impact on how long they can operate in protective equipment.

MOPP

MOPP	PROTECTIVE EQUIPMENT			
	OVERGARMENT	BOOTIES	MASK/HOOD	GLOVES
1 (POSSIBLE)	Worn, opened or closed based on temperature	Carried	Carried	Carried
2 (POSSIBLE)	Same as MOPP 1	Worn	Carried	Carried
3 (LIKELY)	Same as MOPP 1	Worn	Worn, hood opened or closed based on temperature	Carried
4 (IMMINENT)	Worn, closed	Worn	Worn	Worn

Assuming troops are acclimatized, work rates can be exceeded for short periods if adequate rest and water are provided. In all cases leaders must balance the chances of exposing their units to heat exhaustion or chemical agents to the need to accomplish the mission. The ideal MOPP is the category that provides the highest degree of protection from chemical effects and still permits the mission to be accomplished.

The following measures can improve unit efficiency at any MOPP level:

- Rotating heavy work among elements or individuals
- Authorizing longer and more frequent rest periods
- Providing adequate drinking water
- Using truck or air transport when possible

NOTE: For further information, see FM 21-40.

APPENDIX G

Communication Techniques in a Jungle Environment

Section I. GENERAL

Rapid, reliable communications are essential in jungle operations. Command, control, fire support, resupply, and evacuation are all dependent on effective communications. The importance of establishing and maintaining reliable communications cannot be overemphasized. For this reason, a commander must give communications high priority in his planning and supervision.

Jungle operations place additional demands on the resources of communications units and personnel. Tactical and signal units may require augmentation of signal personnel and equipment to accomplish the command and control functions. Besides providing normal tactical communications networks, signal units may have to operate long distance radio sets. Army aircraft will have to be used in many instances to provide a radio relay capability as well as to assist wire-laying operations.

Section II. COMMUNICATING IN THE JUNGLE.

TACTICAL COMMUNICATIONS MEANS

With the range of FM communications in a jungle environment significantly reduced due to the dense undergrowth, heavy rains, and hilly terrain, all means of communications should be used in a manner that complements one another. For all operations, backup means of communications must be planned. The common means of communication available are visual, sound, messenger, wire, and radio.

CONTENTS

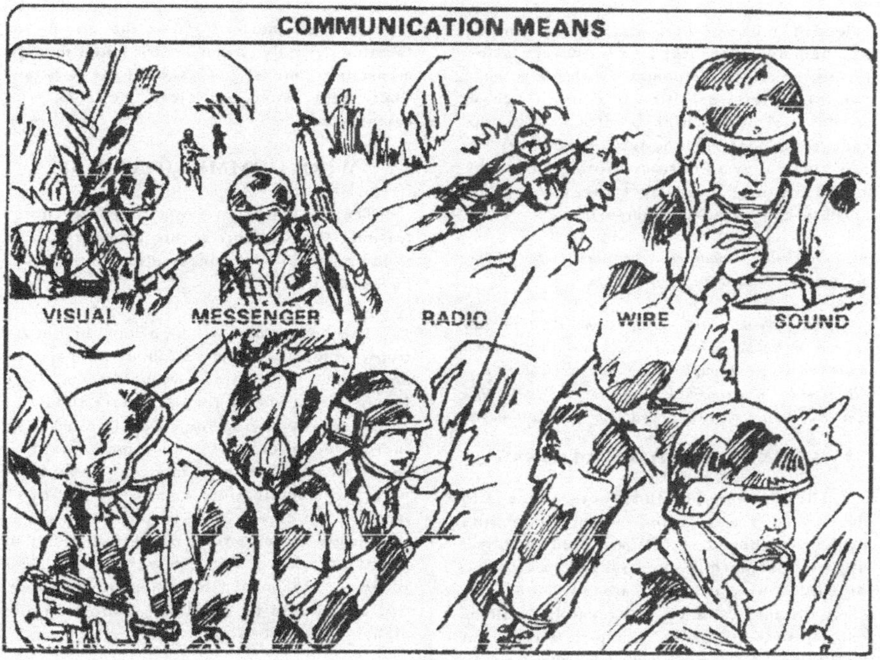

VISUAL COMMUNICATIONS

Such means as arm-and-hand signals, pyrotechnics, flashlights, headlights, smoke grenades, mirrors, and panels normally allow quick transmission of messages and instructions. However, visual means of communication in the jungle are restricted by the dense vegetation. Commanders will rarely have visual contact with all elements or members of their unit.

Arm-and-hand signals are used in all types of operations. The effectiveness of arm-and-hand signals can be improved by insuring that each soldier understands the meaning of the signal and passes it on to others in his vicinity.

Pyrotechnics can be used in most conditions of visibility. Certain standard colors of smoke or flare signals have limited use because they blend in with vegetation (green) or offer little contrast with fog or haze (white). Careful selection of colors (such as red or yellow) which contrast with the background color of the jungle increases the effective range of pyrotechnic signals. The jungle canopy can affect the use of star clusters and parachute flares in two ways.

Overhead clearance required for firing flares, is limited.

Once clear of the canopy, the flare or star cluster may be hard to see by other ground units looking through the canopy. The disadvantage of pyrotechnic signals is that they can be easily seen and imitated by the enemy.

SOUND COMMUNICATIONS

These means include both voice and devices such as whistles, horns, gongs, and explosives. Sound signals are used to attract attention, transmit prearranged messages, or spread alarms. They are good only for short distances. Range and reliability will be reduced by battle noise, weather, terrain, and vegetation. As they may also be heard by the enemy, they should be restricted for security reasons. Sound signals must be simple to avoid misunderstandings. The means for sound signals are usually prescribed by the unit SOP and communications-electronics operation instructions (CEOI).

MESSENGER COMMUNICATIONS

These types are most secure and a good way to send long messages and documents. However, the speed of surface messengers is severely limited by jungle terrain and vegetation. The lack of roads restricts full use of motor vehicles. Foot messengers may have difficulties with land navigation and natural obstacles. The use of air messengers maybe a better alternative, but one that is limited by availability of aircraft, scarcity of good landing zones, weather conditions, and the fact that air activity aids the enemy in locating friendly troop areas. When using a messenger, messages should be written. Their text must be clear, concise, and complete.

WIRE COMMUNICATIONS

These methods are established whenever feasible. Wire is more secure than radio, hard to jam, and allows conversation with break-in capability.

The heavy rain and high humidity of the tropics reduce the range (about 20 percent) and reliability of wire communications. Wire laying in the jungle requires more time and suitable w-ire laying routes are limited. The increased cover and concealment make it easier for the enemy to intercept or interrupt wire communications and to ambush maintenance crews. Wire laid by air on top of the jungle canopy is extremely difficult to maintain and recover; however, it is less likely to be damaged by vehicles or weather. Splices should be carefully made to avoid signal losses when wet.

RADIO COMMUNICATIONS

Radio communications are normally the fastest means available but the least secure.

Effective radio communications in the jungle require emphasis on:

- Use of antennas and field expedients
- Use of retransmissions and radio relays
- Continuous preventive maintenance
- Communications security (CO MS EC)
- Trained operators

The usefulness of radio communications is reduced in jungle operations. VHF and UHF radios (FM radios included) are particularly limited by the jungle growth and terrain that absorb and obstruct the transmissions. It is not unusual for the range of a set operated in the jungle to be reduced by 10 to 25 percent of the normal range.

ANTENNAS AND FIELD EXPEDIENT ANTENNAS

In order to overcome environmental conditions in the jungle, existing antennas must be used properly and field expedient antennas produced to extend the capacity of current equipment.

> **To use existing antennas properly, the following items must be accomplished:**
>
> - **Keep whip antenna vertical when transmitting**
> - **Insure the antenna is not grounded by being in contact with foreign object**
> - **Position antennas to achieve the best line-of-sight possible between stations (such as on top of a hill)**

The problems of line-of-sight antenna locations and operations security are best solved by remoting the transmitter from the command post.

The AN/GRA-39 radio set control group allows the operator to remote the radio up to 3.2 kilometers away from the observation post. It should be noted that there is no practical advantage to be gained by installing the radio and antenna away from the command post (remoting) unless the distance is 1 kilometer or more. The error inherent in radio direction finding equipment does not provide for a consistent accuracy less than 1 kilometer.

REMOTE SETUP

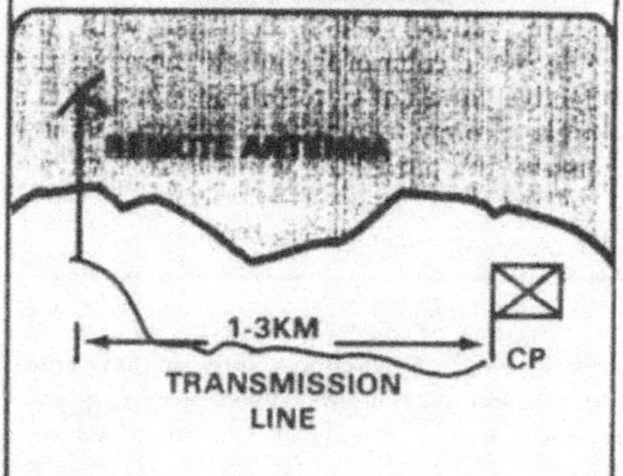

Field expedient antennas will enhance the communications capacity of a unit operating in the jungle if they are properly constructed. WD-1 or portions of the RC-292 antenna can be used as the radiating element (antenna). Wood or plastic spoons can be used as insulators when insulators are not available. Resistors should be obtained through communications maintenance channels.

> **Useful field expedient antennas in the jungle are:**
>
> **Field expedient RC-292**
>
> **Patrol antenna**

The standard issue RC-292 is a highly-effective, omnidirectional antenna. It is usually more effective than a whip antenna, and is particularly effective in the VHF range. With all its components, it weighs 48 pounds and two soldiers can erect it in 15 minutes.

Leaders should consider using field expedient versions of this antenna in the jungle for these reasons:

- Excessive load for dismounted soldiers
- Awkward to assemble in restricted terrain
- Increased chance of losing parts

Portions of the RC-292 may be used to construct a field expedient antenna. Only the vertical sections, antenna base, ground plane sections, and transmission line are needed.

FIELD EXPEDIENT ANTENNA USING PORTIONS OF RC-292

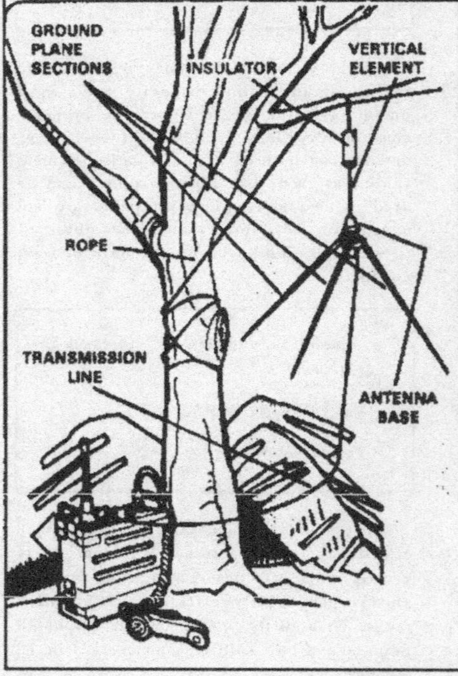

To construct this antenna:

Attach an insulator to the vertical element and add a piece of rope to it.

Tie a weight on the rope and throw it over a tree limb.

Pull the antenna up as high as necessary.

Raise or lower the antenna to change element lengths each time the frequency is changed.

Attach the coaxial lead-in to the radio connector as with the standard RC-292.

Add a rope from the bottom of the head to a stake in the ground. This prevents swaying and damage to the equipment when there are strong winds.

Another field expedient antenna can be constructed without using any of the standard RC-292 components. This antenna has various names such as the field expedient 292, the jungle antenna, and the field expedient ground plane antenna.

FIELD EXPEDIENT 292 ANTENNA

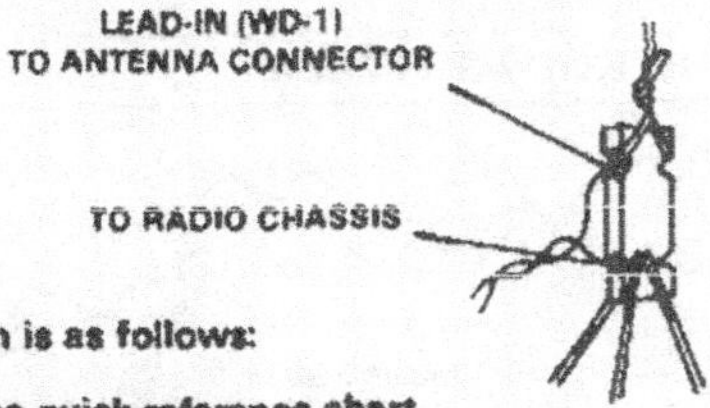

Construction is as follows:

Step 1. Using the quick reference chart at the end of this appendix, cut all four wires for a quarter wave antenna. Connect them as in the following illustration.

Step 2. Two insulators are needed; one at each end to separate the vertical element. Attach a rope with a rock tied to it to throw the rope over a tree limb.

Step 3. Connect the WD-1 as shown, before pulling the antenna up in the air.

Step 4. At the radio, connect the vertical element wire of WD-1 to the antenna connector and the ground plane wire to the radio set chassis.

If frequencies are known before deployment, soldiers can cut wires to the appropriate lengths and carry them on the operation.

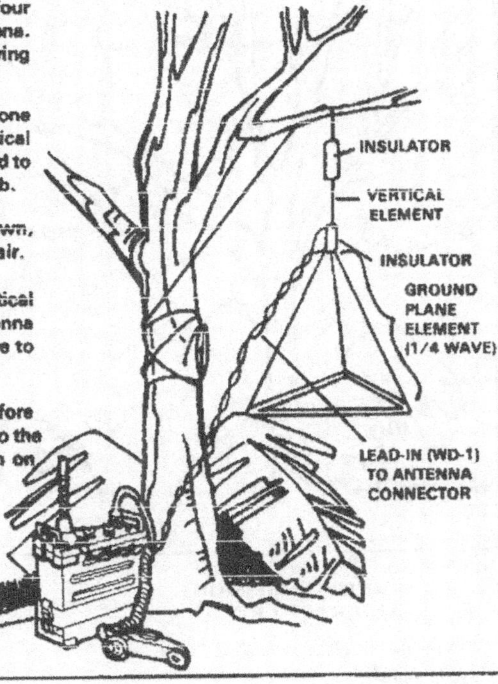

A third type of expedient antenna used in the jungle is the *patrol antenna.* It is the same as the jungle antenna minus the ground plane element. Insure that the length is determined by using the 1/2 wavelength instead of the 1/4 wavelength column of the chart. The range of the set will be increased 2 to 3 times its normal range. This antenna is easier to handle and construct.

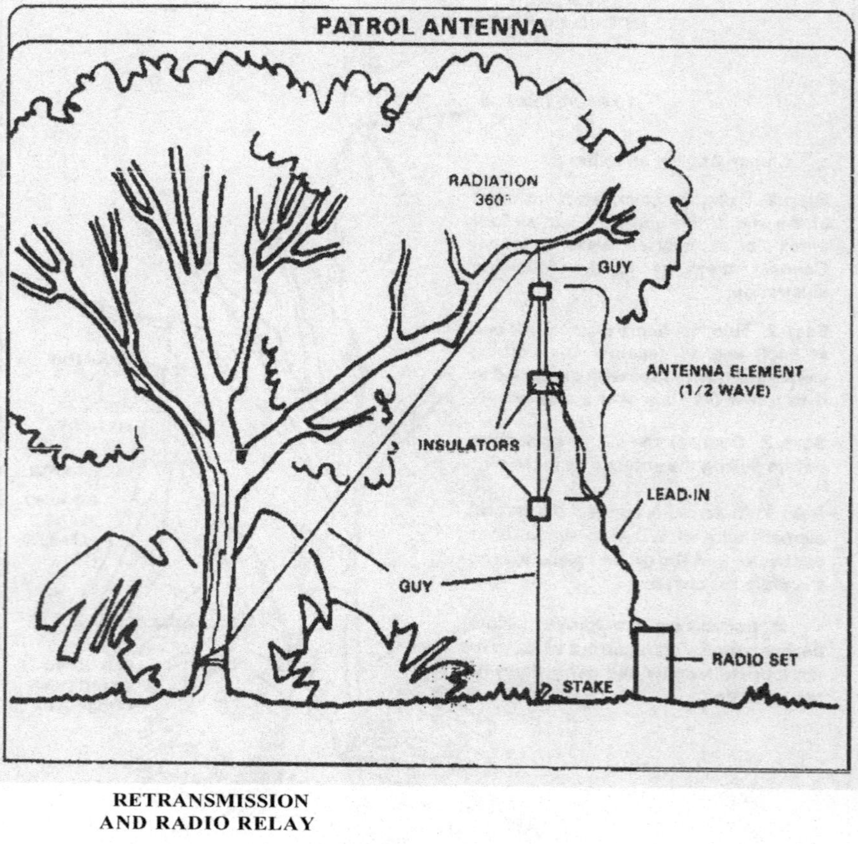

RETRANSMISSION AND RADIO RELAY

With the decrease in range of most radios used in jungle operations, retransmission will be a good method to improve communications. Commanders should consider establishing retransmission sites and using additional signal personnel. Radio relay sites, either airborne or on high terrain, will also improve radio communications between stations.

EQUIPMENT MAINTENANCE IN THE JUNGLE

Moisture commonly found in the jungle environment can result in numerous outages of communications-electronic equipment.

> The following measures, combined with normal maintenance, will help counteract the problems.
>
> - Seal the radios with silicone compound.
> - Pack radios in waterproof containers when not in use (remove batteries).
> - Protect handsets and microphones with plastic bags. (Batteries and rations are packed in plastic bags.) Check frequently for moisture build-up in the bags.

When radios are protected by plastic bags, make sure the battery vent is not obstructed. Radios protected in this manner must be constantly checked to insure moisture does not build up in the bag. Clean radios as frequently as individual and crew-served weapons are cleaned.

Section III. AVOIDING ENEMY INTERFERENCE

COMMUNICATIONS SECURITY

In the past, jungle enemies have relied heavily on friendly radio messages as primary intelligence sources. Communications security denies or delays unauthorized persons from gaining information from telecommunications.

ROLE OF COMSEC

It includes:

- Using authentication to insure that the other communicating station is a friendly one.
- Using only approved codes.
- Designating periods when all radio equipment is turned off.
- Restricting the use of radio transmitters (monitoring radio receivers/listening silence).
- Enforcing net discipline and radiotelephone procedures (all stations in a net must use authorized callsigns and prowords; and they must limit transmissions to official traffic).
- Selecting radio sites with a hill as a shield between them and the enemy.
- Using directional antennas when possible.

GUIDELINES FOR BETTER COMMUNICATIONS SECURITY

- Enforce proper use of authentication systems.
- Use only authorized authentication systems.
- Assume that your radio transmissions are being intercepted by the enemy.
- Use proper site selection, minimum power, and minimum transmission times.
- Do not use electrical communications when messengers will do as well. Messengers are the most secure means of communications, other than personal contact.
- Do not paraphrase or try to talk around the subject. Keep transmissions short and concise. Use authorized codes to send information.

ANTIJAMMING

Radio operators should use the following antijamming procedures to thwart enemy jamming efforts.

Recognition. The first thing an operator must do when there is interference on his radio is to try to find its cause. As symptoms ofjamming are often similar to other types of interference, he should not assume that it is jamming. If the interference decreases when the receiver antenna is removed, the interference is jamming; if it does not, the interference is generated inside the receiver.

Continued Operation. Normal radio operation is continued once jamming has been identified so that the enemy cannot determine the effect of his jamming. The rule is: during jamming, operators continue operating unless ordered to shut down.

Reporting. All operators must report jamming to their next higher headquarters, by another means of communications; for example, wire or messenger. A typical jamming report tells *date and time of jamming, frequencies affected, type and strength of jamming signal, and designa tion of the unit making the report.*

The exact format of the report is found in the unit's CEOI.

FORMULAS FOR COMPUTING WAVELENGTH

TO FIGURE:	FORMULA:	EXAMPLE:
A quarter wave-length in feet:	Divide 234 (constant) by the operating frequency in MHz.	234 ÷ 44.8 = 5.22' or 5'3".
A half wave-length in feet:	Divide 468 (constant) by the operating frequency in MHz.	468÷56=8.36' or 8'5".
A full wave-length in feet:	Divide 936 (constant) by the operating frequency in MHz.	936÷45=20.8' or 20'10".
TO CONVERT:		
Feet to meters:	Multiply by .3048 (constant).	110' x .3048 = 33.5 meters.
Meters to feet:	Multiply by 3.28 (constant).	100 meters x 3.28 = 328 feet.

QUICK REFERENCE CHART EXPEDIENT ANTENNAS

HIGH FREQUENCY (HF) ANTENNA LENGTH IN FEET AND INCHES				VERY HIGH FREQUENCY (VHF) ANTENNA LENGTH IN FEET AND INCHES			
Op freq in MHz	1/4 wave	1/2 wave	1 wave	Op freq in MHz	1/4 wave	1/2 wave	1 wave
2	117′	234′	468′	30	7′10″	15′7″	31′2″
3	78′	156′	312′	33	7′1″	14′2″	28′4″
4	58′6″	117′	234′	35	6′9″	13′5″	26′10″
5	46′9″	93′7″	187′4″	37	6′4″	12′7″	25′2″
6	39′	78′	156′	40	5′10″	11′8″	23′4″
7	33′5″	66′10″	133′8″	43	5′5″	10′10″	21′8″
8	29′3″	58′6″	117′	45	5′3″	10′5″	20′10″
9	26′	52′	104′	48	4′10″	9′8″	19′4″
10	23′5″	46′10″	93′8″	50	4′9″	9′5″	18′10″
11	21′3″	42′6″	85′	55	4′3″	8′6″	17′
12	19′6″	39′	78′	57	4′1″	8′2″	16′4″
13	18′	36′	72′	60	3′11″	7′10″	15′8″
14	16′9″	33′5″	66′10″	65	3′7″	7′2″	14′4″
15	15′7″	31′2″	62′4″	68	3′5″	6′10″	13′8″
16	14′7″	29′2″	58′4″	70	3′4″	6′7″	13′2″
17	13′9″	27′6″	55′	75	3′1″	6′2″	12′4″
18	13′	26′	52′	80	3′	5′11″	11′10″

APPENDIX H

The Individual Soldier's Combat Load

Section I. GENERAL

All equipment considered essential to mission completion should be carried by the individual at all times. Rucksacks should be packed in such a way that they can be dropped or hidden, for recovery later, to allow troops to move freely upon contact or on patrol. In an emergency, they may be discarded completely. Those essential items that must be placed in the rucksack, due to the extended nature of operations, should be removed when the situation dictates. Combat-essential items should not be left in rucksacks that are cached or dropped. Unit SOP should dictate the items that are removed from rucksacks and included in equipment carried by each soldier when on short-duration missions where rucksacks are left behind. Packing lists should also be designed for squads and platoons with careful attention given to balancing loads. (For example, machinegunners should not be given extra equipment to pack until the fighting and existence loads of the other members of the platoon are equal to that of the machinegunners.)

Section II. PREPARING THE LOAD

WEIGHT CARRIED

The problem of excessive weight in existence loads needs to be closely monitored by commanders at all levels. Specific company and platoon SOPs need to be reinforced with inspections prior to deployment, demanding strict adherence to packing lists. Unusable equipment should not be carried.

WATER CARRIED

Water will constitute the second greatest part of each soldier's load. Jungle fighters must be issued more canteens than normal, and they may have to carry 5 to 10 quarts of water attached to their rucksacks, especially in the dry season. Water purification tablets should be carried so that stream water may be used.

AMMUNITION CARRIED

Units should be told how long they can expect to operate without resupply. This figure becomes the basis for planning the equipment, supplies, and ammunition to be carried. Ammunition will normally make up the greatest part of each soldier's load, and ammunition supplies must be planned to last until troops can be resupplied. Rifle and machinegun ammunition and grenades should be loaded on pallets in a rear area so that units can easily be resupplied by helicopter, if required. Every soldier should carry colored smoke grenades and flares for signaling, and tripflares and Claymore mines for security. Special ammunition may be carried, if required by the mission. The most common types of special ammunition

CONTENTS

and C4 explosives, concussion grenades, and CS munitions.

The following description of a soldier's combat load should be used *only as a guide* in developing a unit SOP. Fighting and existence loads will vary according to each unit's special needs and missions.

FIGHTING LOAD FOR AN INFANTRY PLATOON CONDUCTING OPERATIONS IN JUNGLE TERRAIN

	INDIVIDUAL (Carried at all times)				SQUAD AND TEAM LEADERS (Additional)	ADDED FROM RUCKSACK (When rucksacks are left in patrol base or cached, or when on a mission where no rucksacks are carried.)	
	CLOTHING	EQUIPMENT/BASIC LOAD					
	Helmet, with liner, camouflage cover and band. Trousers and jacket (camouflage, jungle fatigue). Belt. Socks. Jungle boots. ID tags.	M16 rifle with sling and one magazine (30 rounds). Ammunition, 180 rounds and six magazines. The following weapon and basic loads, substituted as appropriate: M203 with one round of 40mm HE and one magazine of 5.56mm (30 rounds) Basic load 40mm: 36 rounds HE Basic load 5.56mm 140 rounds and magazine	M60 machine-gun with sling Basic load 200 rounds per gun. Caliber .45 pistol for gunner with magazine, seven rounds Two magazines (14 rounds) with ammunition pouch. Ammunition cases (two for magazines). Hand grenades (two fragmentation). Smoke grenades (two). Canteens (two 1-quart canteens—filled). (Amount depends on availability of local water.)	Water purification tablets. Individual equipment belt/first-aid packet case and suspenders. penders. Snaplink. Twelve-foot nylon rope (Swiss seat). Bayonet knife and scabbard. Ammunition case containing items such as: insect repellent, camouflage stick, foot powder, toilet paper, bandages, matches, notebook and pencil, aspirin, emergency rations, penlight. Individual cleaning equipment. Rations (one meal, C-rations).	Compass. Strobe light. Map and case. Flashlight with batteries. Whistle. First-aid kit, general purpose. Watch.	One poncho per man. One entrenching tool with carrier. Two machetes with case (two per squad). Mask, CBR M17A1 with carrier (optional depending on threat).	NOTE: Critical items such as Claymores, LAWs, M60 ammunition, etc., should never be left with rucksacks.
APPROX WEIGHT (POUNDS) TOTAL	9.36	+	42.63	= 51.99	5.06	8.17	

EXISTENCE LOAD FOR AN INFANTRY PLATOON CONDUCTING OPERATIONS IN JUNGLE TERRAIN

	ON OR IN RUCKSACK			ADDITIONAL EQUIPMENT/BASIC LOADS (To be distributed through platoon during sustained operations. This equipment should first be broken down by squads, then distributed to individuals based on balancing loads across the board.)	
	EQUIPMENT		RATIONS		
	One frame including lower back, shoulder, and waist straps. Pack, medium, combat, field. Bag, waterproof. Poncho (prerigged for shelter). Poncho liner. Socks (four pairs). Trousers (camouflage). Tennis shoes. One sleeping shirt or tee shirt. Health items: soap, toothbrush, razor and blades (use insect repellent for shaving). Insect bar.	Air mattress. Extra insect repellent. Extra water purification tablets. Two quarts water. (Amount depends on availability of local water.) Twenty-foot suspension line. Sewing kit. Mask, CBR M17A1 with carrier (optional depending on threat). Entrenching tool. Sterno/heat tablets/butane stove. Life preserver, inflatable (optional).	C-rations (dry rations are lighter, but more water is needed). 1 day: 1.75 per meal x 3 number of meals = 5.25 TOTAL 3 days: 1.75 per meal x 9 number of meals = 15.75 TOTAL 5 days: 1.75 per meal x 15 number of meals = 26.25 TOTAL	Two PRC-77 radios, complete with four batteries. Five TA1 telephones, w/200 meters wire (one for plt HQ, one per squad). Twelve Claymore mines, four per squad. Twelve LAWs, four per squad. Two M60 machineguns with 600 rounds (200 rounds for gunner, 200 rounds for assistant gunner, 200 rounds for ammunition bearer). Fifteen parachute flares, five per squad.	Fifteen trip-flares, five per squad. Three 120-foot ropes (optional), one per squad. Eight machetes w/carrying case (two per squad, two for plt HQ). NOTE: This weight will increase as unit strength decreases. Radio telephone operators, M60 machinegunners, and grenadiers are left out to determine the average extra weight per soldier.

	WITHOUT OPTIONS	WITH OPTIONS	DAYS	1	3	5		
APPROX WEIGHT (POUNDS) TOTAL	29.07	38.54	-	34.32	44.82	55.32	195	(Approximately 6 pounds per rifleman.)

EXISTENCE ITEMS—RUCKSACK				
	RIFLEMAN	APPROX WEIGHT (POUNDS)	M60 MACHINEGUNNER	APPROX WEIGHT (POUNDS)
RUCKSACK	As listed with options—air mattress and life preserver only.	34.57	As listed with options—air mattress and life preserver only.	34.57
RATION	Three days (note 1).	15.75	Three days (note 1).	15.75
ADDITIONAL BASIC LOAD	One Claymore mine (plt/sqd items—approximately 6 pounds per rifleman). One parachute flare. One trip flare.	4.00 1.00 1.00		
EXISTENCE LOAD		50.32		50.32
EXISTENCE LOAD TOTAL		56.32		50.32
COMBAT LOAD (NOTE 2) TOTAL		99.01		106.98

NOTE 1: Dry rations are lighter and more compact than C-rations. Water may be a problem, however.

NOTE 2: By adding total fighting load and total existence load, to obtain the total combat load, a commander can determine who can receive any additional items to carry.

SAMPLE INDIVIDUAL FIGHTING LOAD FOR 5-DAY OPERATION IN A TROPICAL ENVIRONMENT WITH NO NBC THREAT

RIFLEMAN	APPROX WEIGHT (POUNDS)	M60 MACHINEGUNNER	APPROX WEIGHT (POUNDS)
CLOTHING As listed.	9.36	CLOTHING: As listed.	9.36
EQUIPMENT		EQUIPMENT	
M16 w/mag (30 rounds)	7.91	M60	23.00
Ammunition (180 rounds)	6.21	200 rounds	14.00
Ammunition cases	.86	Cal .45 w/mag (7 rounds)	3.20
Hand grenades (2 frag)	2.00	Ammunition pouch (14 rounds)	.75
Smoke grenades	1.50	Canteens (two 1-quart—filled)	7.20
Canteen (two 1-quart—filled)	7.20	Water purification tablets	.06
Water purification tablets	.06	LBE	1.59
LBE	1.59	Snaplink	.25
Snaplink	.25	Twelve-foot nylon rope	.75
Twelve-foot nylon rope	.75	Ammunition case (subsistence items)	2.00
Ammunition case (subsistence items)	2.00	Cleaning equipment	.75
Gloves	.50	Rations	26.25
Cleaning equipment	.75		
Rations	26.25		
EQUIPMENT	57.83		79.80
CLOTHING	+ 9.36		+ 9.36
TOTAL FIGHTING LOAD	67.19		89.16

Section III. CARRYING THE LOAD

USE OF THE RUCKSACK

Platoon- and company-sized units conducting offensive operations in jungle terrain must emphasize self-reliance. Units must carry rucksacks and be able to operate for extended periods of time with routine resupply being accomplished every 5 to 7 days. This will allow units to develop their operational areas, and preclude compromising their location for frequent resupply operations.

To minimize the weight of rucksacks, commanders must carefully analyze their missions, and allow nothing but mission-essential equipment to be carried. Detailed packing lists and thorough inspection prior to a mission will insure nonessential items are not taken to the field.

When units must carry rucksacks, troops should not be expected to fight with this burden. Unit SOPS must be developed and practiced concerning the disposition of rucksacks in various situations. Rucksacks should contain nothing other than existence items, so the grounding or caching of them should have no effect on the combat effectiveness of an element or unit.

A unit moves from point to point with rucksacks on. After the movement phase of an operation is over, rucksacks should be removed and secured.

DISPOSITION OF RUCKSACKS

Techniques of handling rucksacks will vary according to the situation.

SHORT-DURATION MISSIONS (48 HOURS OR LESS)

Rucksacks should not be carried. Rations can be placed in a sock and securely tied to the back straps on the load-bearing harness. This method will not restrict mobility and noise will be reduced. The poncho, tied to the back of the pistol belt, will be the only other item needed from the existence load. The balance of the equipment to be carried, in excess of a unit's standard fighting load, will be determined by the mission. Bulky items can be carried in rucksacks, but loads should be rotated frequently.

CHANCE CONTACT

Rucksacks should be dropped immediately when contact is made, to allow rapid reaction and maneuverability. The mission to secure these rucksacks should be given to the squad or fire team that is providing rear security. The rucksacks need not be gathered or centralized in this situation, due to the usuall y short duration of this type of contact. If the unit must move because of enemy pressure, or in case of a need for reinforcements, all rucksacks must be left behind. Upon returning to retrieve dropped rucksacks, a unit must be alert for ambushes and care taken to avoid casualties from enemy boobytraps placed in and around the equipment.

PATROL BASE OPERATIONS

Elements or squads moving out of patrol bases to conduct short-duration missions should consolidate and conceal rucksacks prior to moving out. The only time rucksacks should be carried is when the mission does not call for return to the patrol base. If the patrol base is forced to withdraw from its location, all rucksacks will probably be left behind in order to break contact and to move rapidly for linkup with the balance of the unit. Reinforcements to assist a squad or element in contact while away from their patrol base should not take rucksacks when moving to their assistance. If the situation is so critical that the entire unit must move to the assistance of an element or squad, rucksacks should be quickly consolidated, concealed, and left behind.

CACHING RUCKSACKS

This technique is used as a last resort. It is a difficult mission to successfully accomplish. If a situation arises where a commander decides to cache, care must be taken in selecting a cache site. An easily identified area should be used as a reference point. From there, azimuth and pace count should be used to actual cache sites. Natural lines of drift should be avoided and care should be taken not to leave trails into cache sites. Security must be established 360 degrees around the proposed site, and must be far enough out to insure that enemy forces cannot observe the activity. Items that will be needed in the event a unit is unable to return to the cached items should be removed and included in the fighting load. (See short-duration missions.) Natural camouflage (deadfall, thickets, caves, etc.) should be used to conceal rucksacks. Pits can be dug if the area can be returned to its natural state when caching is completed. When returning to cache sites to recover rucksacks, unit leaders must consider the sites danger areas and act accordingly. Each cache site should have the equipment arranged in such a manner that it is easy to determine if anyone has tampered with the equipment. This helps in preventing casualties from boobytraps and mines. Soldiers should not group around the equipment when it is recovered and security must be maintained at all times while redistributing the rucksacks.

In all situations, the commander should remember that rucksacks are expendable. Leaders should tailor the fighting loads of their units to such a degree that the loss of all rucksacks would not hamper the unit's ability to continue the mission.

APPENDIX I

Adjustment of Indirect Fire by Sound

Section I. GENERAL

To effectively use indirect firepower, it is sometimes necessary to adjust fire, and in a jungle environment this requires special techniques. In the limited visibility of the jungle, the only method available for adjustment may be by the sound of the explosion of the round. The techniques described are equally effective in the adjustment of artillery or mortars.

The basic principle of adjustment by sound is that the direction and distance to the target are known, and the direction and distance to the explosion of the round can be calculated from the sound of the explosion. These data are compared to determine the appropriate correction in deviation and in range. The problem encountered in this method is that the heavy foliage distorts sound and makes it difficult to determine range or lateral shift corrections. There is a distinct technique used to determine the correction in each of the two dimensions.

CONTENTS

Section II. TECHNIQUES USED TO DETERMINE CORRECTIONS

CORRECTION IN DEVIATION

The direction from the observer to the target is measured from a map or by compass. The direction to the explosion is measured with a compass. These two azimuths, expressed in roils, form an angle as illustrated.

COMPUTING THE CORRECTION

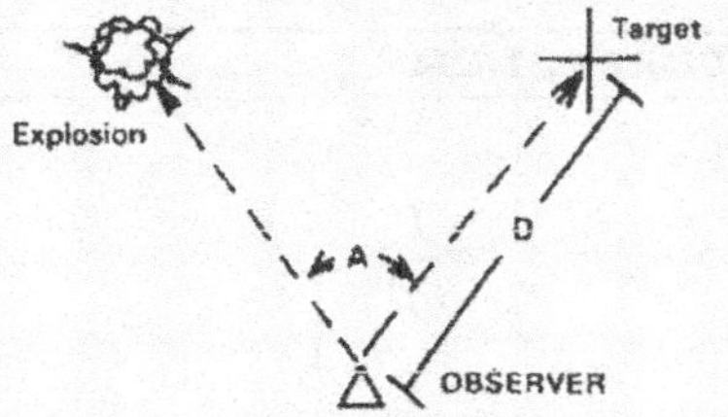

To calculate the correction in deviation, multiply angle A by distance D, with the distance expressed in kilometers, as shown in the example.

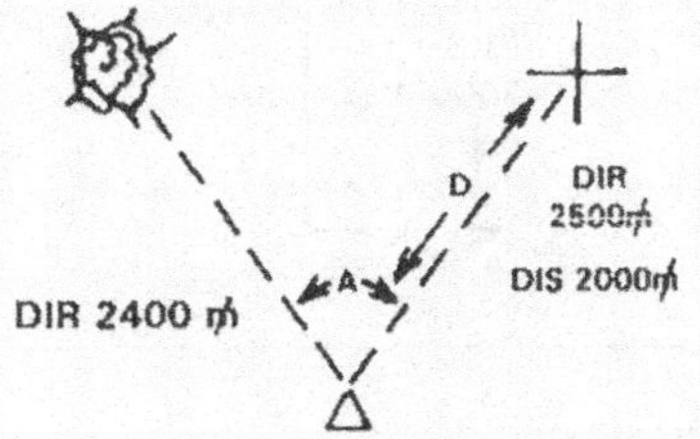

A = Angle in mils (m̸) = 2500m̸ - 2400m̸ = 100m̸

D = Distance to target in kilometers (km) = 2000m = 2 km

C = Correction in meters (m) = ?

A X D = C

100 X 2 - 200 m

The correction is right 200.

CORRECTION IN RANGE

The distance from the observer to the target can be measured on a map or can be estimated. The distance to the explosion of the round is calculated by counting the number of seconds from the impact of the projectile until the observer hears the explosion. Multiply the number of seconds by the speed of sound, 350 meters/second. Add or drop the resulting distance, as appropriate.

To tell the observer when the round impacts, the lire direction center (DFC) provides "splash and count." In splash and count, the FDC sends, "Splash - four, three, two, one-impact." On the word "impact" the round lands, and the observer starts counting the seconds until he hears the explosion.

COMPUTING THE CORRECTION

Example: After hearing "impact" the observer counted 6 seconds before hearing the explosion.

Seconds X Speed of Sound = Distance to the explosion.

6 sec x 350 m/sec = 2100m

2100m - 2000m = 100m

The correction is drop 100.

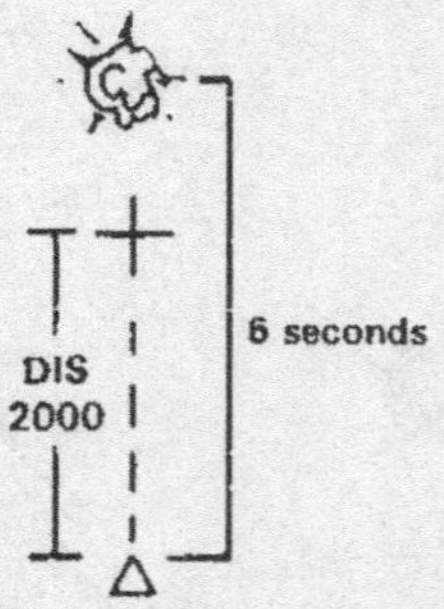

Section III. OTHER CONSIDERATIONS

The FDC normally does not send "splash and count"; rather, it must be requested in the call for fire by saying, "Cannot observe - splash and count."

Many times in dense jungle terrain, a round's effect is lost in the trees. Delay fuze can be used to penetrate the foilage and put the desired effects on the ground. A combination of fuze quick and fuze delay can be effective.

APPENDIX J

Defensive Formation

Section I. GENERAL

Jungle defensive operations are based on the same fundamentals of the defense used in other type operations. Some of the fundamentals may acquire a special significance in the jungle. The basic factors of observation and fields of fire, cover and concealment, obstacles, key terrain, and avenue of approach should always be considered carefully.

Three defensive formations will be presented here: the perimeter, the triangle, and the "Y." Platoon-size elements will be the primary consideration, although the techniques mentioned are adaptable to larger units. It should be noted that all three defensive postures have definite advantages as well as inherent disadvantages. Units should vary their type of defensive posture in order to avoid establishing a pattern.

Section II. FORMATIONS

PERIMETER DEFENSE

The first of the defensive formations is the perimeter. All elements are generally configured in a circular formation, and the terrain is used to the maximum advantage.

PERIMETER FORMATION

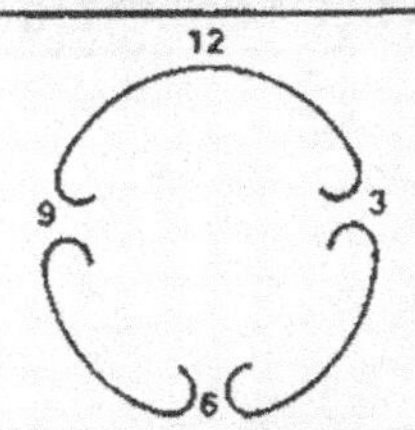

CONTENTS

Advantages are:	It is mutually supported by observation/communication, and patrols are easily coordinated.
■ 360-degree security. ■ Centralized control. ■ Quick emplacement, and it can be executed by any size unit. ■ The frontage of the perimeter can be easily adjusted and internally reinforced.	Disadvantages are: ■ Any penetration creates problems of enfilade, masking, and/or cross fires. ■ It is difficult to achieve final protective fires with the crew-served weapons and, because of the circular configuration, the unit is vulnerable to the entire effects of the bursting radius of an indirect fire weapon.

The basic technique used to establish this defensive formation is the clock system. This involves the platoon using the direction of movement as 12 o'clock, with one squad occupying from 8 to 12, one squad occupying from 8 to 4, and the last squad occupying from 4 to 12. The headquarters element would normally be located in the center of the formation to facilitate control. It is generally suggested that automatic weapons be placed to cover the most likely avenues of approach. At least one machinegun should be kept with the headquarters, under the control of the unit leader, for deployment against a specific threat.

TRIANGLE DEFENSE

The second of the defensive formations is the triangle defense. This formation is a modification of the perimeter.

TRIANGLE FORMATION

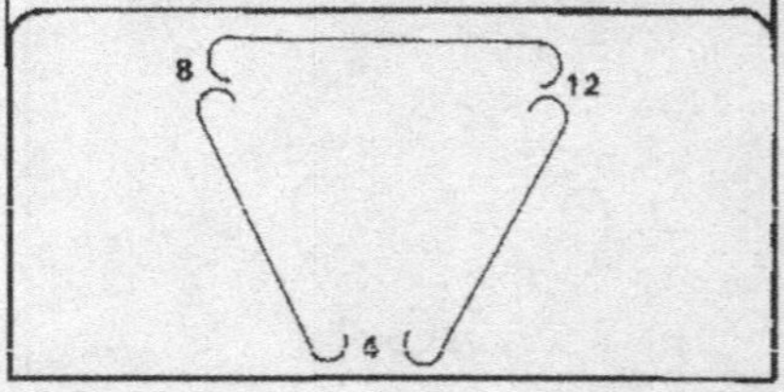

Advantages are:	Disadvantages are:
■ 360-degree security. ■ Quick emplacement, and it can be executed by any size unit. ■ A target approaching perpendicular to any side of the triangle becomes vulnerable to at least one-third of the fighting force and at least two automatic weapons. ■ It is a dual purpose formation in that it may be used as either a defensive or offensive ambush formation.	■ One or more legs of the triangle may be subjected to enfilade fire. ■ Soldiers located at the corners are bunched, thus increasing the danger from indirect fire. ■ Penetration by the enemy creates problems of enfilade, masking, and/or cross fire.

This defensive formation is best established by having the commander move forward with one element and establish a base line. Again, 12 o'clock is used as the direction of movement. The baseline could be established by the first squad, running from 8 to 12. Once this base line is established and the squad is prepared to provide support, the next squad moves forward and occupies the straight line position from 8 to 4; finally, the last squad moves forward and occupies the straight line position from 4 to 12. The headquarters element will normally locate in the center of the formation. If three machineguns are available, one will be placed at each comer in such a manner to allow flexibility in providing final protective fires down either of its two sides. If only two machineguns are available, one may be placed to cover the two sides deemed most vulnerable to attack. The second machinegun should be kept with the headquarters element under control of the unit leader. Firing positions should be pre pared at all three comers.

"Y" DEFENSE

The third defensive formation is the "Y." This formation, like the triangle, may be utilized as an offensive formation, usually in an ambush. Depending upon the situation, it can be a very effective defensive formation.

"Y" FORMATION

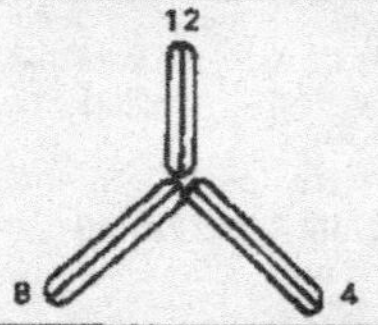

Advantages are:

- 360-degree security.
- Centralized control.
- Maximum firepower in all directions.
- Each penetration by the enemy places him in a new kill zone.

Disadvantages are:

- It must be emplaced during daylight to insure that aiming stakes are accurately emplaced, to preclude firing into adjacent positions.
- The central point of the "Y" creates bunching, and it increases the danger from indirect fire.

The recommended technique for emplacement of this formation is for the unit leader to take one element and the headquarters forward and establish the center of the sector. This establishes the 12 o'clock position. The remaining elements stay in security positions. Since all legs of the "Y" are mutually supporting, the commander has the option of having all soldiers on each leg face the same direction, or he may alternate fire teams. Once the decision is made regarding which method is to be used, subsequent legs of the "Y" must conform to the initial one. Once the lead squad is emplaced, the next squad moves forward to the center point and is closely directed into the 8 o'clock position. After the second element is in position, the third element moves forward to the center point and is closely directed into the 4 o'clock position. If three machineguns are available, one is positioned at each leg, with possible final protective fires across the fronts of adjacent legs. If only two are available, they are positioned at the center point to provide fire down the legs. This positioning of two is not preferred because it does not allow for dispersion; plus, the machineguns are bunched and extremely vulnerable to indirect fire.

Glossary

ADA	air defense artillery
ACP	armored personnel carrier
ASP	ammunition supply point
AVLB	armored vehicle launched bridge
CBR	chemical, biological, and radiological
C&C	command and control
CEOI	communications-electronics operation instructions
COMSEC	communications security
CP	command post
DZ	drop zone
FAC	forward air controller
FDC	fire direction center
FIST	fire support team
FO	forward observer
FPF	final protective fire
FRAGO	fragmentary order
FSCOORD	fire support coordinator
HE	high explosive
HEAT	high explosive antitank
HF	high frequency
ICM	improved conventional munitions
JOES	jungle operations extraction system
LAPES	low altitude parachute extraction system
LBE	load-bearing equipment
LCM	landing craft, mechanized
LZ	landing zone
METT	mission, enemy, terrain and weather, and troops and time available
MO GAS	motor gasoline
MOPP	mission oriented protection posture
NBC	nuclear, biological, chemical

OP	observation post
OPORD	operation order
ORP	objective rally point
PL	phaseline
PLL	prescribed load list
POL	petroleum, oils, and lubricants
PW	prisoner of war
PZ	pickup zone
RCLR	recoilless rifle
SLAR	side-looking airborne radar
STANO	surveillance, target acquisition, and night observation
TACAIR	tactical aircraft
TOC	tactical operations center
TOW	tube-launched, optically-tracked, wire-guided missile
VTR	vehicle track retriever

FM 90-5

By Order of the Secretary of the Army:

16 AUGUST 1982

E. C. MEYER
General, United States Army
Chief of Staff

Official:

ROBERT M. JOYCE
Major General, United States Army
The Adjutant General

DISTRIBUTION:

Active Army and USAR: To be distributed in accordance with DA Form 12-11B, Requirements for Jungle Operations (Qty rqr block no. 307), and Operations of Army Forces in the Field (Qty rqr block no. 405).

ARNG: To be distributed in accordance with DA Form 12-11B, Requirements for Jungle Operations (Qty rqr block no. 307).

Additional copies may be requisitioned from the US Army Adjutant General Publications Center, 2800 Eastern Boulevard Baltimore, MD 21220.

☆U.S. GOVERNMENT PRINTING OFFICE : 1987 - 189-575 : QL 3

PIN: 051485-000

Made in United States
North Haven, CT
04 November 2026

21773571R00225